Classic and Antique Fly-Fishing Tackle

Classic and Antique Fly-Fishing Tackle

A Guide for Collectors and Anglers

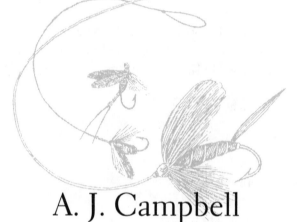

A. J. Campbell

LYONS & BURFORD, PUBLISHERS

10 9 8 7 6 5 4 3 2 1

Design by Joel Friedlander Publishing Services

Library of Congress Cataloging-in-Publication Data
   Campbell, A.J.
      Classic and antique fly-fishing tackle: a guide for collectors and anglers / A.J. Campbell.
         p. cm.
      Includes bibliographical references (p.) and index.
      ISBN 1-55821-400-3
         1. Fishing tackle—Collectors and collecting. 2. Fly fishing—Equipment and supplies.
      I. Title.
      SH447.C36 1997
      688.7'9124–dc21   96-53509
                                 CIP

To English roots,
"I would rather catch a fish . . ."

—Colloquy of Aelfric, 995 A.D.

And American pluck,
". . . a whale is a fish."

—Genio C. Scott, 1869

Contents

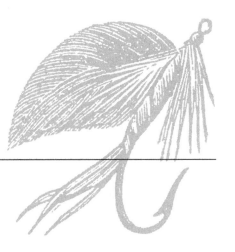

Acknowledgments

Putting together this volume was an undertaking. When possible, I went directly to primary sources, often sitting in front of worn-out microfilm machines until half blind. To the many librarians and curators who gladly gave their aid and personal time, I am indebted.

A hearty thanks to Don Johnson, Alana Fisher, and especially Jon C. Mathewson, all at the American Museum of Fly Fishing in Manchester, Vermont. They were gracious hosts while I photographed a selection of the museum's tackle. Thanks to Patrick Yott, conservator of documents, and William E. Ross, curator of the Milne Angling Collection—both avid fly fishers at the University of New Hampshire Library in Durham. At the State of Maine Museum in Augusta, I appreciate the aid of Ed Churchill and Brian Sykes, who took time from their busy schedules to allow me access to original documents. And in Massachusetts, to Karen Smith, president of the Pelham Historical Society; and Dan Lombardo, research librarian at the Jones Library in Amherst.

Kudos to the many other librarians who aided in finding materials, often without being asked, helping me uncover the production years of many fly-tackle makers. To the staffs at the Boston Public Library; the Farmington, Maine, Public Library; the University of Maine Manter Library in Farmington; the State of Maine Library in Augusta; the Bangor, Maine, Public Library; the Portland, Maine, Public Library; and the Saint John, New Brunswick, Public Library. I also extend thanks to the folks at the State of Maine Research Center for helping me locate vital statistics.

Several avid collectors helped with additional photos and old letterheads, while others lent fly tackle for photographing from their private collections. A special thanks to Brian and Lora McGrath, of *Fishing Collectibles Magazine*, for opening up their Kennebunkport home and allowing me to photograph Brian's side-mounted reel collection; also Bob Corsetti, one of the fairest tackle dealers I've met, for lending classic reel photos; Bob Lang, auctionmaster extrordinaire, for letting me borrow and photograph his Canadian rods; Michael E. Nogay, for primary sources on Charles Orvis; Ken Reback, for authentic maker-seller letterheads; my editor, Jim Babb, for letting me photograph some of his classic rods; and Mark Aroner, for information on Amherst rods.

I also thank David Allen for his beautiful large-format transparencies, as well as Steve and Chris Rubicam, who aided in the "photo finish." To Sally, who put up with all this fly-fishing stuff for years until it finally got photographed. All are commended and appreciated. Without their help, this volume would be a lesser one.

Introduction

The fish were here first. We anglers came along much later and invented the trappings to fool them—simple tools, really: a long rod carved from the limb of a swamp tree, a hank of bustard feathers tied to a hook, and some braided horsehair line. For millennia the branch that tossed hackle also dunked worms, and in the 19th century tackle merchants still offered products that performed both functions.

Just before the American Civil War fly tackle made a leap in popularity and entered the machine-production stage. Carved on milling, lathe, and shaper beds, it reflected a pride in workmanship rarely seen today. In 1858 the first production rods were turned out on lathes by Calvin Gray's crew in Pelham, Massachusetts—direct ancestors of the Montague line of affordable rods. A decade later another Massachusetts native, Thomas Chubb, began a whirring rod factory in Post Mills, Vermont. These men were the Fathers of Popular Sportfishing, and their legacy remains as affordable fly tackle.

Much of this wonderful old angling stuff has exceeded its original popularity, as collectors pay dearly for sporting apparatus of our past. At any well-attended auction, exquisite creations by Leonard or Payne bring four-figure prices—tags so robust we don't dare actually fish with these valuable pieces. Even Goodwin Granger's products are reaching prices that make them impractical for everyday angling.

But collectors will also discover high-quality, eminently fishable classics at reasonable prices—perhaps a Thomas, Granger, or Edwards rod with one short tip. In the same vein, consider the genius of Wes Jordan, last of the great innovators. His classic rods, the impregnated Orvis models, are perhaps the most popular and durable split-bamboo creations available. Fish them in an ice-out snowstorm on the Allagash and quote Alfred E. Newman: "What, me worry?"

A more than cursory glance at Granddad's fly tackle offers a choice, metaphorically a picnic fork in the road. Many post-1900 split-bamboo rods are readily affordable, becoming collectible, and are eminently fishable. Matching reels built within this production era are perfect mates for modern applications. When it comes to stuff too good to actually use, there's a gold mine of underpriced fly-fishing antiques, a lot rarer than once thought and great links to the historic age of Daniel Webster and "Uncle Thad" Norris. What marvelous examples of antiquity.

Think of this book as a ramble down an old woods road, overgrown by generations of puckerbrush. Under the brambles we find the elderly switch, the original spliced wooden rod. Early chapters delve into the birth of split bamboo and original American fly reels. Middle sections follow the careers and production years of major and obscure makers—the grit that pinpoints the origin of so many collectible pieces. Like the tackle industry itself, the book starts in the East and works west; makers are introduced by seniority. There are important hints on ways to identify and date most tackle items you might find, with information on various period stampings and decals, approximate years of production, and key features. If something is uncommon or rare, I'll let you know. And let's peer into the old cellar hole of our retail tackle house that so often signed wares with a die or private label.

To an allowable depth, we wade into the various styles of flies, landing nets, fly containers, and angling accoutrements that have become so popular with collectors. British and Canadian products are covered, arming enthusiasts with a working knowledge beyond the provincial picture. Finally, you're invited to join me in the joy of using the classic rod and reel on today's waters. To release a fish taken in this most traditional way is a never-ending thrill.

This book extends from decades of research and fun in the field. It covers collectibility, and even describes a few repairs that can bring a classic rod up to snuff so that you can take it fishing. For those who are not collectors but have an interest in fly fishing's history, the following pages are intended to entertain.

I hope you enjoy.

                  —A. J. Campbell
                  East Boothbay, Maine

Classic and Antique
Fly-Fishing Tackle

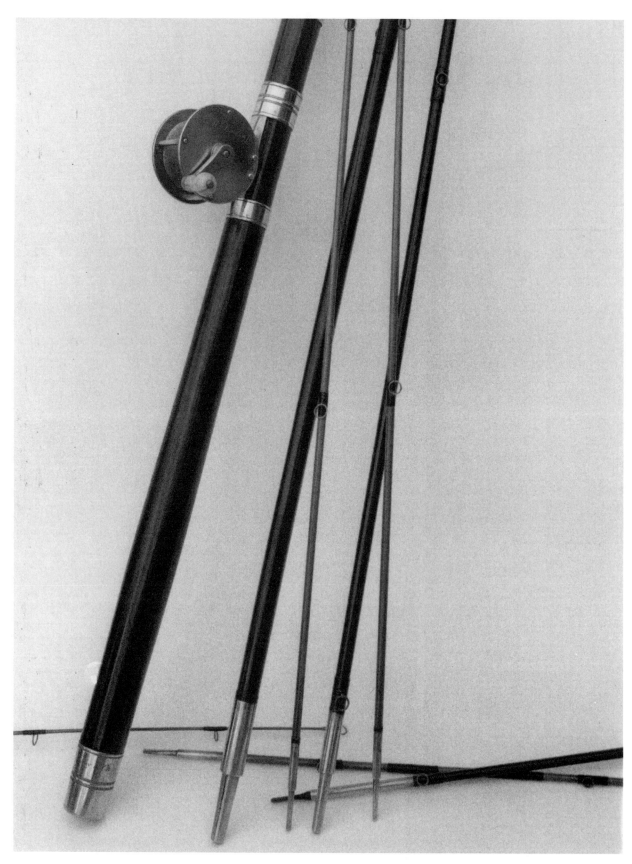

Two-handed trout rods are ultra-rare, and this dated example by Conroy has extra upper midsections and three extra tips. Engraved "1864," this commemorative piece is 15 feet long and was built as a reproduction of an earlier fly rod.

1 English and American Tackle Origins

To say that fly fishing has a long tradition is an understatement. The rod and fly predate the birth of Christ, perhaps going back to the late Iron Age, and appear to have been used from ancient Europe to China. The Macedonians were fly fishers, and so were the Greeks; the poet Theocritus mentioned the "bait fallacious suspended from the rod" in the third century B.C. Five hundred-odd years later, the Emperor Diocletian, after a 20-year career of putting the Roman house in order and persecuting Christians, retired to a life of fly fishing on his private trout streams.

One of the first known illustrations of a rod appears on the floor of the Cathedral of Constantine,

built around the year 330 by Diocletian's successor. This mosaic shows a rod with a liberal amount of spare line between the rod tip and angler. At least two more illustrations show coils of spare line, including a Chinese drawing by Tu Shu Chi Ch'ing and the Mount Nebo mosaic, both dating to the sixth century. The first English mention of the fly rod, by the mysterious Dame Juliana Berners, came in *The Treatyse of Fysshynge Wyth an Angle*, published by Wynkyn de Worde four years after Columbus discovered America. The gizmo that would eventually become the reel had been used in China for 1,000 years, but was unknown in Britain until the 1640s or '50s.

It's in the British Isles that the earliest antique fly tackle gathers dust and that the story of the classic rod begins. First constructed with a yew, juniper, or blackthorn butt section (the *staff*) and a hazel tip (the *crop*), the ancient fly rod made do without a reel. The typical 15-foot rod equipped with a line of equal length could make a 30-foot cast, however—long enough for a lot of modern trout angling.

While the earliest rods had no guides, they featured a loop, or "noose," of line wrapped to the very tip. This very old ingenuity, dating back to the well-trodden illustration on Constantine's church floor, allowed the angler to do two things: actually "shoot" a small amount of line on the initial cast, and give extra line to the occasional large fish. This rig needed little improvement, except for the addition of the upstream cast, the dry fly, and something to hold the excess line.

These hand-fashioned items were the direct ancestors of the fly rod as we know it, and remained little changed from 1550 until the industrial revolution spawned the modern tackle industry after the Civil War. Then, in the space of one generation, the classic fly rods and reels were born—a supernova of tackle.

Early British Tackle

Unfortunately, the few remaining examples of early British tackle were built in the early 19th century, and only by the most prolific of makers. Originally, rod-making was a cottage industry conducted by lance-makers and bowyers. Charles Cotton, Izaak Walton's coauthor on *The Compleat Angler*, owned rods fashioned by an arrowmaker in Yorkshire. These 18-foot-

The development of fly fishing paralleled that of the hook, first seen in a crude yet barbed version from the Bronze Age (illustrated). Better hooks came along in the Iron Age, but it was the advent of steel that spawned the style of hooks we know today. The Knights Templars learned the secret art of forging steel in Toledo, Spain, and brought it to England in the 12th century. From that point on, fly hooks were strong enough to handle robust fish.

Tempered-steel hooks were refined in London by experiments of the Royal Society around 1680. Prince Rupert, a fellow of that society, taught the process to hookmaker Charles Kirby of Shoe Lane, and four generations of the Kirby family dominated the hook industry. After the Kirbys went out of business, around 1775, the superiority of the tempered Irish fly hook—as made in Limerick by the O'Shaughnessy family—came to the fore. By the early 1800s, hookmakers were established in Kendal and Redditch, England. Collectors will occasionally find 19th-century hooks packaged by familiar Redditch makers, such as Allcock and Bartlett. Antique hooks are in great demand by tyers who make reproductions of period fly patterns.

ers featured a whalebone section at the extreme tip, and were spliced together for the entire season.

As early as 1612, Londoner Gervase Markham remarked that rods were available at "every haberdashers store." The fly reel appeared by 1651, when the "wind" was mentioned in Thomas Barker's *Art of Angling*. A couple of years later, Walton's second edition made sparse mention of the "wheel," as if Izaak was unsure of its appearance, location, and function.

In 1662, Colonel Robert Venables mentioned the "winch" in his *The Experienc'd Angler* and illustrated it, rather poorly, on the book's frontispiece. The Venables reel was a crude single-action winch of brass and iron, fitted to the rod by either a spike or a clamp—the illustrator was vague on this feature. I don't think this was an Anglicized version of the earlier Chinese style.

Unlike pizza and pasta, the reel didn't return to Europe with Marco Polo. In the 1400s, the Ottoman prince Muhammed II captured Constantinople, the gateway to the East. The ancient Silk Road was thereafter closed to the British Empire until modern times, necessitating a new method of reaching the Orient—a sea route, which lead to the discovery of the New World.

It's unlikely that the reel arrived in Britain by ship, considering the state of transoceanic commerce in the mid-1600s. It's likely that the development of the English brass-and-iron winch was independent of the oriental reel, which had rotating wooden pegs. It's unknown where the first British winches were made, although the early brass and iron foundries in Birmingham may have been the source—the very same foundries that sparked the industrial revolution a century later.

In Walton's and Venables's time, brass also replaced the rawhide crop as the rod's tiptop. The little brass fairleads for the line—the guides—came in the late 1600s or even later.

Around this same time the earliest known rodmakers were active in London. Walton mentioned that "if you mean to be a fisher, you must store yourself; and to that purpose I will go with you, either to Mr. Margrave, who dwells amongst the book-sellers in St. Paul's Church-yard, or to Charles Brandon's, near to the Swan in Golding-lane: they are both honest men, and will fit an angler with what tackling he lacks." (From the first edition of *The Compleat Angler*. The fifth edition noted that "John Hobs" had acquired Brandon's business or address, and the Lang version of that same edition (London, 1896) mistakenly changed "Hobs" to "Stubs.")[1]

The frontispiece of Robert Venables's 1662 book, The Experienc'd Angler, *depicts an interesting array of tackle, including a spike or clamp winch, a couple three-piece rods, a circular leader case, two bait horns, and a center-hole wicker creel. Then of course there's the other stuff: bobbers, a newt, a maggot, a caterpillar, and an extremely dead pike. The flies resemble early versions of Louis Rhead's bugs.*

Certainly as the 17th century came to a close, the fly rod and reel were readily available in the London marketplace. The first tackle houses were concentrated on each side of the Thames, at Crooked Lane and at Bell Yard Court, where the massive gate known as Temple Bar displayed the heads of executed criminals until 1772. In the early 18th century rodmaker Robert Hopkins displayed his handiwork to gentlemen and titled clientele at 48 Temple Bar.

Unfortunately, fly rods of this era have not surfaced; perhaps the last one was used centuries ago to start the fireplace at a hunting lodge in Hampshire. We do know what they looked like, however. The two-handed models for salmon and trout were 18 and 15 feet long, respectively. And the lighter one-handed versions ranged from 11-something to close to 13 feet. Fly rods built in two sections were replaced by models with three pieces: a butt, a mid, and a tip section. Eventually, the rods were constructed in four sections with two mids, setting the standard as traveling or coach models. John Herro succeeded Hopkins at 48 Temple Bar around 1730, and three decades later the business went to Onesimus Ustonson and his family. At last we have surviving examples of antique tackle—perhaps not to fondle, but at least to view from a safe distance.

The Ustonsons remained tacklesmiths for generations, and were so prolific that some of their wares have survived. The most famous Ustonson

One of the earliest known signed reels, this tiny multiplying brass winch was built around 1810 by James Haywood. It sports a clamp for rod attachment and Birmingham steel face-plate screws; it's marked, "HAYWOOD, MAKER."

rod is an engraved four-piecer build for King George IV (who reigned from 1820 to 1830). A matching 2¼-inch-diameter multiplying fly reel, as well as the original case, net, "casts" (leaders), and flies, are all displayed with the rod a Kew Palace in London.

Average anglers of the time could own similar tackle, albeit less finely embellished; there was a profusion of makers and even a burgeoning export market. In the 1750s John Souch, of the Golden Salmon on London Bridge, started exporting tackle to Barbados, Virginia, New England, and even Newfoundland. This was a two-way voyage; rods built from woods hewn in the Colonies returned to the hands of the woodcutter.

With the Ustonsons, James Haywood, and others entering the colonial market, imported fly tackle decorated hardware shop windows in Saint John, Boston, New York, and Philadelphia. A 1773 advertisement by Jeremiah Allen in the *Boston Newsletter* touted, "Bamboo (solid), Dogwood, Solid Joint, and Hazel Angling Rods, Hair & Silk Lines, Best Kirby & common Hooks, Kirby hooks on Gimp . . . and Brass Winches." London wares traveled as far west as Kentucky, where local watchmakers refined an Ustonson winch into the famous Kentucky reel.

Ustonson's became an even more dominant force in the burgeoning tackle industry upon its appointment as supplier to three successive British monarchs, beginning in 1824. At this time the old London tackle house was operated by one of the most famous ladies in the business—Maria Ustonson. She ran the Temple Bar firm with a deft hand until 1848, when it became Ustonson & Peters.

Maria Ustonson was just one of several British makers to export the many reels now surfacing in America. The London families of Chavalier and

Bowness, separately and together, also played an important role, based on the current availability of their products.

Birmingham's James Haywood began making fly reels around 1802, and his products are actually the earliest surviving English imports. At least four Haywood reels have been found in the United States, several being sold at prices far below their actual value. After Haywood's death in 1828 his widow, Mary, continued making brass fly reels and rod ferrules.[2]

All this smith-age tackle, handmade by artisans—smiths—until the industrial revolution, brought leisure sport from Great Britain to the gentlemen fly fishers of the Eastern American Frontier. Some accompanied Lewis and Clark through the wilds of the Pacific Northwest. The British manufacturers also spawned a new way of doing things that would influence the budding American tackle industry's ability to produce wares for the workingman. The most important revelation came from James Watt, another citizen of Birmingham, England, who finally put a steam engine together correctly in 1769, at precisely the time Onesimus Ustonson was planing wood into eight-sided lengths and scraping them round. Our good Mr. Watt also produced a new substance from coal tar a few years later—a kind of graphite-carbon stuff that might someday find a use in the tackle industry.

American Tacklesmiths

It's natural to assume that the first American rod- and reelmakers used British tackle designs as their prototypes. And upon examining our earliest works, from Kentucky or New York, the similarities are striking. A few refinements and minor changes were made, but the tackle looked very English. Still, if we take the opportunity to view a number of artifacts built on each side of the Atlantic, the differences are subtle but apparent.

In rods, the woods of choice differed—the American butt being ash, with either ash, hickory, or ironwood (hornbeam) for the midsection. Like those in the Mother Country, delicate tips were fashioned from lancewood (sometimes called lemonwood). English rods carried massive butt caps, sometimes fitted with a "spear" to keep the butt from sliding when the piece was leaned against a tree for lunch. Many British reel seats had a wraparound brass plate or sleeve to prevent the sliding ring from binding, and to allow a winch to be removed after the wood

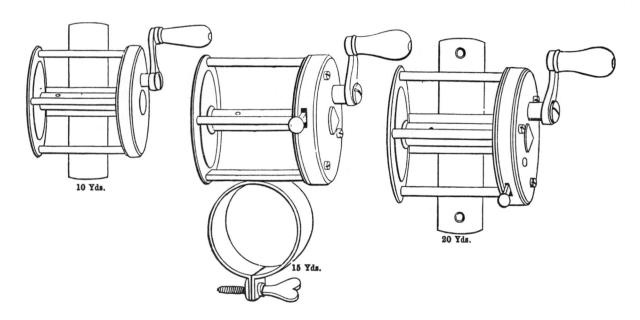

The multiplying fly reel was popular from the mid-1700s until 1850 or later. These are British models, sold in 1867 through Thomas H. Bate & Company, New York.

swelled from dampness. Some American rods sported a similar arrangement, with an additional metal plate mortised into the reel-foot area.

American furnishings followed the simplest British versions, although the Yankee butt cap was a lighter rolled-and-soldered affair, and the seat area usually featured a lower fixed band with a sliding band above it. The sliding unit was retained by a thin band similar to a giant wedding ring. English ferrules were usually doweled and occasionally loose fitting—in some cases so loose that wire loops wrapped to male and female were lashed together so the rod wouldn't "throw apart" during the cast. American ferrules were just tight enough to stay together, and the retaining loops were often omitted.

Collecting early American or British wooden rods is an esoteric occupation, indulged in by those patient enough to *wait* for the next addition to their collection, much like watching paint dry. Most fly rods are dated to the post-1850 era, although American makers of hooks, snells, and so on, were established as far back as the 1770s. Our first American reels may have been built around 1815-26, when George Snyder (Paris, Kentucky) and John Conroy (New York City) made the earliest known examples. Fly rods of American origin don't appear until 1825 (Cape Cod, Massachusetts) and 1836 (New York City).

Within a short period, however, American tackle came to equal or surpass in quality the British originals. In 1849, John J. Brown, a New York tackle dealer at 103 Fulton, wrote in *The American Angler's Guide* that "the best rods were formerly imported from England . . . but they were little adopted to our modes of fishing, and have consequently grown into disuse. American rod makers have introduced great improvements in the article *within the last ten years* [my italics], and can now turn out rods which, for workmanship and beauty of finish, cannot be surpassed." He also urged his readers to "buy American." In 1849, too, Frank Forester (William Henry Herbert), one of the best-known sportswriters of his day, specifically referred to fly fishing: "The art piscatorial is, comparatively speaking, little understood or practised in the United States, bait fishing being all in vogue, even for trout, and an accomplished fly-

fisher—though the number of them is now increasing, *rara avis in terris nigroque simillima cycno*."

At present, we know of only three first-generation American tacklemakers: John Conroy and Ben Welch, both of Brooklyn, New York; and John Dennison, who lived on Cape Cod. By far the bulk of their work—if indeed there was any bulk—has been lost through breakage, misuse, ignorance, and anonymity; much of it was never signed by the maker.

Like the chicken and the egg, we don't know which New Yorker came first, Conroy or Welch. Of the two, Ben Welch built the oldest "almost surviving" American rod. This rather robust implement is missing its tip section or, more likely, tip sections. It was one of three rods built for Daniel Webster of Marshfield, Massachusetts. Our other chap, Dennison, was obviously active by the mid-1820s.

THE WEBSTERIAN LEGEND

Webster met John Dennison before he actually moved to Marshfield. At the time, Dennison was a famed trout guide who made lines and rods as a sideline during the off-season. The Great Orator was well acquainted with Dennison by 1825; during this time the guide built Webster's favorite fly rod—old Kill-all. This is the rod with which he presumably took the legendary Great Trout, at Carman's Brook, Long Island, in 1827. Arthur F. Tait's famous 1854 painting shows Webster at Carman's with tackle in hand. The rendition is reasonably accurate because the tackle had not changed during the quarter-century time lapse.

The three Ben Welch rods were built for the statesman on commission from Timothy Hedges of East Hampton, New York. Constructed in 1847 at Welch's stand on Cherry Street, the rods, along with a case of other goodies (reels, lines, flies), were waiting at Marshfield when Webster arrived after a Senate session. We know from Daniel Webster's letter to the maker that the rods were well appointed, at least by American standards. With that in mind, those who visit the American Museum of Fly Fishing, in Manchester, Vermont, can view the only Ben Welch rod found to date.

Most likely, Benjamin D. Welch became active in New York between the mid-1830s and 1840, the

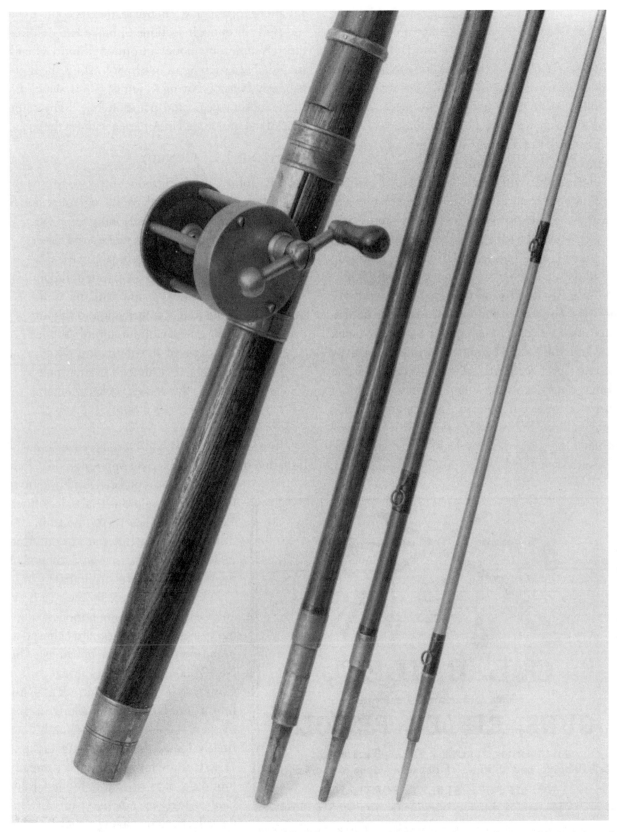

A "Concord coach" model, this four-piece light trout rod came with two-piece ferrules, extra lower and upper mids, and three tip joints. With its light reel bands and reverse taper, it appears to be American. The reel is a small J. C. Conroy #6 multiplier.

same period in which John Conroy started making fly tackle. They were true competitors in every sense of the word. In 1849, through William Trotter Porter's *Spirit of the Times*, the two New York makers went head-to-head in a contest to build the perfect rod, submitted anonymously, judged by merit, cost no object.[3] Welch won the honors, although Conroy was patronized by Porter, whose *Spirit* offices were directly across the street on Fulton.

The lone surviving almost-rod by Welch exhibits clean, almost austere hardware, and sports a solid welt at the ferrule tip of each of the three remaining sections. The rod narrows to a delicate butt cap, and one seat-area ring has a dent. This is an American rod, perhaps not actually born on the Fourth of July, but obviously a participant in many celebrations. And it's a tangible departure from English design, with its diminished butt cap sporting a reverse taper.

While the pastoral Brits might find the activity soothing, sportive Americans had no time to waste lashing splices or wiring floppy ferrules. Their fish and waters were different. In America and Canada shorter salmon rods evolved, along with improved ferrules. However, on both sides of the ocean similar rods were developed for various "general" angling situations. British and American anglers could use a rod built in enough sections to make two or three entirely different models. Londoner John Cheek included two such items in his 1839 listing, the more expensive version having "5 joints, 5 tops, winch fittings, socket, spear, and partition bag." The same style of rod was offered in America without the spear.

JOHN AND J.C. CONROY

In 1843, John Conroy introduced his own version, called Porter's General Rod. William Trotter Porter described this multifaceted creation:

> It can be put together as to make a rod either ten or sixteen feet in length; you may make out of it a light-hand [single-handed] rod for fly-fishing, or a heavy, powerful rod, sufficiently strong to play a thirty-pound salmon or bass, at the end of a hundred yards of line! . . . The sockets of the joints are double instead of single, that the end of the [male] joint fitting into the double sockets [female] having double ferrules around it [two-piece ferrules].

In addition, the rod had German silver furnishings, fluted guides (doubled on opposing sides), four joints for bass or pickerel, and five joints for trout and salmon. It also had three extra tips. Quite a production. It's unlikely that such a rod has survived intact, but collectors should know that the model did exist—just in case.

Conroy also built fly reels in brass and German silver, and a modest number have survived. These are believed to be among his earliest products. The New York City Directory first recorded Conroy as a "machinist" in 1825, giving him a threshold to reelmaking during the period before 1837-38, when he was finally listed under "fishing tackle." (The Conroy folks claimed a company founding date of 1830.) Signed John Conroy reels are rare and expensive little gadgets, some having a click housing on the rear plate that sticks out in a "wedding-cake" design. Copied from

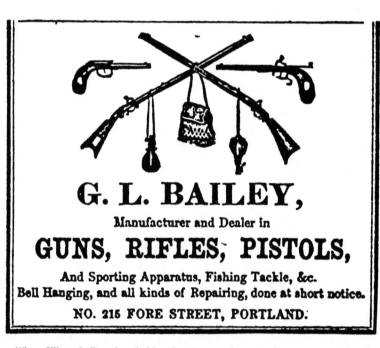

When Gilbert Bailey placed this advertisement in 1853, he was one of the first two recorded tackle dealers north of Boston. The other dealership, belonging to William Neal, was founded in 1843 in Bangor, Maine.

the British version, this New York click reel is as popular now as it was originally.

A great many Conroy rods and reels were unmarked by their maker. Those built during the early years, if marked, are stamped "J. Conroy, Maker, N.Y." In 1840, Conroy took on his son as a full partner, and the firm, now having 10 employees, was renamed "J. & J. C. Conroy Co." This latter stamping, as far as we know, has never shown up on a Conroy product, but articles produced during the 1840s and '50s often carried the marking "Conroy's Makers." By 1864, the business was restructured as "J. C. Conroy & Co., N.Y."; this is the most common of the early stampings. The company also sold tackle through several retailers, including Andrew Clerk and T. H. Bate in the same city. Bradford's of Boston and Gilbert Bailey in Portland, Maine, also carried the J. C. Conroy products.

J. C. Conroy reels and rods were little changed from those of the previous two decades. I have flexed wooden J. C. Conroy trout rods that ranged from 12 feet, 4 inches to 15 feet in length. The 15-footer, a wicked two-handed tool, worked all the way to my lower hand, even though the butt was heavy enough to bludgeon a beaver to death. A veil of mystery surrounds the 15-foot Conroy model. Its 1864 pedigree is a rather late date for a two-handed trout rod—a fly-fishing tool so rare that I've seen only this one. It may well be a commemorative reproduction of his father's early work, having antiquated features and fancy German silver hardware. The rod's 1864 date could have symbolized the passing of the torch.

The 12-foot, 4-inch, or one-handed, J. C. Conroy was, at 25 ounces, a mere wisp of a rod, more appropriate for clubbing muskrats. Typical rods of the time, like this one-handed Conroy, featured a design change that was probably introduced by the end of the 18th century: The reel seat was moved to within a few inches of the butt cap, and the grip area moved up to the "over-the-reel" position. In form and function, it was the first major design change in the fly rod since Diocletian was a pup. This new style would be offered alongside the traditional lower-gripped mod-

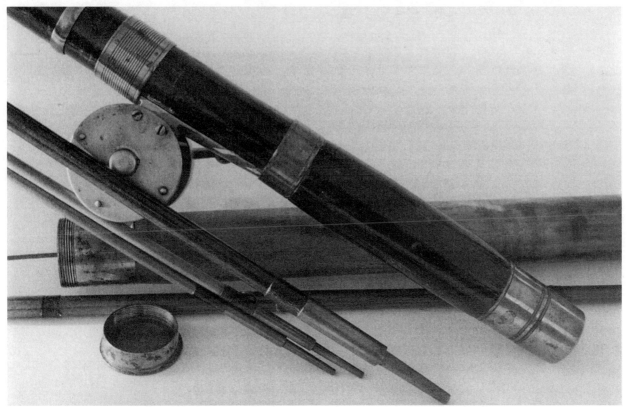

A 12-foot, 4-inch wooden rod exhibiting Conroy ferrules and butt cap. It was reportedly built in Maine, and has the rare 3- to 4-inch extension below the reel seat found on rods built prior to 1860. The bamboo tip case appeared during this early period.

els for the next half century. The original reel-above-the-hand version then continued on as the bait rod.

The Conroys, as well as the establishments of Bate and Clerk, also offered rod furnishings and woods to the amateur maker. And, no doubt, some of these lesser rodmakers built a few wares for the trade. A number of unsigned rods have been found carrying Conroy's butt caps and other furnishings; several examples were built by craftsmen as far from New York as Maine. By midcentury, others had entered New York's rodmaking trade. George Karr, at 143 Grand, started to make his excellent rods. And Jabez B. Crook, at 50 Fulton, also fashioned rods and made reels, several of which have survived into the present century.

J. B. Crook was one of New York's earliest fly-tackle dealer-makers. He began work at the same time as John Conroy and eventually moved into Conroy's original shop space.

During the 1840s and '50s, the craft of rodmaking spread from the cosmopolitan Knickerbocker area to Manchester, Vermont, rural Pennsylvania, and as far away as Saint John, New Brunswick. Of immediate interest, the craftsmen of Manchester exemplified the new geographical approach to rodmaking: a reliance upon a big-city clientele passing through their little burg on the way to the clear streams of the Green Mountains. Others, principally in Maine, would eventually prosper by selling their wares to the passing tourist-angler.

THE MANCHESTER RODS OF SWIFT, EGGLESTON, AND ORVIS

Nestled in the western foothills of the Green Mountains, the Vermont town of Manchester featured the last "civilized" inns and guest houses before travelers moved into the puckerbrush. And Manchester Station was centrally located by railroad from the nearby states of Massachusetts, Connecticut, and New York. Here, three rodmakers—Swift, Eggleston, and Orvis—were active by 1855, building the famous (if your name was Orvis) or not-so-famous (if you were Swift or Eggleston) Manchester rods.

Steeped in obscurity, the partnership of Swift and Eggleston probably made a number of wooden rods from around 1855 until after the Civil War. We know little about this association, proof of the fragility of the lancewood tip. In general, once the tips of a wooden rod were broken to the point where it no longer would function as a casting instrument, the owner chucked it. This seems the sad fate of most wooden rods, and I would hate to guess at the number of Conroy, Welch, Swift, Eggleston, and Orvis rods that ended their days propping up tomatoes.

In 1987, the Oliver tackle auction featured an example marked "Manchester Fishing Rod—Manufactured By—Swift and Eggleston, Manchester, Vermont." It carried a paper label, perhaps the earliest use of this quick and easy method to mark a rod to its maker. It's sobering to think that obscurity can often be dictated by something as simple as a pasted label, so easily removed from one's product. I was lucky enough to examine this first known Swift and Eggleston piece—and stupid enough to pass it by. It was built in the late 1850s or early '60s and had the features of the time, including a stained "pool-stick" butt section, a sliding band, and a fixed upper band, all starting above a ring located about 2 inches over the long butt cap.

The rod was finished with a cord-wrapped handle and simple ferrules. Prior to its use on fishing rods, cord wrapping was common on equestrian crops, but it didn't appear on wooden rods until around 1860. This striking three-piece model also had a midsection that was 7 inches longer than both the butt and the single remaining tip. At about 10 feet long, it may have been built that way or, more likely, shortened from an original length of 11-something. Not really in prime condition and also missing wraps and loose-ring guides, the rod sold for $300.

The obscure partnership of Swift and Eggleston may have ended with the retirement or death of Swift. Our other Manchester rodmaker, Charles F.

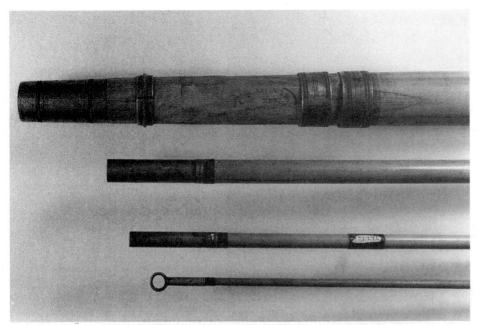

An early Charles F. Orvis rod, signed on the reel seat with a branding stamp. Orvis believed in plain ferrules with no dowels; the rod has the diminutive butt cap that became an American rodmaking trademark. (The American Museum of Fly Fishing Collection)

section as well as on the metal handle fittings. It was made at a later date than the first Manchester rod, probably the late 1860s or early 1870s. The rod had a finely wrapped cord handle (black), very large ornate hardware, and a butt cap marked "C. F. Orvis, Maker, Manchester, VT." It was a three-piecer with two tips, one of which was 6 inches short, supposedly by design.

Orvis built lancewood rods for about 40 years; he continued to make the older models long after he introduced a split-bamboo line in the late 1870s. By this time, his rod-shop foreman is believed to have been Hiram Eggleston, perhaps the son of or the original Eggleston once associated with Swift. All antique Manchester rods, scarce yet occasionally available, are links to the wooden-rod era of fly fishing.

Orvis, was destined to achieve a modicum of fame. Keenly aware of the tourist dollar, Orvis helped his brother Franklin build the Equinox House in 1853. In 1861, he built his own tourist trap just down the street. Adept at mechanical engineering, the 25-year-old Vermonter started his professional tacklemaking business in 1856. Known as the C. F. Orvis Company, the shop was located in a small stone building next to his brother's hotel.

At first Orvis products included wooden rods, built by Charles alone, and fancy stream and lake flies. The flies were probably fashioned by other local talent, usually having a woman's touch. The rods, of course, are the subject of our scrutiny, since wooden Orvis wands were built from 1856 until the 1880s. This means that sufficient early ash and lancewood Orvis products are available to divert collectors awaiting the *next* visit from the ghosts of Swift and Eggleston.

One of the earliest known Orvis rods, a 9-foot, 3-inch, all-lancewood model, appeared at the same 1987 auction that offered the Swift and Eggleston antique. Built in the old reel-over-grip style, the rod had "Serial No. 80" stamped on each

UNCLE THAD'S RODS

Another early maker associated with wooden fly wands, Thaddeus Norris of Philadelphia, may have built rods as early as the 1840s. Born in Warrenton,

By the late 1850s, "Uncle Thad" was an important Philadelphia tacklemaker and dealer. He made wooden trout and salmon models, moving into split bamboo during the 1860s and '70s. Only the wooden rods have surfaced so far.

Virginia, in 1811, he was an accomplished banjo player and smoked a large crooked pipe. Norris moved to Philadelphia around 1835, at first selling tackle and later adding his own rods to the offerings. At least four Norris models have been found, all built entirely of wood. Two of these are in the Anglers' Club of New York collection, one being "snakewood" built in three pieces, the other lancewood. The rods have simple ferrules and reduced shafts above the grip, indicating that they were built in the very late 1850s or after 1860.

Two rods that were similar but had spliced joints appeared at Oliver's Fourth Annual Summer Auction. Only one was marked "Norris" on the top of the reel-seat area. Both butts were made from bird's-eye maple; the remaining two sections were lancewood. One rod was an 11-footer, and the other was a rugged 12½ feet long. Like the Anglers' Club rods, these examples had a reduced butt shaft.

Norris wrote his popular work, *The American Angler's Book*, in 1865, probably at about the same time he was an active rodmaker. Uncle Thad had definite ferrule preferences and, like Charles Orvis, dispensed with the dowel. "I make each joint of my fly rods . . . without any portion of the wood below the end of the male ferrule." Norris also believed in short ferrules, which would not hinder the action of the rod. These two design quirks were sound ideas for his time and were adopted by later makers. Norris ran his retail tackle shop at West Logan Square and probably built many of his rods there. A Norris greenheart salmon rod was mentioned in 1876 by angler-writer Doctor A. G. Wilkinson, who sang its praises. In advertisements, Norris listed "Fine Trout and Salmon Rods, of Iron Wood, Lancewood, Greenheart, and Rent and Glued Bamboo." (Norris was a pioneer in the use of the latter material.)

Angling author Fred Mather met the angling sage in 1873, describing him as "past middle age and not strong nor active. . . . Filled to overflowing with humor, Uncle Thad was as charming a man as one could wish for on a month's trip."[4] The author of the first American angling classic and an inveterate fly fisher, Norris died on April 11, 1877. The Norris Case is not closed, however; and a lucky collector will find a signed "Norris" bamboo rod in the future. Finding out just how

Forest and Stream, *April 24, 1879.*

many sections are split bamboo and how many strips were used will enhance a wonderful discovery.

 Recognizing Early Reels and Wooden Rods

The fly-fishing tools of the pre–Civil War period represent the most historic of angling tackle. Later period pieces are available but rare. A number of English multiplying fly reels came to the early United States, including those by Maria Ustonson and by James Haywood, the Birmingham reelmaker who died in 1828. His brass winches continued to be made by his widow, Mary, until about 1840. These

British reels were copied by our earliest makers, and collectors find American brass winches marked by Conroy, Crook, Bradford's, T. H. Bate & Company, or Andrew Clerk & Company. All signed reels built within this time frame are quite recognizable—and expensive. Even small, unsigned trout reels bring good prices; the most expensive were the rare #6 New York ball-handled single-action models.

The brass winch was not exactly a refined reel, having only one moving part and no bearings. As late as the 1840s sportsmen living in the country's interior had difficulty obtaining British or American winches at all; crude tin-knocker's specials were fashioned by smiths of lesser talent to make do for a season or two.

Many rods of the period were unmarked by the maker, although a few carried commemorative or presentation dates. This small percentage was also engraved, in script, with the original owner's or maker's name. Thus we have the "Furman Rod" (1832, English), the famed "Webster Rod" (1847, American), the "God Bout River Rod" (pre-1857, English, see figure 15-8), and the "J. C. Conroy Rod" (1864, American), most of which are in the care of the American Museum of Fly Fishing. At present, a dated, signed, or presentation rod, intact, should sell—regardless of size—for between $600 and $1,200. Much rarer than the signed reels, these implements are priced rather low. Many nameless complete rods, having at least one tip left, sell for peanuts—less than $100. Occasionally I fish with one in ponds—if I'm sure they contain only very small trout.

WOODEN ROD SECTIONS

Rods of the mid-1800s exhibited extreme variation; the only common link may have been their lancewood tips. The standard package, or "put up," usually produced a three- or four-piece rod. Sometimes the mids were shorter than the butt, but rarely. The tips, however, may have been to length, or built shorter or longer than the butt and mids. Thad Norris explained that the four-piece rod was popular during the old days of stagecoach travel, but the advent of the more spacious steam train meant a rod could use longer sections—thus the popularization of three-piece models. The first long-haul passenger rail line began in 1839, between Boston and New

The Nottingham reel began showing up in America by the 1860s. It was a functional model, no frills, with a moderate price tag. Collectors can acquire these reels today, unsigned, for $100 or less.

York, but it would be another two decades before sportsmen could travel by rail into the interior.

Rods built during the first half of the 19th century were packaged as follows: for four-piece rods, one butt, one lower mid, two second mids, and up to four tips, although three tips was standard. In other words, if the rod now has seven pieces, it's intact. Such a complete rod is a rare and momentous find. Three-piece rods of the time started to approximate the classic formula. The earliest were constructed, like the four-piecer, with extra mids and tips to allow for puckerbrush accidents. A complete early three-piece rod had as many as six sections: the butt, two mids, and three tips. You'll notice that later three-piece rods from the split-bamboo era dropped one mid and one tip, reaching the classic 3/2 formula (three sections, two tips).

Some rods were also made with from five to seven sections, not counting extra tips. Dubbed "valise" or "traveler's" rods, these multipiece instruments were built on both sides of the Atlantic, although they were most popular in Great Britain, as made by the Aldreds, John Cheek, W. H. Alfred, and others. Cheek, also an umbrellamaker, operated his "Golden Perch" at the Strand from 1837 until 1842

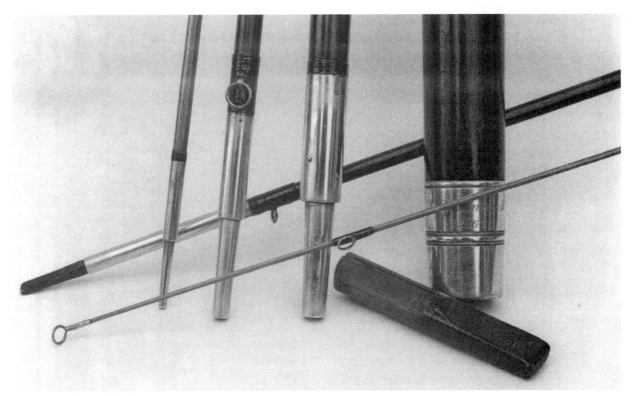

Details of the fine German silver fittings applied to the 1864 J. C. Conroy rod. Large covered dowels and one-piece ferrules were standard in the trade until after the Civil War.

and offered rods in up to 10 sections, "for Pocket or Portmanteau."

Various parts of early wooden rods hold the keys to actual age. By recognizing a few clues to extreme antiquity, you can be certain that the rod was built during or prior to the Civil War. Examine a rod with a mental checklist, going from the shafts and grip area to the ferrules, bands, and butt cap. The following guidelines will help you identify a rod from this period.

GRIPS AND SHAFTS

Wooden rods were often built in odd lengths—not the standard 10 feet, 10 feet, 6 inches, and so on. Rods of this era may instead be 11 feet, 3 inches or 12 feet, 7 inches, as each angler had a favored weight. Altering the length was the only way to produce lighter or heavier rods from a caster's standpoint, as illustrated by Genio C. Scott's (*Fishing in American Waters*, 1869) preference for a rod of "twelve feet and two inches long."

As mentioned, butts were usually built from ash stained black, brown, or verdigris, then varnished;

some were even painted black, or "Japanned." The mids were made from any of several woods, including ironwood, lancewood, mahoe, ash, and greenheart. Greenheart was not popular with early American rodmakers, but it was and is a mainstay in Britain, and finally became popular here around 1880. Mids were stained to match the butt section. American tips were fashioned from lancewood, always left unstained, and coated with either lacquer or coach varnish. British tips were either of lancewood or greenheart.

The oldest British antiques were shaped in one continuous taper, butt to tip, sometimes called the "pool cue" look. Early American models tended to have no taper, or even a reverse taper, in the area below the reel grip, with the pool-cue shape starting just above the reel seat. Rods produced from the late 1850s onward had a distinct hollowing, or reduced area, just above the grip. Around this time, the most popular grip moved from behind the reel to in front of it—the same configuration as the modern fly rod. The grip had no covering, such as cord or rattan, and could only be defined as an area located to the rear

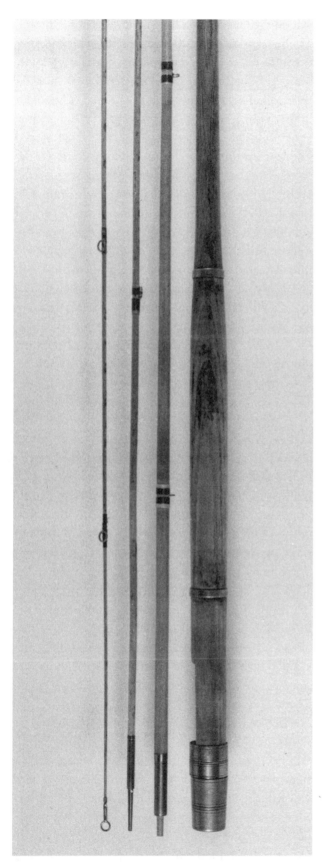

and forward of the reel bands. I have never found a rod built before 1860 that had an actual band denoting the top of the grip. These bands, however, showed up during the Civil War era. At that time, the reel seat was also lowered to the very butt region of the rod—the beginnings of the classic design.

FURNISHINGS AND METALWORK

British furnishing design was static through the mid-19th century; examples made by Maria Ustonson, the last of the clan to remain in the tacklemaking trade, were little changed from those made by her grandfather. Rods by the Bowness and Chavalier families

Plain Butt Ferrules.

Butt-cap designs from Thomas H. Bate & Company, 1867.

This 1841 engraving shows an angler using a "spear" screwed into the butt cap. The rod was spiked into the ground, leaving the angler's hands free to land a large fish.

The little band denoting the top of the "handle" appeared around 1860. Later rods sported a rattan grip.

were similar in design. I recently restored a Bowness & Bowness, circa 1855, that had the old brass spear fitted to the butt cap. Talk about Waltonian! This rod had a short, small male ferrule on the top of the butt with a machine-threaded hole drilled in it. The female ferrule, located on the midsection, had an internal screw that tightened down over the butt. The two tips and the top of the mid were cut to a splice and required a bit of time to assemble. The entire rod was built from light-colored greenheart.

Most English rods were fitted with plain brass furnishings, although you may occasionally see bluing. More expensive rods had German silver accoutrements. Early-19th-century British ferrules were reasonably tight, and it's not easy to distinguish them from American ones. Ferrules of the day were rolled with no trim rings. At this point the tackle trade was going mechanical, with turning engines (lathes) and rolling forms in use by the largest houses, but ferrules remained plain until the 1860s. The reinforcing welt at the top of the female ferrule was introduced prior to 1830, contrary to a much later date given by some modern writers. Many British examples from the 1800-1850s era sported a butt cap affixed to the bottom of the rod by a round-head screw. Two classic examples, the Furman and God Bout River rods, had this retaining screw.

American furnishings were made in the same manner, each piece hand rolled and soldered into a tubular shape. Again, ferrules lacked trim bands, with the possible exception of one near the juncture with the wood. As a rule, the butt cap had a flush-soldered end disk, and its trim bands were often as plain as one or two simple rings, as seen on the Ben Welch rod; in the case of Conroy, there were several grouped and finely knurled trims. The fixed and sliding bands were light for their size, and the reel-foot area was recessed into the wood. On a salmon rod, this area was located as much as a foot or more above the butt; on one-handed (trout) models, the reel seat could be about 9 inches above the butt on lower-grip models, or 3 inches from the butt on those gripped above the reel.

Most brass ferrules were plain with an exposed dowel. Better ferrules, usually made from German silver, had a covered dowel. Some makers, such as Norris and Orvis, preferred a simple ferrule with no extending dowel, since they believed that the dowel, when wet, would cause the ferrule to swell and jam. If you find a two-piece ferrule, it's usually a female, but occasionally such males are seen. Early ferrules were rolled over a mandrel and soldered. The solder line is visible when the ferrule has an old patina. When this style of ferrule is polished, the patina and solder seam disappear; always attempt to retain the patina. Welts at the female ferrule tip, as seen on the Webster, Furman, and Mack rods, were similar to British examples. Some rod furniture was Japanned with black lacquer to keep streamside flash to a minimum. British tiptops were usually wrapped with thread or copper wire, the latter showing reddish with age. Early on, American tops were constructed from a ring soldered to an appropriate tube. The Furman rod has the improved tops.

Pre-1800 rods had the earliest form of standing guides; a tube or double rings soldered to a flat base. In the 1850s, Ainge & Aldred described a newer style: "The eyes or conductors of the line are small brass rings secured to the rod by a small slip of metal and tied or wrapped with silk thread: this arrangement allows the eye to fold to the side of the rod when not in use and is a vast improvement on the old permanently projecting eye." The newer loose-ring guide progressed in popularity throughout the 1800s, but salmon rods are sometimes found now with tube or standing-ring guides, which evidently never went out of vogue on larger sticks.

Smith-age tackle is a rare and valuable find. If the rod looks old yet has no soldered seams on the metalwork, it was built during a later era. Rolled-and-seamed furnishings, the plainer being the elder, are the true key to determining extreme age in a wooden fly rod.

Before leaving the early to mid-19th century, we return to Great Britain, and to James Watt and his steam engine. As it turned out, the steam engine spawned an industry—Machine Begat Machine, if you will. In 1800 we had Henry Maudslay developing the screw-cutting machine. Then in 1836 James Nasmyth came along with the shaper, a belt-driven gadget that was modified into the beveler and milling machine. If that wasn't enough, Nasmyth makes an encore in 1842 with a huge, noisy contraption called the steam hammer, which was miniaturized into the belt-driven drop hammer. And what, you may ask, did these cutting and clanking machines have to do with the tackle biz? Everything! Especially in the not-so-distant future.

Notes

[1]Walton, Izaak. *The Compleat Angler*. Andrew Lang edition. London: F. M. Dent & Company, 1896.

[2]Turner, Graham. *Fishing Tackle: A Collector's Guide*. London: Ward Lock, 1989.

[3]Schullery, Paul. *American Fly Fishing: A History*. New York: Lyons & Burford, 1989.

[4]Mather, Fred. *My Angling Friends*. New York: Forest and Stream, 1901.

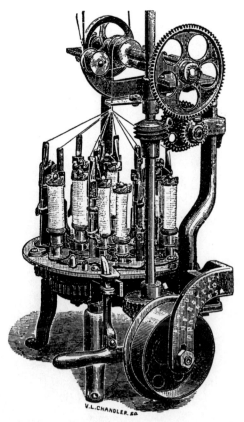

The machines are coming. The first American line braider, similar to this Rhode Island model, was probably in operation by 1860.

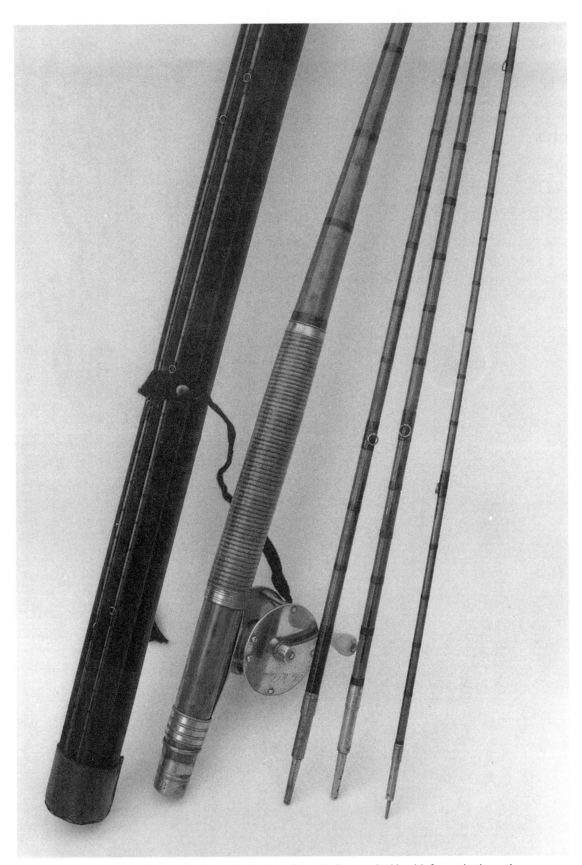

This early complete split-bamboo rod, built by Charles F. Murphy, has a six-strip butt, and mids with four-strip tip sections.

2 A Search for the Marvelous Sliver

In two decades pioneering tacklesmiths such as Dennison, Conroy, and Welch had made American tackle directly competitive with English products. In particular, the second Brooklyn generation would help establish the reputation of American rods and reels. Today we are left with a few rare and exquisite examples made by John C. Conroy and perhaps Robert Welch (believed to have been the son of Benjamin Welch). These two Brooklyn makers were joined by others, establishing New York City as the heart of the American craft. Early New York products became rather stylized, exemplified by the Conroy rod illustrated in the previous chapter; this design remained static for three decades or more. For collectors, the distinct look of these items aids in identification even when the piece is unmarked.

Matching the young American rods was the New York click reel, perhaps the most collectible of the era's cosmopolitan tackle. Click reels were built in an array of sizes, from the tiny #6 upward to the #1, and they were the finest American-made winches available in the prepatent period. The Conroys, Crook, and others now anonymous fashioned the best of the raised-housing click reels from German silver. Many were stamped by the seller, and even these retailer's examples are coveted. During the decades before the Civil War New York click reels were the gentleman's choice, the first winches to be mounted on the emerging split-bamboo rods.

Between 1845 and 1868 the popularity of American fly fishing grew. Tacklemaking spread into rural areas, where gunsmiths began fashioning sporting apparatus for their clientele. During this period more than a dozen gunsmiths in Pennsylvania, New York State, and Maine started to design rods and reels that looked very different from "City" tackle. The gunsmiths learned their trade through the old bonding system, with a young man working under a master craftsman for two or more years. As an underpaid journeyman, the aspiring smith formed every part of a rifle, fowling-piece, or rod until he could make the finished product unaided.

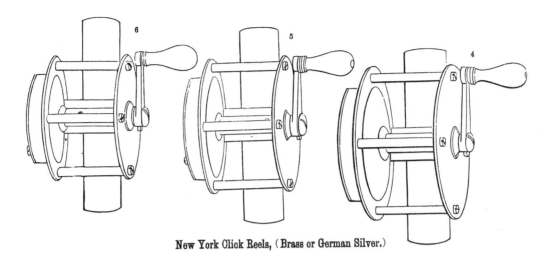

New York Click Reels, (Brass or German Silver.)

The New York click reel, shown from Bate's 1867 catalog, was a popular match for the expensive early split-bamboo rod. Moderately rare, these reels now bring a healthy price.

By this process, and over a period of years, one master craftsman training several new gunsmiths passed along his sense of design and aesthetic excellence to his former apprentices. This led to local versions of sporting apparatus quite different in style from those made in Brooklyn or even other regions of the country. This was very apparent in the Cult of the Side-Mounted Reel, headed by Morgan James and William Billinghurst in New York State. Farther to the east, in Maine, the Multishot Smiths designed tackle that looked entirely different from the upstate New York products. The earliest of these American sects, the Split-Bamboo Clan, can be traced to a single gunmaker who was active in Easton, Pennsylvania. Various tackle designs produced by this regional bonding system spawned some of the most interesting rods and reels ever built in the New World.

The beautifully delicate surviving tackle of these gunsmiths and the original New York makers explains the rapid growth of American tacklemaking—an industry that had been controlled almost completely by the British only two decades before. For the remainder of the century British tackle would remain encumbered with heavy fittings and bayonet ferrules.

Nothing illustrates the English way of making tackle better than the search for an effective split-bamboo rod. Born in Britain, the bamboo shaft would become stuck in a time warp that finally led to the futuristic Scottish expression, "I doon't think wee can take much mour o' this, Captain Kirk!"

The advent of the split-bamboo rod was perhaps the most romantic and controversial occasion since the birth of (insert favored deity here). From the British Isles, the split-bamboo idea journeyed to America, where it was eventually refined by enter-

A J. C. Conroy & Company fly reel with a raised click housing. With its offset foot, the reel is typical of those built during the Civil War era.

prising smiths. The first rudimentary cutting of the culm probably occurred in the late 1760s, when Britisher Robert Clive stabilized trading conditions in India. Thereafter, Calcutta cane, actually a member of the grass family, was exported to England by the East India Company. In turn, the Indian product was imported into America from the British Isles via New York packet ships. After 1834 large American vessels ventured to China and returned with Tonkin cane, a stronger species of bamboo first incorporated into a rod in 1869.

As the leading American tackle producers of their time, John Conroy and his son started to import "east India cane" (the Calcutta strain) in the spring of 1845, as advertised in Porter's *Spirit of the Times*. The Conroy import was sold to the trade in large bundles, and although the stuff could have been used straight, this was also the year that Samuel Phillippe started to build his split-bamboo rod tips. No doubt Conroy and other New York makers were also fooling around with cane as an alternative tip material, just as the contemporary British were doing.

Wooden tips were very prone to breakage. In the beginning of the 17th century the first 6 to 18 inches of rod tips were often fashioned from "whalebone" (actually whale baleen). Whalebone was more elastic than wooden tips, and achieved some popularity with fly casters. Baleen was a common by-product of the whale fishery and adopted to various uses, including coach whips and ladies' girdle busks. Every seaport from Liverpool to Boston had a dealer in whalebone, and the English used this material for fly-rod tops well into the 19th century.[1] Perhaps some American makers did likewise, since whalebone was very easy to procure.

Another method of improving the tip section involved the rending and gluing of lancewood strips. This rather obtuse idea was espoused as late as 1859, by author "Frank Forester" (Henry William Herbert), who also liked British rods built in "the country." Both materials, whale baleen and rent

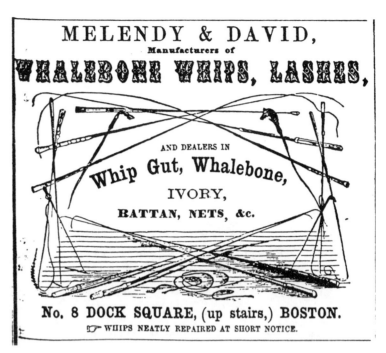

Whalebone, featured in this 1857 ad, was used for refined rod tips from Charles Cotton's time until the emergence of split bamboo. Whalebone showed up again—in the form of a wood-whalebone laminate—in Thomas Tout's September 26, 1871, rod patent.

lancewood, were unsatisfactory solutions to the early fly fisher's nemesis: the broken tip.

Around 1800 pioneering rodmakers canned whalebone and begain experimenting with split bamboo. The earliest rods had a short Calcutta cane top section made from two or three glued bamboo strips spliced to a tip's lower wooden shaft. For the next four decades costly spliced-in "rented" tips increased in length until the section's lower wooden area was eliminated entirely. This is a handy collector's yardstick for determining the age of pre-1840 spliced-tip fly rods.

The Ministry of Silly Bamboo

The original idea of splitting and gluing bamboo tips can be found if we return to England in the year 1801, when a rodmaker named Charles Snart made the outrageous claim that India cane would become the wave of the rodmaking future.

Following Snart's lead, bamboo strips started showing up on rods, often in the silliest of combinations—two-strip rods and strips glued together pith side out being two of the sillier. Almost four decades would pass before a moderately successful four-strip rod would be produced. Such august tackle firms as Ustonson's and Giles Little's had ample opportunity to beat the Yanks to the split-bamboo punch, but just couldn't seem to come up with the right formula.

During their 40 years of wandering through the bamboo wilderness, the British makers must have tried some exotic approaches to the rending and gluing of their strips. British angler-author Edward Fitzgibbon ("Ephemera"), in his 1844 Handbook of Angling, recommended the rent-and-glued rod. In the second edition of this work, Fitzgibbon revised his position, stating that the design was good for salmon-rod tips only: "I have changed my opinion with respect to rods made entirely of rent cane or any other wood rent. Their defects will always more than counterbalance their merits." Obviously, the horse-hoof and vegetable glues of the time were unrending.

In 1851 a rodmaker named Irwin repaired one of the Earl of Craven's split-bamboo rods, an example then 20 years old made by "Usterson [Maria Ustonson] Temple Bar." Giles Little of 15 Fetter Lane stated that, in all of England, only three or four makers could successfully create a split-bamboo rod, yet in the same year a slew of such instruments was exhibited in the Great Crystal Palace Exposition in London. Ainge & Aldred won top honors with a three-strip rod, evidently with all sections made of split cane. Most items exhibited had rent-and-glued tips only. Rods by other makers included those by Ustonson & Peters (1847-55), J. Bernard & Son, and J. K. Farlow.

The 1856 Bohn edition of Walton's Compleat Angler noted that "the split or glued up rod is difficult to make well and very expensive. It is made of three pieces of split cane, which some say should have the bark (enamel) inside, some outside nicely rounded." The British bamboo-tipped rod appeared as early as the 1830s, yet by the mid-1850s it was still a three-stripper, and the builders were still vacillating about where to place the enameled side of the cane.

THOMAS ALDRED'S CORRECT THREE STRIPS

In defense of Thomas Aldred, of the tackle firm Ainge & Aldred, we note that William Mitchell, an English expatriate living in New York City, wrote in an 1883 edition of American Angler that "Thomas Aldred of London, claims, and I have never seen it disputed, to be the inventor of the three-section [strip] glued up bamboo rod. It was . . . previous to the Crystal Palace exhibition in 1851."[2] I agree with Mitchell: Aldred probably produced the first complete three-stripper, and in so doing became preeminent on both sides of the pond. There is a good chance that his early work of the late 1840s included the "correct" method of fashioning the tips—enamel outside, which formed a stronger rented shaft. This success gave him the confidence to make additional rod sections in the three-strip design. As far as we know, Aldred never took the next logical step to a four-strip layout.

My own search for a three-strip rod, either whole or tips only, finally ended at the American Museum of Fly Fishing and its 12-foot Furman rod. This ancient piece, which must have been examined by many researchers, was apparently cataloged as a totally wooden rod, yet close examination reveals it was actually built with three-strip Calcutta tips. The fixed reel-seat ring inscribed "G. C. Furman, New York, 1832" makes it the oldest known rod in the museum (as well the New World). All the lower sections of this five-piecer were built from various woods stained very dark, starting with a butt fashioned from either maple or beech, moving up to hickory and lancewood on the second through fourth sections.

The elusive three-strip tips, all four of them, were a real rig, each being spliced into an 8-inch lower area of lancewood. These 75 percent split-bamboo tips were longer than the rest of the rod, which was common for the period. A multitude of fine intermediate wraps bound the delicate rents; this departure from the rest of the rod's construction was the real tip-off for me. Strange that nobody had noticed them before. The high order of workmanship needed in making tips so fragile that they tapered to mere thousandths of an inch was most impressive.

The Furman rod tips also appeared to have been spliced to remove the bamboo's nodes, standard

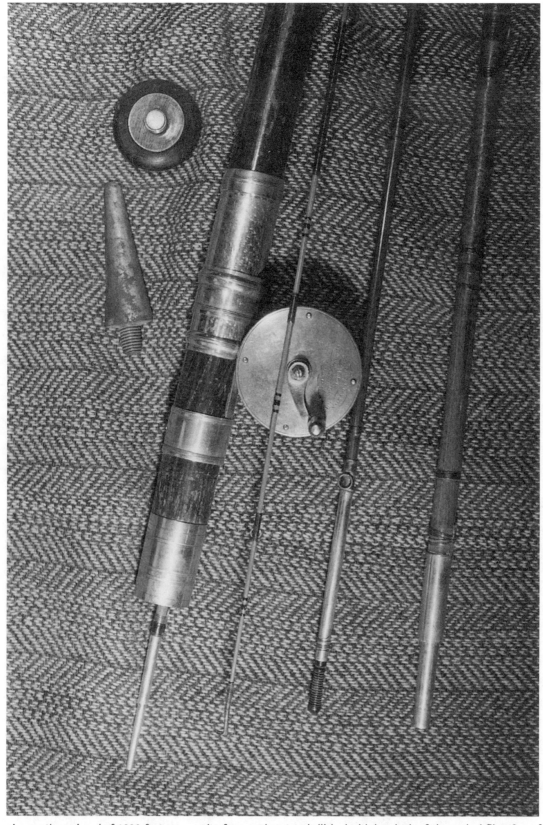

The expensive gentleman's rod of 1830 features an ultrafancy reel seat embellished with hundreds of tiny etched flies. One of the oldest fly rods found to date in America, it also sports German silver-covered dowel ferrules; expensive 12-inch, two-strip bamboo sections are spliced into the wooden tips. The extra two-strip tip is stored in the hollow butt section. The rod originally came with an iron spear. The scripted, curved crank reel was made by John Bernard of London.

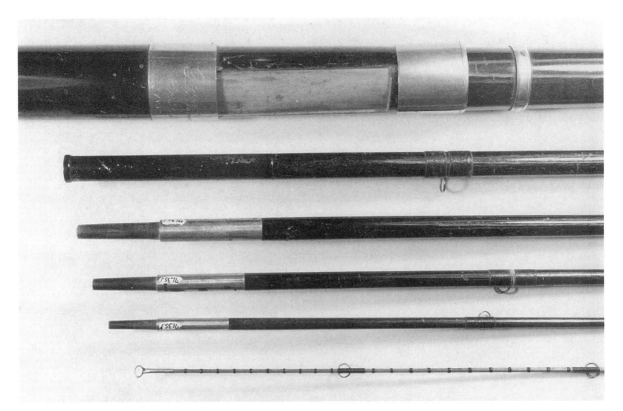

Engraved "1832," the Furman rod is the earliest dated, spliced split-bamboo article known to have been used by an American sportsman. Notice the delicacy of the partial three-strip bamboo tip joint—especially compared to the size of the butt section. (The American Museum of Fly Fishing Collection)

practice until around the 1840s. I couldn't be certain of this, but no nodes were showing and the wraps were lengthy—as long as the splice wrap where bamboo met wood. Surely these marvelous slivers had been fashioned by someone as talented as Thomas Aldred, yet these *were* built by another maker before Aldred's time. And, most important, they were the missing link between the millennia of wood rods and our less than two centuries of split bamboo.

The Four-Strip Rod in America

No one knows for sure when the idea came along of employing four strips of cane instead of three, but it appears to have been an American phenomenon. Three-strip tip sections certainly came to the Americas by 1832, and 20 years later William Mitchell repaired and altered a split-bamboo rod for James Stevens of Hoboken, New Jersey, that had been made by William Blacker, 54 Dean Street, Soho,

London (also a noted fly tyer). The following year the magnificent Thomas Aldred three-strip rod, having been the angler's darling of the Crystal Palace, was displayed at the New York Exposition. Mitchell pointed out that, in earlier times, three-strip sections had been built by removing the nodes and splicing along each strip, probably with a staggered pattern. The rods shown at the Crystal Palace and the New York Exposition were built with three strips to length, with nodes left on and probably filed before wrapping.

Although the next improvement, the four-strip tip, may have been British in origin, we can only go by the evidence of surviving examples and documented testimony, which indicates that the additional strip was added by Samuel Phillippe, Robert Welch, or a yet anonymous maker. Of the two known possibilities, Phillippe is the more likely candidate. Various sources, including Mitchell, Doctor James Henshall, and Charles F. Murphy, credit him with the four-strip tip and midsection.

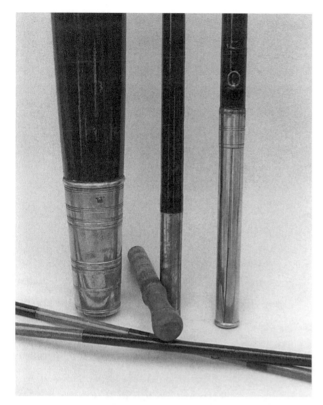

The Mack rod exhibits the diminutive butt cap and plain male dowel favored by American makers. The four-strip split-bamboo tips were spliced after the nodes were removed.

Other pre–Civil War rodmakers are known to have built four-strip tips as a standard angler's option. By 1865, when Thaddeus Norris wrote *The American Angler's Book*, he stated that "the butt of a fly-rod should be of well-seasoned white ash, the middle joint of ironwood, and the tip of quartered and spliced bamboo." Norris then mentioned salmon rods; "the tip of the best Malacca cane, rent and glued." Uncle Thad was not alone in his opinions.

Genio C. Scott listed his salmon-rod preferences in *Fishing In American Waters*, published in 1869. "An excellent salmon-rod is made with hickory butt, next joint of ash, a third of lancewood, and top of split bamboo. . . ." He then listed two salmon-rod styles built entirely from wood, those of Martin Kelly of Ireland ("preferable to any that I have seen except the split bamboo") and the traditional Celtic "Castle Connell" spliced rod. Then Scott stated, "Mr. Johnson of Boston, makes an excellent salmon rod, and so does Robert Welch, of New York." We may gather that either Welch or Johnson was making a salmon

rod with the "top of split bamboo"; perhaps they both were.

The copyright date of his book and the author's copious knowledge of rod trends make Scott very important to collectors. A columnist for Porter's *Spirit of the Times*, he was familiar with the best in tackle and was the only descriptive writer about rods prior to the tackle "boom" that arrived right after *American Waters* was published. Genio C. Scott wrote thus of his personal 12-foot, 2-inch trout wand: "A Robert Welch rod, of ash for the butt and second joint, lancewood for the third, and split bamboo for the fourth or top joint, is the best rod that I have ever owned for general fly fishing." The trout rods of this era built from such materials ranged in weight from 9 to 15 ounces.

Robert Welch was, as mentioned, believed to have been the son of Ben Welch and the probable successor to that Brooklyn tackle business. The identity of our rodmaker, "Mr. Johnson" of Boston, remains elusive, and attempts to find a Johnson listed as a rodmaker in the *Boston Business Register* from 1843 to 1868 have proved futile. There was a listing for A. J. Johnson, who ran a popular hardware and import stand. At this time, Boston's leading tackle dealer, Martin Bradford, carried the split-bamboo rods of William Henry Alfred & Son, London. The Alfred rods were built with the enamel on the inside.

At Faneuil Hall Square, William Read & Sons sold top-of-the-line rods made by Thomas Tout. The area's earliest known rodmaker, Tout lived in nearby Cambridge and may have supplied his handiwork to Johnson as well. Scott's contemporary, William H. H. Murray, wasn't impressed by Bradford's 1869 fly-rod offerings. He wrote, "If I was buying in Boston, for my rod I should go to Read's."

The Thomas Mack rod, an almost exact duplicate of Scott's preferred rod and constructed within this time frame, exhibits early four-strip tips. Although built for a Bostonian, it could have been made by any one of these three known American makers, though Samuel Phillippe is an unlikely candidate.

This salmon rod, engraved "Thos. Mack, Boston, Mass.," was built in four sections and marked "Heavy Salmon" in writing on the original bag. It is only 13 feet, 3 inches long, with the Welch progression of

WILLIAM READ & SONS,

13 Faneuil Hall Square, Boston,

DEALERS IN

FINE FISHING-RODS AND TACKLE,

KEEP CONSTANTLY ON HAND

The celebrated "Tout" make fine-spliced Bamboo Trout and Salmon Rods, for fly and bait fishing, of very best quality.

Medium Quality Fly Rods of Hornbeam, with spliced bamboo tips, in great variety.

Medium and Common Fly-bait, and General Rods of all prices and kinds.

ALSO, REELS, FLY-BOOKS, BASKETS, HOOKS, FLIES, in great variety, and every article in the line.

Adventures in The Wilderness, *Murray, 1869*

woods in the butt and mids; both tips are four-strip Calcutta bamboo, neatly wrapped with black thread at the guides and well over the nodes. Scott didn't care for the reel seat of this rod, with its metal inlaid sleeve to help prevent the reel from binding. To him, the style was not artistic and looked "gaudy."

The advent of the bamboo tip was an important step in fly-rod evolution, one that had a direct relationship to the complete split-bamboo rod of the not-too-distant future. A wooden rod built with a three- or four-strip tip can serve as a cornerstone of either a historic- or a bamboo-era collection, since it is the transitional piece between the two. These pivotal antiques are among the rarest, yet a number of such rods, such as the Furman piece or our Mr. Mack's salmon killer, must still exist, unrecognized as what they really are.

Four-strip tip construction would continue into the post–Civil War years; any tip so built is rounded, regardless of the number of strips. Good workmanship made these sections look like their solid-wood predecessors, and you may miss tight glue lines with a cursory inspection. On wooden rods, be suspicious

of any model that has intermediate wraps on the tips only. It could be a three- or four-stripper.

FOUR- AND SIX-STRIP TIPS AND MIDS

For a further look at the evolution of the American fly rod we need to follow up on the man who probably developed the four-strip tip. Born in Reading, Pennsylvania, on May 25, 1801, Samuel C. Phillippe moved to Easton in 1817–18 and apprenticed with gunmaker Peter Young. Around 1830 he established his own shop, as both a gunsmith and a violinmaker; one of his musical instruments won a medal at the Franklin Institute Fair in Philadelphia. A consummate craftsman, Phillippe also became a hookmaker, and with sometime fishing partner Phineus Kinsey of Easton he developed the "Kinsey" hook patterns. Other angling friends included Ebeneezer A. Green, George W. Stout, and Thad Norris, all of whom later made rods employing split bamboo. His son, Solon, followed Samuel's gunsmithing and rodmaking trades and became another of the clan to use the new material.

The elder Phillippe reportedly made his first rods in 1845, the earliest of which had three-strip tips, also known as British style. By 1848 the Pennsylvanian had produced versions with four-strip tips and mids. The example in the Pennsylvania Historical Museum in Harrisburg shows this style of construction, the butt section mortised of light and dark hardwoods.

Another early Phillippe rod, designed as a reel-above-the-hand model, became the pride and joy of Dr. James Henshall, the man who dragged every other early split-bamboo artisan through the muck to the favor of his rod's maker. The opinionated Henshall claimed that Phillippe was the "inventor" of the six-strip rod, yet described his personal 11-foot, 4-inch Phillippe model in enough detail to discount his own assertion: "It is made of four sections of bamboo, except the butt, which is of stained ash, and is neatly wrapped with black silk on the bamboo joints. The ferrules and reel bands are brass."[3] This Henshall-

This circa-1865 Thomas Mack "Boston" salmon rod was fashioned with exceptional German silver fittings, upper and lower mids, and a couple four-strip bamboo tip sections. It was probably built by Thomas Tout of Cambridge, the only local rodmaker at the time.

owned Phillippe is believed to be in the custodianship of the New York Angler's Club.

Whatever style of rod Phillippe made—most had butts crafted of various hardwoods—remaining examples are coveted treasures. There's no doubt that Samuel Phillippe influenced his acquaintances Norris, Green, and Murphy, and these gentlemen also built fly rods of exceptional quality. In turn, these three men swayed others as the four-to-six-strip cult spread. According to Solon Phillippe, his father built a rod with a six-strip mid and tip sometime prior to 1859. In that year the younger Phillippe made a similar rod in the six-strip style. At this time the elder Phillippe was selling his handiwork through Andrew Clerk's, the famous old New York tackle house. Samuel Phillippe's limited production appears to have ended prior to 1865, and on May 25, 1877, he died at his home in Easton.

Influenced by Phillippe, George Stout, also of Easton, once stated, "I came to this town in 1851. I made my first split bamboo [rod] in 1860, and got my idea from Phillippe's rods. I was an amateur only, and never made more than a dozen in all." So much for Stout, since we have nothing more to go on. Sam's angling friend, Thad Norris, was sufficiently impressed that he made a number of rods incorporating split bamboo and advertised them nationally.

Ebeneezer A. Green, most likely the "W. Green" remembered by Solon Phillippe, was a sporting gentleman of mechanical talent living in Newark, New Jersey. Solon obviously had trouble with names, referring to Bloomsbury, New Jersey, as "Bloomsburg," the spot that his father helped Norris select for the site of a new fish hatchery. Another Phillippe groupie, Green had the notion of making a rod entirely of split bamboo; around 1850 he produced such a piece, which evidently followed the strip style of rod used by Samuel Phillippe during their trouting forays.

Green's masterpieces, believed to be "a couple," were built entirely of wood, as found in the Angler's Club rod or in the later four-strip sections. Known examples also may have sported rattan-wrapped handles. A venerable 12-footer in the Jay Vergas collection, attributed to Green, has this grip configuration and was built entirely from four-strip sections. This Green rod, or one similar to it, encouraged another Newark craftsman, who would carry the ideas farther.

GEORGE BURGESS, NORRISTOWN, PENNSYLVANIA

On occasion, but not often, we collectors discover some great maker who quietly languished in the wings; and so George Burgess comes to us late, over a century after his passing. As another exponent of the emerging split bamboo fly rod, Burgess was an active tacklemaker from the 1850s until his death in 1882. An eastern Pennsylvanian and a contemporary of Samuel Phillippe, he incorporated some noteworthy ideas into his products.

I examined the first two Burgess fly rods retrieved: the original piece as found and researched by Steve Vernon in mid-1996, plus a second rod discovered a few months later by Harold "Doc" Herr. Both were built around 1860 and signed by the maker, establishing the rods as the oldest known collectibles built by an American split bamboo artisan.

Steve's rod is an 11½-footer, somewhat diminutive for the period, made with what appears to be a 4-strip ash butt section and a 2-strip lancewood mid. The tip sections are built nodeless, spliced from three pieces of 4-strip split bamboo. The 4 strips are of unequal width, 2 narrow ones opposing 2 wider ones, aligned to keep the glue lines from sheering while casting, and they are glued in the "center enamel" British style.

Although the handpiece is fashioned from walnut, it is covered by an intricate weave of linen cord—one of the earliest American examples of a cord-wrapped grip, an extension of Burgess's other vocation as a maker of buggy whips and riding crops.

The second rod is a robust 12-footer for larger trout. It has a solid ash butt and mid, with a single surviving 4-strip bamboo tip built in the same style as Vernon's example. The handpiece below the ash butt is turned from maple or birch.

Both of these historically important George Burgess fly rods have their wooden handpieces connected to the upper butt section by a splice hidden under a 2¼ inch metal band—the same type of construction as used by Phillippe in Easton. The rods are stamped, "BURGESS, MAKER, NORRISTOWN, PA.," in three lines, and they are packed in a round wooden form with brass reinforcing bands at each end.

With the appearance of a third marked rod by this maker (a bait model), Burgess is ranked as a major pioneer in split-bamboo construction. There is a similarity between George Burgess rods and early models sold through Krider's Sportsman's Depot in Philadelphia. Quite possibly, the Norristown craftsman may have built some of the rods marked to John Krider. Collectors will find Burgess in extremely good company—and he stands with Phillippe, Nor-

The Six-Strip Rod Emerges

Although no one can be sure who truly originated the six-strip rod, most collectors agree that C. F. Murphy probably built the first complete six-stripper and was the first maker to offer complete split-bamboo rods to the American market. With no reference to its maker, Genio Scott mentioned several times a rod built totally of bamboo, weighing 7 to 10 ounces, as if it were a commonplace trouting tool in 1869. Murphy versions have survived in reasonable numbers and, though expensive, are within reach of collectors. Because they are eminently important collectibles as well as historic fly-fishing tools, we shall describe the man, his production, and his rod features in detail. From this, you should be able to date and identify most remaining Murphys.

Charles F. Murphy was a wiry New York City Irishman of small stature, "a little given to brag of his

Forest and Stream, *December 7, 1876*

ris, and a very few others, as one of the first Americans to split the Calcutta culm.

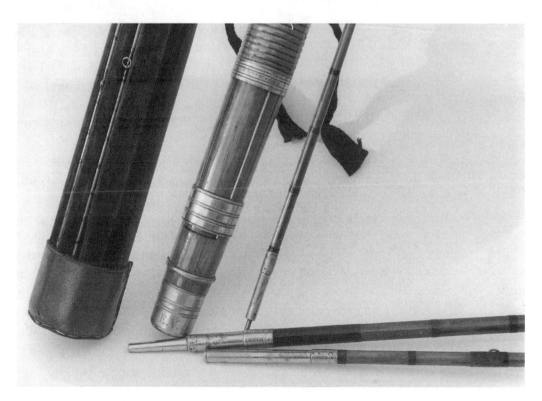

This early C. F. Murphy split-bamboo fly rod features the excellent craftsmanship of its Newark, New Jersey, maker. Its number of sections conforms to the "old formula": one butt, two mids, and three tips.

exploits," a professional fireman, and a gifted wood-worker. In the outdoor sports, Murphy was an avid wingshot and a noted fly caster. He moved to Newark, New Jersey, at some point prior to 1860, but often revisited Andrew Clerk's and J. C. Conroy's. Influenced by Samuel Phillippe rods, with their four-strip mids and tips (as sold through Clerk's), and encouraged in the endeavour by E. A. Green, Murphy built his first complete four-strip fly rod in 1863.

In his own words, "I soon found that four strips left too much pulp on the inside—for the strength is all in the enamel—and I [then] made rods of six and eight strips." The eight-strip rod was abandoned, but his complete six-strippers became stylized and were soon being promoted by such notable anglers as Francis Endicott, Genio C. Scott, and Robert B. Roosevelt. Murphy continued, "Two years later I made a salmon rod and Mr. Andrew Clerk took it to Scotland where it attracted much attention [no doubt]. Clerk then gave the rod to Genio C. Scott, who fished it up on the St. Lawrence and took some big pike and mascalonge."

We can pick up the story from Scott, who wrote the Canadian section to his *American Waters* while the thrill of using the Murphy salmon rod was still fresh. Writing in 1869, he understated the Scottish incident: "Dr. Clerk . . . carried with him a split-bamboo rod *made by their house*." Scott concluded his discourse on the Murphy rod by saying, "This is the *fourth season* it has been used, and, though it has . . . killed many salmon [upward of] thirty-five pounds each . . . it appears as good as new." (My italics.)

Going back to Murphy's own recollections, he stated, "Then, in 1866, I made a split-bamboo bait-rod for black bass, and arranged with the firm of Andrew Clerk & Co. to sell my rods, which they did for some years, and then they began to make them on a larger scale." I am inclined to take the Scott version as the more accurate; it places Murphy products—marketable trout, salmon, and bass rods—at Clerk's in the beginning of 1865. Fittingly, the January 21, 1865, issue of Wilkes's *Spirit of the Times* carried an Andrew Clerk & Company advertisement espousing "The EXCELSIOR FLY ROD, manufactured from split bamboo, stronger, lighter, and more elastic than any other."

MURPHY'S COMPLETE SIX-STRIP FLY ROD

Collectors should be aware that Murphy's early products had four-strip tips. Early 1863–65 survivors built in that style have unusual, tiny cupped areas in the shaft above the grip, first noticed by Martin Keane, author of *Classic Rods and Rodmakers*. Other rods have been found with staggered pinholes in this same region. The earliest six-strip Murphy tip sections were introduced at the end of his pinhole period and at about the time he began his association with Andrew Clerk.

Murphy's association with Clerk & Company lasted until 1875, when the old Maiden Lane tackle house replaced his superb craftsmanship with the work of H. L. Leonard, who had started his split-bamboo career, ironically enough, by examining a four-strip Murphy in 1870–71.[4] Remember also that Murphy's super-salmon rod, which so impressed Scott and the Scots, created a steady market for Leonard's Abbey & Imbrie salmon sticks in the British Isles. By March 1876 Murphy was on his own and advertising his wares in *Forest and Stream*. His "fine trout rods" sold for $35 each, and as always he was ready to "fill orders to a limited number." Just how limited? When active, Murphy probably finished a couple rods a month. Today, Murphy's hand-planed artifacts are appreciated by the few collectors lucky enough to own one. There aren't many complete Murphy rods kicking around, so if you have one, send it to me. The American Museum of Fly Fishing has four Murphys, so it doesn't need any more!

Judging from its taper, the oldest Murphy I've examined was quite early. Only the butt section was original, and it had the strange pinholes. At one time it was probably an 11-footer built in four sections. Another early Murphy example, in the North Wing of the East Boothbay Tackle Museum, is a 12-foot model with two six-strip mids and a six-strip butt. The three tip sections are of four-strip construction. The rod remains as one of the best examples of Charles Murphy's small but superb output. With precision German silver metalwork, this 1864–65 model was built in a classic early-19th-century configuration: one extra mid and two extra tips, similar to the wooden rods built by Conroy from 1840 to 1865. The original price of this rod would have been about

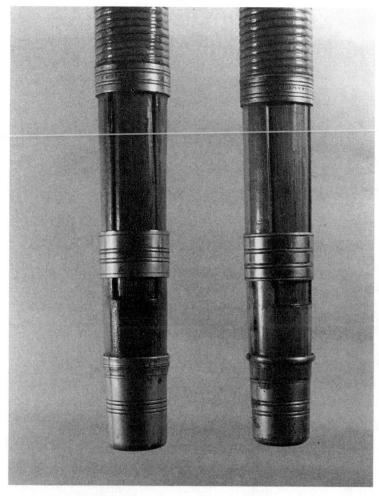

Two Murphy reel seats. The circa-1863 version (left) has a larger butt cap. The circa-1865 model (right) shows the standard style of Murphy's work.

rod over the wooden versions of his day. These were major accomplishments for a man whose products Leonard would refer to as being crude—yet were built while the Mainer was still stuffing squirrels into cute poses.

Dr. James A. Henshall also slighted Murphy. In the 1881 edition of *Book of the Black Bass*, Henshall cited the Newark maker as the first to build a split-bamboo bass and salmon rod (no mention of the number of strips). Henshall credited Samuel Phillippe as the originator of the split-bamboo rod (no mention of completeness or lack thereof), however, and claimed that Leonard was the first to put the six-strip model on the market. Of course, Henshall owned a Leonard—not to mention believing that handmade fishing line was better than machine-made stuff, and that rounded split bamboo was superior to hexagonal shafts.

Despite a contemporary lack of recognition, Murphy's lifetime production was impressive, judging from the number of existing rods. This is in contrast to the ultra-rare work of Phillippe, who made so few rods that they are now almost never seen. Quite a feat considering that the Newark craftsman worked alone. In a letter to Harry Buckley dated 1908, Charles F. Orvis wrote the following about Murphy: "I knew briefly the maker of the rods. He was a good hearted fellow but 'crooked his elbow' quite too much—he was a good mechanic . . . he was the first man that made split bamboo *for the trade* [Orvis's italics] but making only a few. He only made what he could make himself—would not have help."

Charles F. Murphy, perhaps never really appreciated in his somewhat humble lifetime, continued to build rods at 40 Halsey Street until his death. Fred Mather wrote about Murphy in his 1901 book, *My Angling Friends*: "Brought up by the fire laddies, where the only qualities recognized in a man were honesty, pluck and muscle, he was an entertaining companion. He died at his home in Newark, N. J., in 1883, at the age of sixty-five."

$60—a fantastic sum to pay for a fly rod in the Civil War era, when a good horse might sell for $30.

Murphy never really made a killing in the trade. Late in life, Fred Mather, who knew "Murph" personally, claimed that "he seemed to be unfortunate in usually meeting with an accident of some kind."5 T. S. Morrell, writing in the March 13, 1880, issue of the *Chicago Field*, stated that "Murphy is a very conscientious, careful workman . . . spending more time on his work than is profitable, and consequently is poor."

Indeed, Murphy tried to make his rods as light and delicate as his time in history would allow. The tip sections of the previously mentioned 12-footer were tapered down to 0.065 inch. Through his excellent workmanship, the Newark maker was instrumental in convincing American anglers to buy American and to prefer the complete split-bamboo

An 1853 advertisement for Philadelphia gunmaker John Krider. Within the decade, Krider would join the Split-Bamboo Clan and make a number of exceptional fly rods.

I estimate Murphy's output at about 400 rods, rods so good that they ensured the popularity of split bamboo and paved the way for the next logical step in the progression of the American tackle industry: machine production of the classic six-strip rod.

One of the design changes of the 1860s, perhaps not of Murphy origin but certainly seen on his rods, was the new "spike ferrule." Similar to the covered-dowel ferrule, but more delicate, the spike ferrule would be used by various makers for the next three decades. Other rod craftsmen, often home hobbyists, moved to the newer spike ferrules as well. Known tackle-house owners and pioneer bamboo rodmakers of the spike-ferrule era included J. B. Crook of New York and John Krider of Philadelphia.

THE TALENTS OF JOHN KRIDER

Of the several rodmakers influenced by Phillippe or Murphy, John Krider remains a relative unknown. Krider was a talented Philadelphia gunsmith, machinist, and author who was active by 1836, when he was first listed at 2 Walnut Street as a maker of guns and fishing tackle. In 1853, when his *Krider's Sporting Anecdotes* was published, he was running a

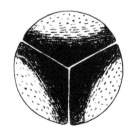 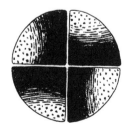 

From right to left—the common British or William Henry Alfred & Son method of placing bamboo enamel in a fly rod shaft; J. B. Crook's and John Krider's early configuration for a "center enamel" four-stripper; and the enamel-outside shaft as built by Phillippe, Green, and Murphy. The dark areas are the "power" fibers. Note the dramatic increase in power obtained by placing them on the outside of the shaft.

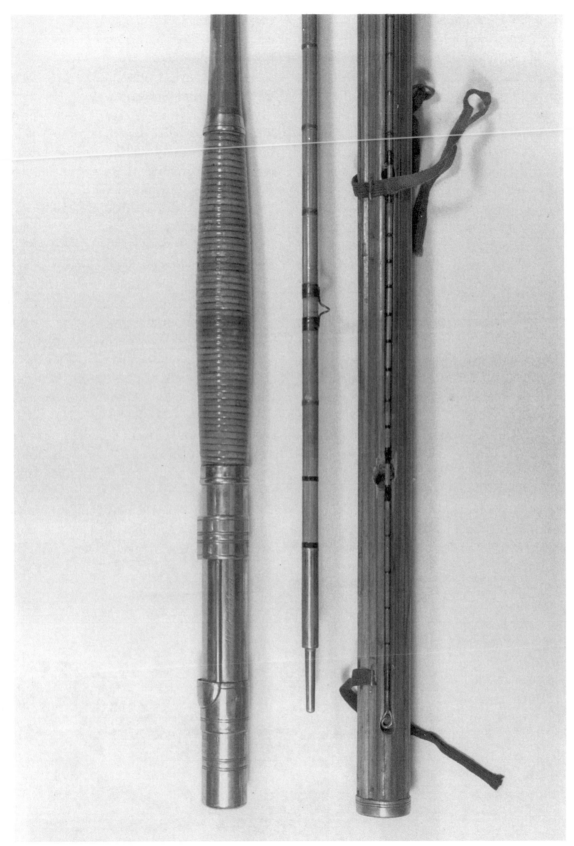

A John Krider rod, dated 1878, shows the rugged German silver reel fixture that followed the early open split-bamboo seat. This piece was built in the classic smith-age style, including a wooden form tipped with German silver. (The American Museum of Fly Fishing Collection)

popular retail-wholesale business at the northeast corner of Second and Walnut Streets, where he carried a good selection of imported fishing tackle and made guns and wooden rods. By the late 1860s, Krider was selling split-bamboo rods of his "own make," and the earliest model I've seen smacked of Murphy's influence. I have not found a signed wooden Krider rod, but there's good reason to believe that enough of them were made for one to appear someday.

His split-bamboo production, in four- and six-strip rods, is a different story; a number of examples are in collections. A John Krider 10-foot, 6-piece "trunk" rod, fashioned in about 1870–75, had male spikes similar to Newark ferrules, and the reel-seat area looked amazingly like the work of Charles Murphy. The rod also showed the workmanship of a man practiced in precision gunmaking; even the wooden form case exuded fine craftsmanship. Any sportsman who glanced into Krider's street-corner shop window would have been captivated by this beautiful casting instrument.

By 1877, Krider's shop was the popular "Sportsmen's Depot," where avid anglers met to mull over subjects ranging from artificial flies to artificial fish hatcheries. Selling everything from guns to dog biscuits, he also touted his newer "Krider's Celebrated Center Enamel Split and Glued Bamboo Rods." Possibly Krider built early rods with the bamboo aligned sideways, so that the enamel of each strip met in the center of the shaft, as advocated by contemporaneous maker James B. Crook of New York.[6] By 1878 his catalog listed about 20 wooden rods of American as well as British origin, along with "KRIDER'S CELEBRATED SPLICED BAMBOO RODS," dropping the mysterious phrase "center enamel." The basic late-1870s version, which sold for $30, was a three-piece rod with an extra tip. This same model with two mids and three tips sold for $40; a four-piece fly rod having the extra mid and tips was priced at an astronomical $60.

An example has surfaced of Krider's 1878 "basic" model, fitted into a wonderful old German silver-capped wooden form, that gives us an idea of what these rods looked like. By this time the shaft area is a bit more defined; the reel seat is rather heavy and of full metal construction, similar to those found on the rods of other gunsmiths, such as Wheeler and Leonard. The rod is wrapped in red silk, much like those of Murphy and Leonard, and its workmanship is on a par with their rods. Like the earlier John Krider rod, this one was built rounded, in a six-strip Calcutta style. Krider also experimented with other configurations, including some 10-strip "splices."

I rate Krider's split-bamboo rods very highly and feel that he was one of the first successful designers of the emerging split-bamboo rod. Known examples are marked "John Krider, Maker, Philadelphia." Unlike the products of Murphy and Leonard, these rods were marketed by the maker only, and there's a good chance that Krider's total bench-built output may carry his mark. Collectors will find Krider rods appreciating in value even when additional examples come forward for sale.

There was a real connection between Samuel Phillippe's work and the subsequent examples by Green, Murphy, and Krider. Solon Phillippe mentioned that Charles Murphy had visited his father's

A THING OF BEAUTY WITHOUT ALLOY.

THE BROOK TROUT.—*Salmo fontinalis.*

Fishing in American Waters, G. Scott, 1869

shop in Easton, "to learn something of his method of making split-bamboo rods." As a contemporaneous gunsmith and angler, Krider may have done the same, or he could have been influenced by Murphy's rods alone. The aesthetic styles of all five makers were very similar, and Phillippe gets the nod as the Sagamore of the Split-Bamboo Clan. The rods built within the Easton gunmaker's parameter were not like the products of Conroy and the New York makers. Instead, they reflected a charm all their own.

The early split-bamboo rods, as built by the Phillippes, Mr. Green, the "Fire Laddie," and Krider, gained popularity on the "woods highway," as sportsman used the new shafts in their search for native brook trout from the streams of Pennsylvania through the Adirondack lakes of New York; in Vermont's tumbling Green Mountain brooks, across the White Mountain waters of New Hampshire, and, finally, to the Maine frontier. The Big Trout Mother Lode was discovered in the Rangeley Lakes, and the bamboo rod was there in 1868. In the space of three to four years we have come from the meager output of a few craftsman to the beginnings of the industrial revolution in split-bamboo tackle—a great testimony to the influence of Murphy and the Phillippes.

Notes

[1]South, Theophilus. *The Fly Fisher's Textbook.* London: R. Ackerman, 1841.

[2]*American Angler.* May 19, 1883.

[3]*Outing.* May 1902.

[4]Keane, Martin J. *Classic Rods and Rodmakers.* New York: Winchester Press, 1976.

[5]Mather, Fred. *My Angling Friends.* New York: Forest and Stream, 1901.

[6]Kelly, Mary Kefover. *U.S. Fishing Rod Patents and Other Tackle.* Plano, Texas: Thomas B. Reel Company, 1990.

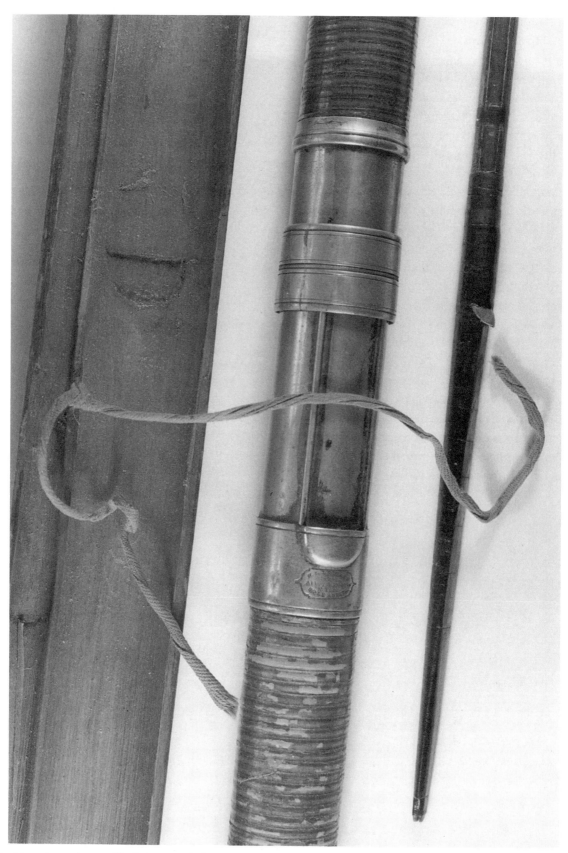

A spliced two-piece salmon rod by H. L. Leonard, showing features commonly found on wooden Maritime models built in Saint John, New Brunswick, by Dingee Scribner. This is one of the earliest production rods, marked by Abbey & Imbrie, and made from rounded six-strip Calcutta cane.

3 Gunsmith Tacklemakers of Maine

The popularity of the split-bamboo rod and of fly fishing increased markedly after the close of the Civil War, as veterans sought the solitude of woods and streams. The postwar revolution in transportation opened large tracts of wilderness to sportsmen, particularly the narrow-gauge railroads then extending into the Maine wilderness. Tacklehouse owners Charles Imbrie and Martin Bradford were also key players in the growth of the new tackle industry. Through their patronage, the country's earliest bamboo craftsmen found an entrée to the sporting market. I've already mentioned the Phillippes' and Charles Murphy's connection to Clerk's, but budding Maine talent also discovered willing wholesale buyers in Boston and New York.

For the most part, the cutting-edge rods built during the waning years of the 1860s were four-strippers that differed from those built during the past decade in only one way: They were made entirely of split bamboo; wood was phased out of the quality rod market. The strictly American rod had come into its own. Between 1865 and 1868, at least 70 Murphy six-strippers were in use (a figure tabulated from his peak production of two rods per month). Add to these the work of the Easton gunsmiths, and it's all but certain that the six-strip rod hooked a few of the largest brook trout in one of the period's most popular angling areas—the Rangeley Lakes region of northwestern Maine.

During these last years of the 1860s the split-bamboo craft spread from its original home between the Delaware and Hudson Rivers to northern New England. Local craftsmen—gunsmiths, patternmakers,

and machinists—were captivated by the new rods, and set to work learning to build them.

The New England business climate was on a postwar roll, entertaining not only the old notion of manufacturing but also the new one of production. Yankee entrepreneurs found ways to crank out wares with speed and accuracy. Imagine a pair of shoes that actually *were* a pair, buggy whips that whipped, and axes that could hack down a cord of timber before they needed sharpening.

These inventive Yankee industrialists stole ideas from our old friends Watt and Nasmyth, building machine upon machine to split and sew and strap and slam. As the northern machine age expanded, New England businessmen such as Isaac Cutler found more time for recreation, including angling.

The First Split Cane in Maine

George Shepard Page, a New Yorker and one of the founders of the Oquossoc Angling Association, owned and used a Murphy fly rod on the Rangeleys. The rod must have impressed Cutler, because he wondered if a pair of Farmington gunsmiths down on the Sandy River could construct something similar.

Farmington was a pastoral community below the Rangeley Lakes—a notable town, the Franklin County seat, and the residence of gunsmiths A. G. Wheeler & Son. Albert Galletin Wheeler, originally a machinist-gunsmith from Chesterville, moved back to his native Farmington to increase his trade,

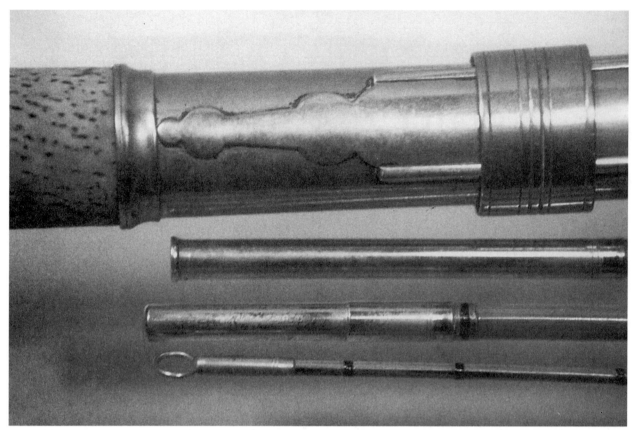

This fancy German silver overlay on an early C. E. Wheeler reel seat is reminiscent of the gunmaker's patch box cover, found on contemporaneous rifles and fowling pieces.

which was listed in the *Maine Business Register* from 1856 until his death in 1883. One of the state's top gunsmiths, he trained a generation in the art of precision manufacture. Most of his fowling pieces and rifles were marked "Wheeler & Lawrence," "Wheeler & Stevens," "A. G. Wheeler & Co.," or "A. G. Wheeler & Son," all with a "Farmington, Me." address.

The "Son," born Charles E. Wheeler on July 13, 1847, produced a number of fowling pieces marked "C. E. Wheeler, Farmington, Me.," but his real interests were in music and angling. Upon hearing Cutler's request, the younger Wheeler agreed to make a six-strip rod that could handle the big Rangeley trout. Cutler was so pleased with Wheeler's split-Calcutta creation that he encouraged the young man to make more on speculation.

This event took place in 1868, establishing Wheeler as the first documented maker of the new style of fly rod east of the Hudson. He was not Maine's first rodmaker, however. That honor probably goes to William Neal, a Bangor smith.

Wheeler moved from guns to the long wand and became the only major rodmaker to confine his total production to within the borders of the Pine Tree State. His business grew on both the wholesale and the retail levels, and he started to sell general tackle as well.

In 1876 Wheeler submitted samples of his work to the International Exposition at Philadelphia and was awarded a Medal and Diploma of Excellence "for the manufacture of split bamboo fishing rods." That same year, a June 22 advertisement in the *Farmington Chronicle* revealed a substantial business: "Chas. E. Wheeler, Manufacturer & Dealer in Fishing Tackle, Fishing Rods, Fish Hooks." The ad also stated, "Any kind of rod made to order at short notice. Split Bamboo Trout and Salmon Rods a specialty." At this time, Wheeler's operation was located at "Broadway"—actually on the corner of High Street and Broadway, right next to a music hall, where the rodmaker served as bandleader. The historic old Wheeler rod shop still stands as one of the few origi-

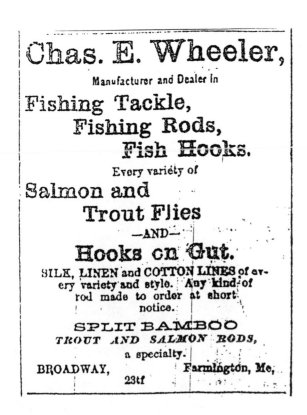

Farmington Chronicle, *June 22, 1876*

nal wooden structures remaining in downtown Farmington.

For many years collectors knew little of Wheeler; perhaps his name and products were confused with those of "W. R. Wheeler," a minor Connecticut maker. The problem was the absence of a maker's address. The first Wheeler rod to appear at auction was marked, "C. E. Wheeler, Maker," with no address. This beautiful 10½-foot early fly rod was built with two mids and two tips, and had a cork-covered handle, loose-ring guides, and "fine silver-soldered German silver fixtures." It appeared at Richard Oliver's premier Antique & Classic Fishing Tackle Auction in July 1985, and sold for $180—a bargain.

Early Wheeler rods can now be distinguished by their soldered-rail reel seats and seamed spike ferrules. They usually have the aforementioned sheet-cork grip, considered an improvement on the rattan grip of the previous decade. A similar 10½-foot C. E. Wheeler fly rod in the Maine State Museum collection exhibits the furnishings of the 1870s; the shaft is six-sided, not round like most rods of the period. After 1880, soldered-rail reel seats were maintained, but the ferrules were changed to a simple one-piece capped style, possibly purchased from Thomas H. Chubb, but most likely hot-swaged right at the Wheeler shop.

Rods built during the first two decades of manufacture were put up in canvas-covered fitted forms. The forms' end caps were made from German silver, round like a "Chinese hat," and were distinctively Wheeler creations. I believe these may have been Chubb forms modified with Wheeler's caps.

By the mid-1880s the shop employed six to eight "competent and careful assistants." The two-story, 25-by-40-foot manufactory was equipped with the "latest improved machinery," designed by Wheeler himself and operated by steam "as a motive power." The rods were described as beautifully finished and richly ornamented, and some retailed for $50. With annual sales exceeding $5,000 in the wholesale trade, it seems fairly obvious that the bulk of C. E. Wheeler's output was marked with the retailer's name rather

Schoverling, Daly & Gales ad in the April 12, 1888, Forest and Stream

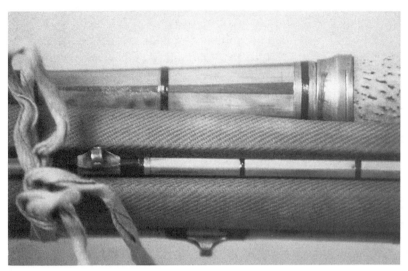

A circa-1880 Wheeler mortised rod, wrapped with dark brown and tan silk, and featuring hand-fashioned salmon guides and a sheet-cork grip.

velvet cases, which retained the German silver Chinese-hat-shaped ends. Throughout his 49-year rodmaking span, Wheeler usually added little special touches to his rods, such as custom-soldered guides and tops or overlaid German silver reel seats. Some top-of-the-line models had six additional tapered, mortised cedar strips spliced into the grip area.

As far as I know, all of C. E. Wheeler's creations were six-strippers, with a conservative production estimate being about 5,500 rods during his most active years, between 1875 and 1915. All known examples are also hexagonal. Was Wheeler the first to deviate from the round rod? Lots of other questions remain unanswered. Was the Wheeler beveler the first such machine in the country? Did he employ one of the first "drop hammers" in the biz? Maybe. One thing is sure. We can identify a Wheeler rod if it's still in the original covered form, regardless of what retailer's name may be stamped on the rod itself. The Chinese-hat end caps were a unique C. E. Wheeler trademark.

than the maker's. Although we know that tackle houses in Boston, New York, Cincinnati, and Chicago sold Wheeler rods, we have few identified outlets. In 1887 New York's John P. Moore's Sons were sole agents for "Acme split bamboo rods manufactured by Charles E. Wheeler." The next year Moore's successors, Shoverling, Daly & Gales, continued to carry the Farmington products. In Chicago, the Wheeler outlet was Joseph Van Uxem, a large establishment on Randolph Street.

Early rods marked by the maker had a rectangular stamp framing the words "C. E. Wheeler, Maker." Later rods, built from the 1890s until 1916, had a stamped, raised rail seat, large fancy grip checks, and ½-inch cork ring grips. Some rods carried an additional stamp that read "Farmington, Me." in one line. These later models were put up in blue

As interest in antique fly tackle increases, these early Maine rods will appreciate in value, especially those few in pristine condition. A Wheeler rod from the 1860s has not yet surfaced, but a few very early models may well survive. Charles E. Wheeler helped

As this advertisement from the 1898–99 Franklin County Directory shows, Charles Wheeler manufactured more than just rods. His Patent Net Staff and Bow is a great collectible today.

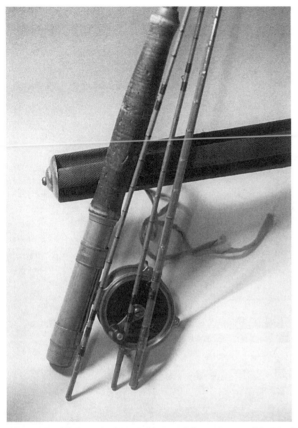

A later Wheeler fly rod, this 8½-foot model was "put up" in his distinctive fitted form, which featured Chinese-hat-shaped German silver end caps.

the first Registered Maine Guide, always used a Charles Wheeler rod for her exhibition casting at David T. Abercrombie's Camp in New York. And without doubt the Wheeler rod was a mainstay on the Rangeley and Kennebago Lakes.

When the April 21, 1916, issue of the *Franklin Journal* hit the parlors of Farmington, the townsfolk learned that the beloved leader of Wheeler's Band had finally succumbed to tuberculosis. The headline also mentioned that the late Charles E. Wheeler had been a "Pioneer in [the] Bamboo Fishing Rod Industry." Any collector lucky enough to find a complete Farmington rod can be assured that it is a treasured part of our fly-fishing heritage, built by one of the first American makers of the six-strip cane rod.

GILBERT BAILEY'S PATENT REEL SEAT

Another Maine pioneer in the tackle industry, Gilbert L. Bailey, chose a slightly different tack. His obituary described him as "one of the most ingenious mechanics who ever lived in Portland," as well as being "the inventor of almost countless articles of more or less value." The son of a farmer and an avid trout fisherman, Bailey was born in West Newbury, Massachusetts, in 1824. By the end of the 1840s the young man had moved to Maine and established his Portland gunsmithing-tackle business.

Bailey retired from gunmaking early on, although he always did repair work for valued customers. His distinctive hobby, "inventing," led to more than 100 improvements in the way of doing things in 19th-century America. Without the aid of

move the business of rodmaking from the smith age into the machine age, and in so doing takes a catbird seat alongside Hiram Leonard and Captain Chubb.

Soft-spoken and somewhat frail, Wheeler had many customers who become close friends. "Flyrod" Crosby, the Annie Oakley of the Maine woods and

1874 Maine Business Directory

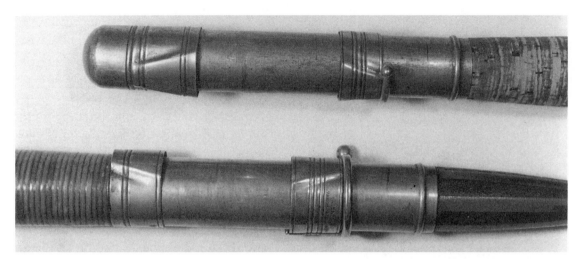

Two T. J. Conroy–marked rods carrying the Gilbert Bailey reel seat. The top rod has a grip made from thin ⅛-inch cork rings, indicating post-1890 construction. The lower rod is a mortised model with a rattan grip, made by Frederick Malleson. (The American Museum of Fly Fishing Collection)

an attorney, Bailey received 21 U.S. patents for such widely varied gizmos as the "first automatic door check," "a dumbell for exercising," "the curtain roller," "a buckle of a new type," and, my favorite, the infamous "ice creepers" that keep today's ice fishermen from falling on their butts.

It was no coincidence that so many of Bailey's inventions "of more or less value" related to angling. Gilbert's "split shot Trout Sinkers," to quote from his obituary, were made prior to 1874 and remain in anglers' vests to this day. Bailey also developed "a new form of landing net," "an improved fishing reel which is still in general use," and "a reel strap for fishing rods."

This last item, the reel strap, is most intriguing, since it has mysteriously shown up on a number of rods sold through T. J. Conroy's in New York. Patented on March 4, 1884, Gilbert Bailey's reel seat featured a swinging locking bar that snapped over a sliding, tapered band to hold the reel foot in place. Quite a piece of work and built for a number of years, it was featured in J. Harrington Keene's tackle section of George Shields's *American Game Fishes* in 1892 as an improvement over Chubb's design. Keene said, "American ingenuity comes to the rescue again."

You may today find the Bailey reel seat on some rods carrying the 1884 patent date and marked "T. J. Conroy, Maker, N.Y." One Conroy piece is a hexagonal six-strip light trout rod, mortised with cedar and featuring a rattan grip below the Bailey seat. Another

example, a heavier fly rod, has rounded four-strip construction and a grip made with narrow, ⅛-inch cork rings. Although the maker of the seats and accompanying rods is unknown, likely candidates would be Fred Malleson or the Varney Brothers. The seats are swaged and the ferrules serrated and drawn, appearing to be of Fred Malleson origin. The shafts look like Malleson's work. The total package exudes quality—gleaming patent reel seats, mortised shafts, fine red wraps, and quality ferrules.

Recently, I purchased one of these beautiful maverick Bailey-Conroy models, a delicate 10-footer built for a 4-weight line. It illustrates the interconnection of makers, dealers, and wholesalers in the tackle trade. Obviously a cooperative effort,

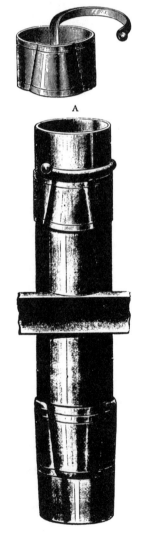

American Game Fishes, Shields, 1892

the rod arrived on the market during the height of the gimmicky "patent era" (1875–1900) that paralleled angling's expansion age.

Gilbert L. Bailey continued as an important eastern tackle dealer-jobber, remaining on Portland's Middle Street from the year he patented his unique seat until 1902. He died two years later at the age of "79 years, 11 months." His only long-term Maine rivals in tackle dealings were the Ramsdell brothers, who lived farther east in Bangor, and C. F. Nason of Lewiston.

CHARLES NASON'S LANDING NET

Before leaving western Maine, we'll travel to the city of Lewiston, on the eastern side of the famed

carried Dame, Stoddard & Kendall, Orvis, and Chubb products. The latter company's rod trimmings were standard on local-built rods, as made by Jim Daniels who, like Nason, lived just across the river in Auburn. The connection to Orvis and Dame, Stoddard & Kendall can be traced to a landing net sold by both companies in the 1880s. Nason's 1883 advertisement in the *Maine Business Register* listed him as a "Gun Maker and dealer in Fishing and Shooting Tackle." It also mentioned that he was the "Patentee and Manufacturer of Nason's Flexible Bow Net Rings."

On August 31, 1875, Charles Nason received a patent for a new style of landing net designed for portability. A search of U.S. patents in the "Nets and

Agent for Dupont's Smokeless Powder.

CHAS. F. NASON,

⟶⟶◦ GUNSMITH ◦⟵⟵

——AND DEALER IN——

Guns, Ammunition, Fishing Tackle, and General Sporting Goods.

14 Main Street,   Lewiston, Me.

Androscoggin River. Here a contemporary of Gilbert Bailey began a gunsmithing-tackle business in 1860 that continued for the next 40 years. Originally, Charles F. Nason was located on Lower Main Street, where he conducted trade until after 1883. He then moved to 3 Main Street for a short time, and finally settled in at 14 Main, where he remained for the rest of his tenure. He made exquisite half-stocked target rifles, marked "C. F. NASON, LEWISTON, ME.," punctuated on each side by an eagle logo. I mention the stamping because Nason also marked a number of angling items. He was a superb shootist and in 1878, at the Androscoggin Club, topped the score of Mr. Ira A. Paine, the "champion shot of the world."

As a Lewiston tackle merchant, Nason apparently had Vermont connections. I believe that Nason

Pounds" category reveals that 39 inventors were granted various rights within the genre, going back to 1795. All were commercial stop seines, fyke nets, and so on, with the exception of Charles De Saxe's "landing net" of 1854. De Saxe's net may have been the first of the collapsible breed, but Nason's was the first one that worked—a spring-steel net produced in quantity by later makers, who generally accepted the credit for it. Collectors recognize the evolved models as produced or sold by Orvis, Harrimac, Meisselbach, and Richardson.

The spring-steel idea was natural to a gunmaker, since the material was used in various firearm parts, most notably the trigger spring. Along with conducting a brisk tackle trade, Nason built his landing nets for more than a decade, until he bowed out of the

The Nason patent landing net, produced from 1875 until the maker's death, was in 1887 wholesaled exclusively through Dame, Stoddard & Kendall. A later device, with the ring modified into a teardrop shape, carried Charles Nason's idea into the 1920s. From Forest and Stream, May 26, 1887.

business in 1887. His successor, John B. Littlefield, continued their production, advertising "Nason's Patent Net Rings" in the Maine and Androscoggin County directories. In the spring of the next year, one of Littlefield's customers shot himself in the side while toying with a loaded revolver and died from the wound. The negative publicity destroyed John Littlefield's retail career.[1]

167,189. Landing Nets. Chas. F. Nason, Auburn, Me. [Filed July 30, 1875.]
Brief.—The flexible hoop is easily detached and packed with the handle and fishing-rod.

This brought Nason out of retirement, and once again he conducted a lively and safe business at the "old stand" until 1901, when he retired for good and moved to a farm in Greene. He died there on October 25, 1909.

Nason's Patent Net Ring was produced for 25 years, and examples surfaced at auctions in 1989 and 1995. Marked by the maker, the net's flexible, one-piece rim straightened and slipped into its handle, forming a unique "walking cane." The Orvis connection is an identical net advertised in Charles Orvis's 1884 catalog. This Orvis net was either an extremely coincidental patent infringement or, more likely, Nason's product. The Charles Nason spring-steel net hoop is the original model from whence all others sprang and is a highly collectible item.

The Bangor Gunsmiths

In eastern Maine, the lumbering city of Bangor (*Ban-gore*, not *Banger*), built where Kenduskeag Stream enters the Penobscot, was the last outpost of civilization before Indian country—or, as local businessman-author Lorenzo Sabine characterized it, "the Eastern Frontier." Bangor had its own top regional gunsmiths. Contemporaries of the Wheelers in Farmington, Bailey in Portland, Leland of Augusta, and Nason in Lewiston, the Bangorians were adherents of the Multishot School, fashioning rifles designed to fire two or more shots in succession. The sage was William Neal, who may have been Maine's first rodmaker and had a shop at East End Kenduskeag Bridge from 1843 to 1853. His apprentices included such future notables as the Ramsdells and Hiram Leonard.

After Neal's death in October 1853, Charles V. Ramsdell and his brother, John W., at East Market Square, became Bangor's top gunmakers and tackle dealers and remained so from 1855 until the 1880s. In their latter years, the Ramsdells were assisted by Francis J. Philbrook and Hiram L. Leonard. Of the two

JOHN W. RAMSDELL,

Manufacturer of and Dealer in

GUNS, RIFLES & PISTOLS,

And all articles connected with the same.

All kinds of

JOBBING

AND

REPAIRING

done to order and warranted.

Constantly on hand the largest assortment of

TABLE & POCKET

CUTLERY

AND

Fishing Tackle

to be found in Maine.

No. 3, East End of Kenduskeag Bridge,

BANGOR, ME.

☞ Blasting Powder in quantities to suit purchasers. ☜

A historic year. In 1871, when J. W. Ramsdell placed his first ad in the Bangor Business Directory, *he was assisted by machinist Frank Philbrook. Also in 1871, Hiram Leonard—an associate maker with the Ramsdell brothers the previous year—started a small rod business over Bernhard Pol's eyeglass and watch store at 28 Main Street, Bangor.*

superbly talented Ramsdell assistants, Hiram Leonard would achieve regional, national, and, finally, international fame. Philbrook would fall into obscurity, head for South America, and eventually retire to a manufacturing town in Massachusetts.

Born in the central Maine village of Sebec on January 23, 1831, H. L. Leonard was the son of an oar-maker and spent his formative years in the forests of New York and Pennsylvania. By his mid-20s Hiram was back in Maine as the East's greatest "white hunter." His prowess and stamina in the puckerbrush were legendary; despite his light frame, Leonard managed to carry a 135-pound moose quarter from Little Spencer Pond to Lobster Lake in 1856, a trek of some 16 miles over rugged terrain. The next spring, Hiram was one of two passengers to swim ashore from a stagecoach that had careened into the Piscataquis River in Foxcroft. The other man went to the nearest house to keep himself from freezing. Leonard remained behind, built a raft with his belt hatchet, and saved the two remaining passengers and

driver. The horses drowned. Through this sort of pluck and his can-do attitude, Leonard had become a backwoods icon before he was 30.

In his work *The Maine Woods*, Henry David Thoreau, the original flower child, mentioned Leonard in July 1857, relating that the hunter-gunmaker was leading a caribou hunt into Canada "across the Restigouche and the Bay of Chaleur, to be gone six weeks." In the summer of 1858, Hiram explored the Tobique River region of New Brunswick, where he also fished for salmon; his love for salmon eventually led him to build his own hatchery. The following year Leonard married Elizabeth Smith Head of Bangor, and in 1860 he took her into New Brunswick on an extended fur-trading sojourn. The couple set up base camp on a branch of the Kedgwick River and remained there for several years.[2]

At some point after the Civil War the Leonards moved back to Bangor, where Hiram became an associate gunmaker at Charles Ramsdell's shop. In

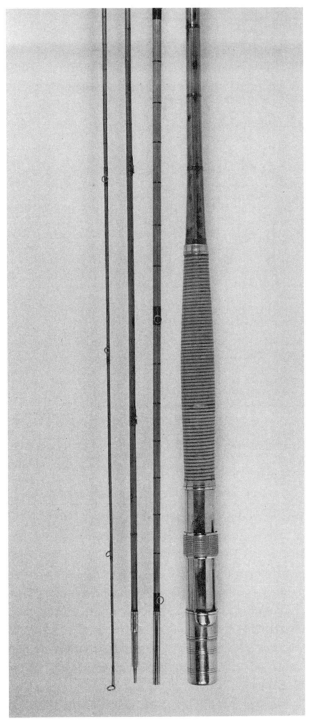

An early H. L. Leonard production fly rod, with a massive German silver seat and female ferrule; it's marked with the 1875 patent date only.

vived as examples of his very earliest work. During the latter part of this period, in 1869–70, Leonard produced his first four wooden fly rods. I believe these earliest models had a Canadian influence, possibly being built in just two sections. The fly rods were for his personal trout and salmon angling, and two were sold to others: the local druggist and the commissioner of fisheries.[3] At that time he and Lizzie lived at a boardinghouse on Thomas Street, and he was listed as a "gunsmith." In his own words, "I made my first rod in Bangor ME. Material used was ash and lancewood. I made it for my own use, not intending to make a business of rod manufacturing. I, however, sent it to Bradford & Anthony . . . being advised to do so by a friend of theirs."

HIRAM LEONARD'S SIX-STRIP RODS

You know the rest of the story: A Bradford & Anthony drummer flashed a pair of four-strip bamboo rods at Hiram, and he would soon improve upon them by going six-strip. In 1871, following the breakup of his associates at the original Ramsdell shop, Leonard opened a business in two rooms on Main Street and began hand fashioning products that would eventually shake the tackle industry to its foundations. The man Frank Philbrook recalled as "one of the greatest hunters who ever lived" had been tamed. Hiram worked alone during this first year, making rods for Bradford & Anthony. They looked remarkably like the rods built by Charles Murphy, but according to a 1905 Leonard letter, these six-strippers were of his own "invention." As yet, no examples have surfaced that bear a stamp linking him to his short-term association with that old Boston firm.

H. L. Leonard then moved "into Strickland's block on the bridge," where he remained from 1872 until 1874. At this shop he first took on an additional helper and gradually increased his work force to "6 or 7" men. You'll find Strickland-period rods obtainable but rare. These prepatent models sport commercial fittings, most notably a large reel seat with soldered raised rails and a long, ornate butt cap, all of German silver. They are marked by the maker, but have no Bangor stamping. Some early male ferrules show exposed bamboo between the junction of the wall and the spike.

the late 1860s, Leonard started a part-time taxidermy shop over Bernhard Pol's watch and eyeglass store. This is probably where his first ash and lancewood rods were built. Leonard's stuffed moose heads are best forgotten, but a few multishot rifles have sur-

In late 1874 or early 1875, Leonard again moved, this time to Dow's Block on Hammond Street. Here he employed 11 men and one woman, and hooked up with the old New York tackle house of Andrew Clerk's, which was restructured that year to become Abbey & Imbrie. With this prestigious company as his "Sole Agents," Leonard's business increased, although A&I-marked rods are extremely scarce. A unique example of a Leonard fly rod dated to this period is a 14-foot grilse (small Atlantic salmon) model. With one butt and one tip, this six-strip curiosity was built in the two-piece "spliced-rod" configuration originated by Saint John, New Brunswick, makers. The rod also sports two small German silver cups, which receive each tip of the splice when lashed together, exactly like the cups found on rods by Dingee Scribner (established 1860), the dean of Saint John's rodmakers. Shades of Walton and Cotton! The construction required 12 strips of Calcutta cane longer than 7½ feet before trimming. To my knowledge, this historic piece is the only spliced Leonard rod found to date.

THE BIRTH OF PRODUCTION BAMBOO

The Leonard spliced salmon model was probably built to Castle Connell specs, producing what is now called a "parametric" action. It may have been designed for the Canadian or British market, and in the chronology of the split-bamboo article may have been one of Hiram's last bench-fashioned products—or one of the first rods made, in part, by a new machine. In 1876, the middle of his A&I period, the ex-gunsmith designed and built the famed Leonard beveler. This wooden machine has been "rumored" to great heights as the first beveling engine to speed up the production and cutting of bamboo strips. For this advancement several modern writers have called Hiram a genius, while others give the credit to Hiram Hawes. Nobody mentions Wheeler, who, the same year, could build a rod in "short order at short notice." Thus in 1876 we find two Maine rodmakers entering the production age. As evidence on behalf of Leonard I'd like to offer the comments of Doctor A. G. Wilkinson in the October issue of *Scribner's Monthly*, and show the accompanying engraving.

A fudged double-stamping of the rare Abbey & Imbrie rod die, found on a two-piece Leonard salmon model.

Wilkinson was fascinated by rod theory and by the new beveler in particular. He explained its abilities:

This beveling is done with a saw or plane if preferred, but more expeditiously by having two rotary saws or cutters set at an angle of 60 degrees to each other, in case the rod is to be of six strands. The strip is fed to the cutters by means of a pattern which, as the small end of the strip approaches, raises it into the apex of the angle formed by the cutters. This preserves a uniform bevel and still narrows each strand toward its tip end so as to produce the regular decrease in size of rod as it approaches the extreme end.

These strips can also if desired be filed to a bevel by placing them in triangular grooves of varying depths in a block of lignum-vitae. The pieces are then filed down to the level of the block which is held in a vise during the operation. In the accompanying illustration, some pieces are being thus worked out by hand, while others are tied ready for gluing, and still others glued and ready for the ferrules. For this sketch I am indebted to Mr. Leonard, to whom every angler in America owes thanks for what he did as the pioneer in this art and for what he is constantly doing in perfecting these excellent rods.

Wilkinson was talking about salmon rods, so there is a link to our spliced grilse model: "The old well known firm of A. Clerk & Co., New York, introduced into the market the Leonard rod of six and twelve strands, and have since been supplying Europeans with all they get of this article." Leonard's export trade is indicative of the heightened output created by the beveler's speed. From this information it seems that a greater portion of H. L. Leonard's Abbey & Imbrie–marked rods were the first models to pass through the blades of the historic machine.

Thus, in what seems to be the mother of all déjà vu, *exactly* 50 years after its invention James Nasmyth's incredible whirring shaper showed up in a different skin in Bangor, Maine. The year 1876 marked the nation's centennial, but for anglers it also marked the birth of the production split-bamboo rod industry. Without the Leonard beveler, and the models that followed, we collectors would have few fly rods to pursue. Without Doctor A. G. Wilkinson, we wouldn't be able to pinpoint a major event, nor would we have his etching of Leonard at the vise using one of Job Collet's locally made files to true up a strip of Calcutta. As great an invention as the shaper was, each strip still needed handwork before it was "right" according to Hiram's high standards.

An 1876 portrait of Leonard. Titled Making Split-Bamboo Rods, *this etching accompanied a Doctor A. G. Wilkinson article on salmon fishing. Wilkinson was a* Big Cheese *at the U.S. Patent Office.*

With the exception of the first six months of production, standard A&I-era Leonards carried the new waterproof ferrules, with Hiram's first patent stamped on the lower female. The Abbey & Imbrie association lasted until the end of 1876, when Leonard resumed his own advertising. At this time, Hiram went back to his original "H. L. Leonard, Maker" stamp, adding a second that read "Bangor, Me." Also in 1876, Hiram Leonard, like fellow Mainer Charles Wheeler, exhibited an award-winning rod at the Centennial Exposition in Philadelphia, and received a Bronze Medal and Diploma. Shades of Aldred and the Crystal Palace, Leonard's

Forest and Stream, *December 7, 1876*

169,181. FISHING-RODS. Hiram L. Leonard. Bangor, Me. [Filed Aug. 30, 1875.]
Brief.—To render access of water to the end of the wood of the rod joints impossible in the case of rent and glued rods—an especial desideratum.

rod was reportedly an ultrafancy $2,500 jobbie with jeweled mountings.

In 1877, Hiram went into partnership with New York tackle-store owner James H. Kidder[4] in an attempt to expand his production facilities and purchase the rights to Frank Philbrook's patent reel. At this point the ex-fur trader and Lizzie moved into their own "new" house on Thomas Street.

FRANK PHILBROOK'S RAISED-PILLAR REEL

During Leonard's tenure, the Bangor gunsmith-machinists played an important role in the tackle trade. One of the Ramsdell brothers, for instance, probably made the famous "H. L. Leonard, Maker" stamping die. The second most famous person to emerge from the multishot ranks, Francis J. Philbrook, became a gifted mechanical engineer. As Leonard's future associate, Philbrook entered the trade in 1868–69 as a journeyman working for C. V. and J. W. Ramsdell. Upon the dissolution of the Ramsdell partnership in 1870, Philbrook remained as a machinist with John Ramsdell, at the original shop on the east end of Kenduskeag Bridge.[5] The next year, the Bangor directory officially listed Philbrook as an employee of J. W. Ramsdell. At that time Ramsdell advertised "Constantly on hand the largest assortment of Table & Pocket Cutlery and Fishing Tackle to be found in Maine." Notice that the left side of the ad concerned "Jobbing and Repairing done to order." In 1873–74, Philbrook severed his relationship with Ramsdell and listed his vocation as a "machinist," working with Tucker & White.

Following his journeyman years, Philbrook formed a partnership with C. G. Staples and opened a shop over Job Collett's file factory on Exchange Street. The *Maine Business Directory* listed "Staples & Philbrook" as "Gunsmiths, Machinists, and Model Makers" in 1874 and 1875. This short-term association ended with the disappearance of Staples, and by 1875–76 Frank Philbrook was on his own at the same shop address as a "Gunsmiths' Machinist

F. J. PHILBROOK,

Gunsmiths' Machinist & Model Maker,

OVER JOB COLLETT'S FILE FACTORY,

Exchange Street,   BANGOR, ME.

☞ *All kinds of Small Machinery made and repaired.* ☜

A rare advertisement posted by Philbrook between his partnerships with C. G. Staples and Edward F. Payne. From the 1875–76 Bangor Business Directory.

and Model Maker." I believe that these small machine firms—the Staples partnership, Philbrook alone, and later with Edward F. Payne—made a good number of Leonard's early rod furnishings.

About a year after the termination of his partnership with Staples, Philbrook designed a new style of fly reel, an invention that ultimately became America's first raised-pillar winch. Philbrook handed a prototype to Leonard, offered to sell the rights, and it was a done deal. Philbrook contracted to manufacture all future reels for Leonard. In a flyer sent to eastern retailers dated May 1, 1877 (a month before Frank's patent was assigned to him), Leonard announced, "The REELS of my make are the lightest, strongest, handsomest, and most durable ever made." Frank Philbrook's reel may not have been the first raised-pillar winch, however. London reelmaker John McGowan, active between 1857 and 1882, designed a brass raised-pillar reel that also incorporated a protective rim to keep line from entering under the crank handle.

So now, in this centennial year, three of Maine's top gunsmiths had moved from the hand fashioning of rifles and fowling pieces to the *manufacture* of fishing tackle, using the country's first two beveling machines—kept tucked away from competitor's prying eyes—along with a complement of stamping drop hammers, specialized lathes, patterned knurling tools, and other finishing gadgets. Along with Captain Chubb in Vermont, Wheeler, Leonard, and

Philbrook fashioned an American fishing-tackle industry, making quality casting instruments available to those who previously could not afford expensive bench-made smith-age offerings. Philbrook's reel would evolve into an American classic.

Philbrook took Edward F. Payne as a junior partner during the early months of 1877–78, the years when he wasn't mentioned in the *Bangor City Directory*. By 1879–80, Frank was boarding at the Franklin House and had moved his machine shop to 18 Broad Street, upstairs. The directory listed the new association as "Philbrook & Payne, fishing reels" and "Philbrook & Payne, fishing rod trimmings." The business moved again in 1881, to 53 Exchange, and the partnership lasted until 1885. The earliest known reel, a marble rubber and German silver model, was marked "Philbrook & Payne, Makers—Pat. Apld For." This 2⅜-inch-diameter click reel had an orange-and-black rubber handle and side plates. A second marbled rubber and German silver reel, identical in every detail, has been sold at auction; more very early Philbrook models may be unearthed.

Another Philbrook & Payne model, made after April 1877 but before the official patent assignment, probably came in two sizes: the original 2⅜-inch winch and a smaller 2-inch version. This second model had the orange-and-black marbled side plates, but the handle knob was solid black hard rubber. Marked "H. L. Leonard, Maker," it was a contract reel signifying the new association. All Philbrook &

A graceful handle, marble hard-rubber plates, and gunsmith screws characterize the beautiful Philbrook & Payne reel. (The American Museum of Fly Fishing Collection)

Payne marbled reels are today rare and expensive—some carry a five-digit price tag; they're perhaps the most exotic and historic of all American reels. The style of screw used in their construction was the gunmaker's "cheese head," used by only a handful of reelmakers. This feature alone separates Philbrook & Payne reels from all other look-alikes.

The next Philbrook & Payne models—a black rubber and German silver version and the famous Leonard "Bi-Metal" raised-pillar reel—are also rare. Both were built with the single- and double-crank handle, as seen on the marbled originals. The bi-metal reel, constructed with a German silver center plate swaged into a bronze pillared rim, is an absolute knockout. Its average current selling price is $1,500. "First model" bi-metal reels had an "H. L. Leonard, Pat. June 12, 1877" stamping as well as a chamfered handle screw, which dates back to the earliest Philbrook prod-

This superb example of Philbrook & Payne's work was owned by Hiram Hawes, and probably by his uncle before him. It's now in the American Museum of Fly Fishing, Manchester, Vermont. (The American Museum of Fly Fishing Collection)

A small bi-metal H. L. Leonard fly reel. This second model was built by Julius vom Hofe in Brooklyn, New York, but has Philbrook's patent.

era bi-metal variations can be found, including a lighter model with a bronze rim and aluminum side plates.

At least one reel, looking remarkably similar to Frank Philbrook's, has been discovered marked "Appleton & Litchfield, Boston, Mass." The partnership of George Appleton and Henry Litchfield was a short-term Washington Street affair lasting from 1885 until 1888. You should be aware that early "Appleton & Litchfield" reels—those built prior to 1886—could be the products of Philbrook, especially if they have gunmaker's screws. Later A&L models were probably manufactured by Julius vom Hofe.

In the end, of course, Philbrook changed vocations. The 1887–88 Bangor directory listed Frank as an "engineer," and a few years later he was working for George W. Parker and Henry Peakes, "boot and shoe manufacturers." The various romantic collector stories—that the Philbrook & Payne partnership dissolved because Frank died, or, even worse, Frank shot himself—are bunk. Philbrook became a well-respected mechanical engineer, eventually moving across the bridge to Brewer. In 1901 he moved to Ecuador, as an equipment designer, and returned to Brewer eight years later. At this time his occupation reverted to "machinist," and in 1923, Francis J. Philbrook and his wife Pauline removed to Rockland, Massachusetts.

While working in the Bangor area, Edward Payne lived in Brewer with his father, Henry, a carpenter. In 1885, the younger Payne left the Philbrook

ucts. Later models may show some metal variations, but the screw heads are a good way of identifying these Bangor products.

"Second model" bi-metals, built in Brooklyn from the latter part of the 1880s until 1894 or later, have an oval-head handle screw and pillar screw and a beefier single-crank handle drop hammer stamped from a new die. Although built on Philbrook's patent and so marked, they were made in Brooklyn by Julius vom Hofe. The real tip-off is the "little round hole," the maker's trademark of Julius vom Hofe. These newer models were usually marked "H. L. Leonard, Pat. No. 191813." These last of the Philbrook-patent Leonards usually sported the optional balanced handle; the style was to become standard on the Leonard-Mills reels, with the exception of the small Fairy models. Several vom Hofe–

Introduced in the spring of 1879, the original 5-ounce Catskill rod was built in Bangor for more than two years. Its production would continue in Central Valley, New York.

partnership and moved to Central Valley, New York, where he would be employed at the new Leonard plant until around 1890. It's possible that Frank Philbrook continued to make the Leonard reels alone for one more year, until 1886. On July 22, 1890, Ed Payne patented his "improved" version of Frank Philbrook's original. He assigned the patent to Thomas Bate Mills,[6] and it became the renowned "Leonard-Mills" reel, which was produced in various forms until 1960. It was built, at first, by Julius vom Hofe and then contracted out to various makers for the next half century. Francis Philbrook, however, remains as the champion of the American raised-pillar reel, a winch so pleasing in design that it never went out of favor.

H. L. LEONARD, MAKER—WM. MILLS, SOLE AGENTS

When Hiram Leonard acquired the rights to Frank Philbrook's reel he was running a full-fledged manufactory. His 1877 broadside advertised a plethora of wands: a 17-foot Salmon rod, 14-foot grilse rod, trout and bass rods, a trout rod with a 12-strand butt, a trout rod with five joints (trunk style), a ladies' rod (4¼ to 6 ounces), and a trout rod of greenheart with split-bamboo tips (shades of Robert Welch). His closing statement read, "Heretofore I have found it difficult to execute all orders promptly, but with the increased facilities now at my command I shall be able to meet the wants of all my patrons."

Collectors have often mentioned the 12-strand "rod," assuming that Leonard actually made such an odd item from 12 strips, tip to butt cap. Leonard's actual words were "12-strand butt," a term for the mortised rod. The Leonard 12-strand butt was nothing more than a 6-strip butt section with 6 additional cedar strips spliced into the grip-and-swell area. The spectacular mortised Leonard in the Gary Howells collection is perhaps the finest of the genre, and it's actually older than this 1877 date.

In 1878 James Kidder sold his Leonard holdings to William Mills & Son. Early in that year, Leonard had been selling through Kidder's upstairs establishment at 19 Beaver Street. Then, in February 1879, Hiram listed as his showroom 7 Warren Street—the address of William Mills & Son.[7] The Millses insisted upon being H. L. Leonard's exclusive sellers,

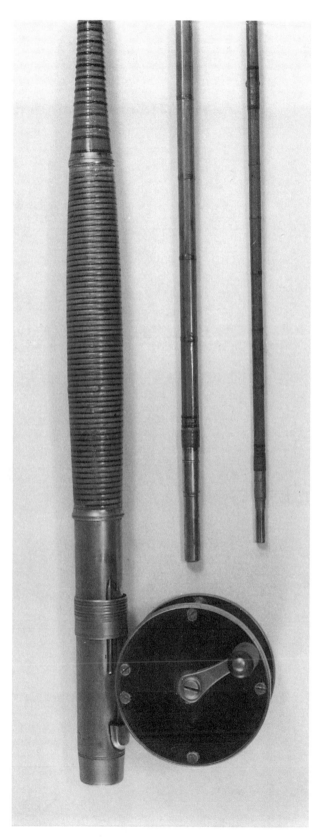

A new-improved Leonard rod built after 1878, designed to cut down on weight, and made with a shorter swell above the grip. This 10-foot, 5-ounce model is probably the original Catskill.

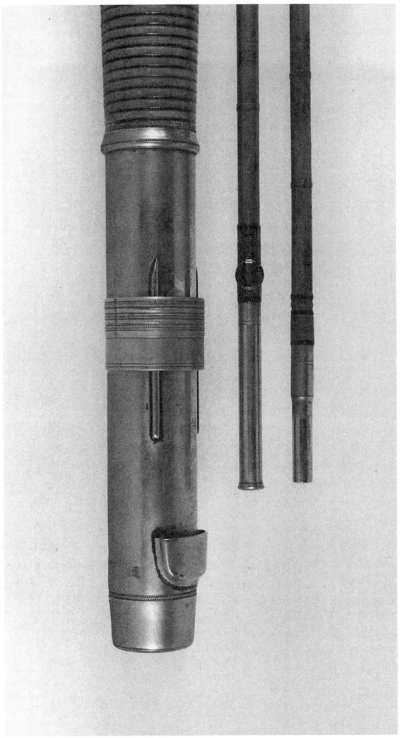

Details of the H. L. Leonard 10-foot Catskill fly rod. It has rather modern ferrules, similar to those shown in the patent drawing, and a delicate George Varney–style reel seat.

hexagonal rods, not rounded at the edges like those Leonard built through the Abbey & Imbrie period. Thomas Bate Mills was a rather pushy partner, suggesting rod refinements that Leonard may have resented. Within two months of the Mills takeover, Leonard introduced a new rod model, the Catskill. The model was first advertised in the April 24, 1879, issue of *Forest and Stream*, listed at a 10-foot length and weighing 5 ounces.

At this time Leonard was assisted by his nephews, Hiram W. and Loman Hawes, who had left the family gunsmithing trade that their father, Dwight, carried on in Honesdale, Pennsylvania. The Hawes brothers were first listed in 1879–80 as "fishrod makers" and boarding at 3 Thomas Street, the Leonard home.

In 1881 the Mills family convinced Hiram Leonard to move his entire operation to the old idle T. H. Bate rod factory, located in Central Valley, just northwest of New York City. The idea behind the move was to reduce shipping costs between the Leonard plant and the New York retail outlet, and to gain tighter controls over production. In Bangor, for instance, less-than-perfect split-bamboo blanks were tossed away; in Central Valley, they were used in the Mills Standard rod.

Leonard soon became disenchanted with this loss of quality control and sold more of his business to the Millses. By 1886, Thomas Bate Mills owned a controlling interest, and the phrase "H. L. Leonard, Maker" was deleted from future rods

and Hiram's last Bangor rods carried the stamping "H. L. Leonard, Maker, Wm. Mills & Son, N. York, Sole Agents."

These Mills-marked wands had the second patent on the lower female ferrules and were true and reels. Like a puppet head of state, the former adventurer and gunsmith remained with the Leonard-Mills Company until his death at Oak Clove on January 30, 1907.

The very last rods marked to the "maker" exhibited a new style of ferrule. Dispensing with the old spike, these surprisingly modern ferrules consisted of a one-piece drawn female and a two-piece male. The only carryovers from the Bangor-style rods were the rattan grip and a pronounced swell that extended about 2 inches above the grip check. The early Catskill rods had a very light soldered-pocket German silver reel seat designed to cut down on weight. These delicate reel seats were designed by Leonard employee George Varney and displayed the design that he would later use in his own rods. Like the new ferrules, these Varney seats were surprisingly modern looking for the mid-1880s.

Leonard's Catskills were exhibited in 1883 at the London Exposition and awarded a "special gold medal." In ensuing years, Leonard-Mills Catskill rods were built with a delicacy remarkabe for the times. Leonard rods would evolve further under the direction of Ruben Leonard and Hiram Hawes, as the founder's influence slowly faded. The original Catskill was also built as a 9½-footer weighing about 4⅝ ounces.

In 1890, the Petite Catskill rod joined the lineup. It was a light-line rod, at 9¼ to 9½ feet, and slimmed down to 3⅛ to 3¾ ounces. Then, four years later, the amazing Fairy Catskill was introduced—a little 2-ounce wonder at "8⅛" feet in length.[8] The Catskill-series fly rods had sliding-band wooden reel seats and were built in three sections, the two lighter models designed by Ruben Leonard as continuations of the six-strip excellence that had begun in his father's little shop in Bangor.

Late Catskills from the Ruben Leonard period (1890–1925) were, until recently, often complemented by the diminutive Leonard-Mills Fairy reel. The Fairy was a miniature and modernized version of Philbrook's reel, with aluminum parts replacing German silver. Like crows, tackle collectors love tiny bright objects, and the Fairy reel is a coveted item. As the most collectible rods of the Hiram Leonard era, the Bangor products, early "lady's rods," and Catskill models are now eagerly sought.

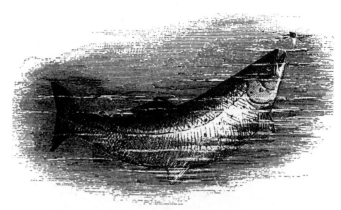

"THE RISE"—ADAPTED FROM BRACKETT'S PICTURE.

The Rise, an 1876 etching from a photo of a painting by Walter M. Brackett, supplied by "the genial Prouty" of Bradford & Anthony, Boston.

When the last bamboo shavings were swept from under the beveler and the last scrids of German silver were cleaned from the milling machine, our pioneering Maine gunsmiths put the labor-intensive handmade tackle industry to bed. For example, in 1876, working in his Bangor shop with his new beveler, Leonard built 200 rods—half of Murphy's lifetime production—in just one year. In so doing he led the way for a newer generation of tackle producers, men like the Heddon, Edwards, and Bartlett brothers. The new six-strip rod and stamped reel had proved themselves as good as or better than tackle made the old-fashioned way. Today, we revere the remaining Wheeler, Leonard, and Philbrook & Payne fly tackle as original examples of true precision production.

Notes

[1, 2, 5]Demeritt, Dwight B., Jr. *Maine Made Guns and Their Makers.* Hallowell, Maine: Maine State Museum and Paul S. Plumer Jr., 1973.

[3, 8]Keane, Martin J. *Classic Rods and Rodmakers,* New York: Winchester Press, 1976.

[4]Brown, Jim. *Fishing Reel Patents of the United States, 1838–1940.* Stamford, Conn.: Trico Press, 1985.

[6]Brown, Jim. *A Treasury of Reels.* Manchester, Vt.: The American Museum of Fly Fishing, 1990.

[7]Kelly, Mary Kefover. *Fishing Collectibles Magazine.* Summer 1991.

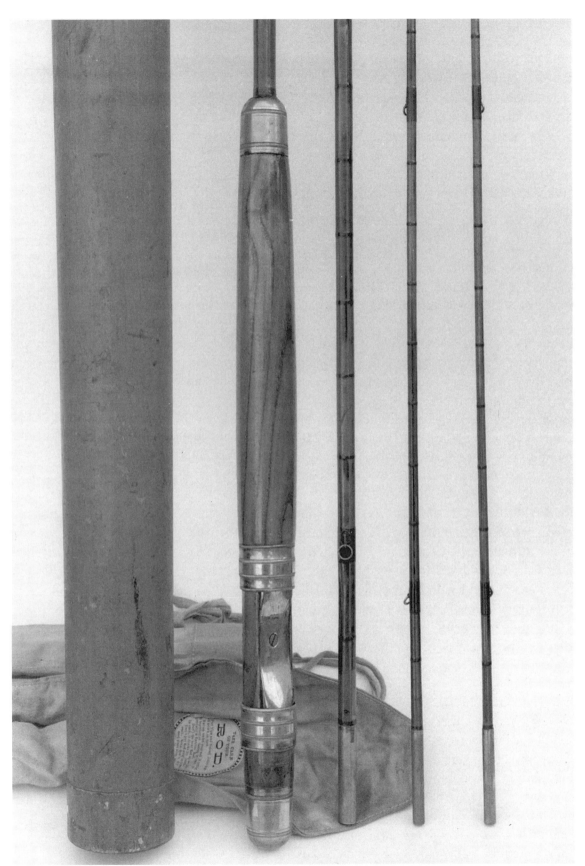

A beautiful C. F. Orvis split-bamboo fly rod, Serial No. 6, featuring Hiram Eggleston's patent reel seat and capped ferrules. The handle is Orvis's "Sumac Hand Piece."

4 New England Pioneers

The earliest New England firm catering to anglers was Rhode Island's Ashaway Line & Twine Manufacturing Company, established in 1824 by Captain Lester Crandall, a former commercial fisherman. At first its product line was confined to the "Cuttyhunk" threads used by fresh- and saltwater bass buffs, but after the Civil War, Ashaway introduced linen and Spanish silk fly lines. Fly lines are not high on the collector's list. Fortunately, other Yankee entrepreneurs turned to making rods and reels. We have looked at the superb early products of the Maine gunsmiths, but other pioneers were active before these artisans.

doing also introduced stamped rod fittings to the trade. These first affordable trimmings were a boon to aspiring rodmakers, and the list of those who used Chubb hardware reads like the *Who's Who* of the tackle world—from rural home-shop makers like Maine guide J. B. Daniels to the biggest names in the field: Nichols, Goodridge, Thomas, Edwards, Fowler, and a score of others. Shortly after Chubb entered the business, the availability of good working ferrules brought a plethora of new makers into the fold, and tacklemaking in the Northeast grew steadily into the beginnings of the Tackle Explosion.

ASHAWAY LINE & TWINE MFG. CO.

In 1858 a young millwright named Calvin Gray modified the bedpost turning lathes in his father's shop in Pelham, Massachusetts, to produce the first American production rods.[1] Gray's enterprise eventually grew into the giant Montague City Rod Company, but the largest early New England rodmaking operation was established in the Green Mountain State.

A decade after Calvin Gray began rodmaking in Pelham, a shrewd mechanical engineer named Thomas H. Chubb started a similar factory, and in so

There was even a minor gaseous shock wave in Beantown, where an obscure little rod shop started off as the "Boston Split Bamboo Fishing Rod Company." Many smaller New England outfits grew into true production companies, such as the growing factories of Nichols and Orvis. Between 1880 to 1910, more than 50 rod innovations were registered at the U.S. Patent Office and went into actual production. Reels, too, sprouted like fiddleheads, and they were increasingly designed by "engineers." And here's the best part: So much of this fly-fishing stuff was made

Through the increased use of machinery, tackle kings such as Chubb could stamp out products at a fast clip. At left is one of Thomas Chubb's famous drop hammers in operation. Such scenes dispel the craftsman-rodmaker stereotype; this chap could have been working in a shoe factory.

that it's still genuinely affordable. To a great extent, we collectors can thank Captain Chubb and his successors.

Thomas H. Chubb, The Machine Man

The Chubb success story is a wonderful tribute to inventiveness, and although the Captain left precious few signed rods behind, his products and influence can be seen in the work of dozens of other post-1870 rodmakers. Chubb was the enabler, the engineer other makers turned to for hardware—the man with the drop hammer.

In the spring of 1869, fresh from the flatlands of Texas, Thomas Chubb stood on the banks of the Ompompanoosuc River, just north of the village of Thetford, Vermont, and decided to build a fishing-rod factory. At this time there were about 10 rodmakers in the entire country. With the exception of

Gray's Pelham products, American rods were built the old-fashioned way—one at a time. Chubb's shop was completed just in time for the "October freshet" of the same year to wash it away. Undaunted, Chubb rebuilt, and within five years this talented diemaker had established the country's foremost machine-age rod factory. The machinery, building, and stock were valued at $28,000. Unfortunately, on the night of February 14, 1875, the manufactory and its contents burned to the ground.

Chubb rebuilt again, although he had "but little insurance." The new three-story, 23-room facility looked like a Lowell shoe factory and was powered by a 75-horsepower waterwheel. If we could go back in time and walk through this rod factory, we would see the machines that revolutionized American fishing tackle. From this plant, rod components were carted to the post office in Post Mills and shipped to the trade.

Between the autumn of 1875 and the winter of 1880, the Captain would develop the prototypes of his patented furnishings and several historically

Has a new Retail Catalogue. It contains a list of the old and reliable

CHUBB RODS, REELS, Etc.,

also many new articles, among which is the

CHUBB FLY BOOK,

and this is the Neatest, Handiest, most Durable and Cheapest Fly Book made; also the

Henshall-Van Antwerp Reel,

which is greatly improved.

Finest quality Split Bamboo and Lancewood Rods, Reels, Lines, Flies, Hooks, etc. Everything that the Angler uses. Write for Catalogue. Address

THOS. H. CHUBB,

The Fishing Rod Manufacturer,

Post Mills, Vt.

(*Mention this paper*).

Forest and Stream, *June 20, 1889*

important machines. Also within this time frame, Chubb built the industry's first large power lathes specifically designed for turning lancewood rod sections, as well as one of the first three split-bamboo beveling machines. (The other two bevelers were built at approximately the same time by Leonard and Wheeler.) Chubb's most important machines, however, were the series of drop hammers, miniature versions of Nasmyth's Scottish steam hammer of 1842. These were used in cold-stamping and annealed swaging of rod furnishings, such as guides, butt caps, reel seats, and ferrules.

Indeed, the machine room held the heart of Chubb's genius. There, to the clanking drone of an overhead belt-driven feed, sheets of brass and German silver were smashed, slammed, and spit out into the furnishings used by 90 percent of Chubb's tackle-building contemporaries. In a very real sense, Chubb was the father of popular sportfishing. Before his assembly-line approach, the angler equipped with tackle more complex than a handline was a well-heeled individual indeed. In the space of a few years, Chubb put more tackle into the hands of the average man than the smith-age makers had during their entire tenure.

Before we canonize Thomas Chubb, we should note that others had early rod- and reelmaking and swaging machines, including C. E. Wheeler, Frank Philbrook, Hiram Leonard, and C. F. Orvis. An examination of a Leonard boat rod of the 1878–80 period (marked "H. L. Leonard, Maker, Bangor,

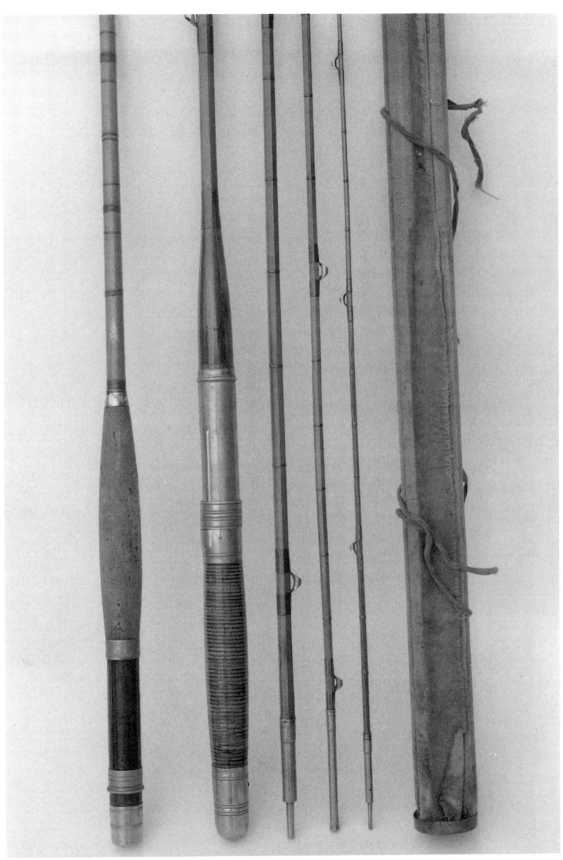

The lancewood rod at left has loose-ring guides and the Chubb "star" logo. At right is a superb early Chubb trout model, with ferrules peened to a hex, hand-fashioned tube guides, and a fitted form case. This little 7½-footer is marked "E. E. Eaton, Chicago."

Chubb's patent reel seat, made from tubular stock and with a swaged butt cap, found its way onto countless small-timer rods.

Me.") showed tiptops looking remarkably similar to the "tube tops" that Chubb would patent on May 8, 1883. Whether Chubb lifted and modified Leonard's two-piece tube top (as built by Philbrook) is a matter of conjecture, but the basic idea appears to have been Leonard's. Chubb's forte was to produce rod fittings that were sound, inexpensive, and accessible. In that he was extremely successful.

In 1883 T. H. Chubb also started to manufacture the "Henshall Van Antwerp reel." Although primarily a multiplying bait model, the winch was used by ouananiche and grilse anglers who wanted a bit of gearing. Patented on May 22 of that year, the "first model" Van Antwerp was a click reel with an additional drag button. Initially it was the brainchild of Doctor William Van Antwerp, who commended Chubb for making "a most perfect Fly Casting reel."

On September 27, 1887, Thomas Chubb patented the lever drag, which was mounted on the right-hand rim of the Van Antwerp reel. This variable drag worked well for fly fishers, who controlled it with their pinkie. Unfortunately, by 1890 Chubb moved the location of the lever to the other side of this (by then) "third model," and the Van Antwerp became a bass reel with one of the first thumb bars. The Henshall Van Antwerp, a squeaking compliment to two doctors, is a great collectible with a great price. For upcountry flavor and reminiscences of

togue and landlocks, the "second model" exhibited period features that few other reels had, including at least three dips in a "nickle bath." Of course, any Van Antwerp is dear to tradition-minded bass anglers.

The Van Antwerp model was not the only reel Chubb built; several single-action models have been found that are takeoffs on the simple New York (and earlier British) fly reel. No doubt contract reels had other tackle-house stampings, but any reel marked by the maker is fairly rare. Obviously, Thomas Chubb made far fewer reels than rods or fittings.

In his September 27, 1884, advertisement in *The American Angler*, we get an overview of the entire Thetford Village operation. Chubb had the "LARGEST FISHING ROD FACTORY IN THE WORLD," with the "BEST OF SEASONED STOCK, Either Turned, Planed or Square, in Ash, Lancewood or Greenheart ... MATERIAL FOR SIX STRIP or HEXAGONAL SPLIT BAMBOO RODS ... Complete Sets of Brass, Nickel-Plated and German Silver Rod Trimmings," and so on. In the full-page ad, Chubb listed dozens of various guides, reel seats, dowel ends, ring tops, three-ring tops, sheet metals, shellacs, gums, cloth cases, you name it. Frankly, I wish he were in business today.

Chubb was "prepared to supply reel-makers with the various parts for making reels, either click or multiplying, such as spools, plates, handles. . . ." The advertisement continued by offering services to the trade, such as Japanning, tinning, nickeling, and polishing. Finished rods were "Supplied to the Jobbing Trade Only," one of the reasons collectors find so few signed Chubbs. Most Chubb rods were sold semi-finished to a retail or wholesale tackle house; the house wrapped the guides, varnished the shafts and wraps, and stamped the rod with the retailer's name.

Thomas Chubb's major production probably began in 1870, at first limited entirely to wooden rods and stamped fittings—until his "patent period." Starting on December 14, 1880, the Vermonter patented a half-dozen ideas that were incorporated into his products and those of other makers. The original 1880 patents were threefold: a metal-sleeve reel seat with a fixed and sliding band; a method of rotating a reel seat to align with a set of double guides; and, most important, the notion of making the reel-seat barrel from one piece of hollow tubing.

Thomas Chubb's famous "star" logo. On rod shafts, it appears as a gold acetate label.

As the original seamless reel seat, Chubb's version had double rails to encase the reel's foot. On fly rods, a swaged one-piece butt cap slid over the lower end of the seat and was pinned in place. These reel seats sometimes have the December 14 patent stamped on the barrel.

Thomas Chubb's next patent, obtained on August 29, 1882, protected the idea of stamping tie guides from a sheet of brass or German silver. These cheap drop-hammered guides were mailed off to countless cost-conscious rodmakers. On September 12 of the same year Chubb patented a ferrule aligned by a longitudinal groove, a patent that evidently never went into production. Chubb's last patent was for the funnel tube top mentioned earlier. Another Chubb innovation was the "reinforced female" ferrule, which had an external sleeve soldered at the area where the end of the male was finally positioned.

Signed Chubb rods are often quality split-bamboo models built during this patent era. Marked with a stamping on the reel seat, the bulk of these rods were built in a hexagonal configuration, although some models, like the Dr. Baxter Salmon Rod, were glued up in eight strips. Lesser rods, often made from lancewood entering a separate grip and reel-seat area, had the traditional Chubb "star" logo affixed as an acetate label just above the grip check. This label does not indicate age, since Montague continued the labeled Chubb line until the Depression.

The Captain met with another disastrous fire in 1891 and, rather than rebuild, decided to sell the company and surviving machinery. A story goes that the Bartlett brothers of Montague arrived in Thetford while the factory's ashes were "still smoldering."[2] Chubb stayed with Montague as a mechanical consultant for about a year. As a result, many Chubb innovations were carried into the 20th-century Montague product line and—in the best Chubb tradition—were sold as anything from an L. L. Bean Salmon Rod to a Bob Smith's Trout Special.

 ## James Barbour Daniels, The Shoe Leather Man

Of the up-and-coming rodmakers to use Chubb's products, none shone in eccentricity of method as

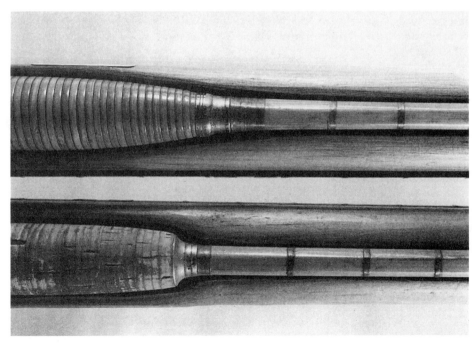

Two of Jim Daniels's rods, showing the shaft and grip styles used on earlier Auburn models (top), and later Scarborough products (below).

brightly as Jim Daniels. Born in Woodstock (near West Paris), Maine, on February 22, 1840, J. B. Daniels spent his youth hunting and fishing in the woods of western Maine. As a young man he established himself as a first-rate house carpenter, but he spent most of his spare time as a guide on the trout and salmon lakes of Oxford County.

By the early 1860s he specialized in fly fishing, using the ash and lancewood rods of the period — which, he said, kept breaking under the strain of Maine's larger trout. Since the nearest rods were sold in Portland, this meant downtime; he'd have to wait until his older cousin, Linc Daniels, brought a new rod up from the city. Linc was a famed early guide and had seen sports using the new and much stronger split-bamboo models built by Mainers Wheeler and Leonard. It was Linc who prompted Jim to build his first six-strip bamboo fly rods. There's no question that the time frame coincided with the mid-1870s availability of Thomas Chubb's rod furnishings. These nickel-plated mountings were probably purchased from Charles F. Nason, who had a gun-and-tackle store on Main Street, Lewiston. Daniels's first rods were built for personal use after he moved to Auburn in 1865. During his angling jaunts, other sportsmen admired Jim's handiwork, and in 1878 he offered his first rods for sale.

The Daniels rod was produced for the next 57 years — perhaps the longest career of any individual American rodmaker. Jim hand built about 1,000 rods, averaging around 20 per year. Shortly after 1880 he added bait and trolling models to his line, but the bulk of his output was the long wand. Daniels never had machinery of any kind, and around 1887 probably had a planing bench at his workshop at 14 Mechanics Row. His original hardwood planing form was built with one of his molding planes, modified to produce a series of 60-degree grooves. The Auburn rods were built during the winter and finished at his home, then 143 High Street.

Daniels never signed or stamped any of his rods, so no one will ever find a marked Daniels example. When the 250 various wrappings were completed, Jim varnished the sections and hung them on a special rack to dry. He then made a soft pine rod form, which was a production in itself. Wooden forms built by the best — Chubb, Leonard, Scribner, Nichols,

and a score of other makers — don't even approach the accuracy of fit found in the Daniels version. Lewiston-Auburn was the shoemaking capital of New England in Jim's time, and he made the end pieces of each rod form from scrap sole leather. This is the one surefire way of identifying a genuine J. B. Daniels rod: Look for the thick shoe leather! His sole supplier was the Pray-Small Company, manufacturer of ladies' boots and shoes, where he worked as a watchman during the 1890s.

Each leather-soled rod form was "signed" with a rinky-dink inked gum-rubber stamp, "MADE BY J. B. DANIELS, AUBURN, MAINE," in three lines. Unfortunately, the ink faded with time, and very few stampings survive. The rods themselves had a multitude of orange and tan (sometimes red) wraps, rattan grips, and reinforced Chubb ferrules that were hex-shaped for the first ¼ inch entering the split-bamboo shaft. Daniels's blanks were often built from Tonkin cane and lacked the telltale black burn marks that typify Calcutta cane.

The most productive period of Daniels's career was from 1904 to 1911, when we find him listed in the *Directory of Androscoggin County* as a "fish rod mnfr." at 13 Walnut Street. Well into his 60s, Jim was building at this, his peak, some 40 rods per year. In 1912, and after the death of his wife Lucinda, Jim packed up his Tonkin culms and shoe leather and removed to Scarborough to live with his daughter. Unique in an already odd trade, Jim Daniels fashioned every Scarborough rod, from the initial six strips to the finished wrapping, while sitting in a chair after dinner in a corner of the kitchen. He built the entire rod in his lap! The Scarborough models had cork-ring grips, German silver snake guides, agate strippers, and red wraps. Jim no longer used a gum-rubber stamp, and as far as we know all these later rods were completely anonymous.

Daniels continued to fly fish until he was 87, and hunted into his 90s. When he was honored as the oldest citizen in Scarborough, Jim claimed that his great longevity was due to his "constant habit of chewing and smoking tobacco and drinking rum for 78 years." Since the 1935 passing of the Grand Old Rodmaker of Maine, a few Daniels rods have come to light, although they've often been sold as "vintage Chubbs." It was said that, around 1930, at least 300

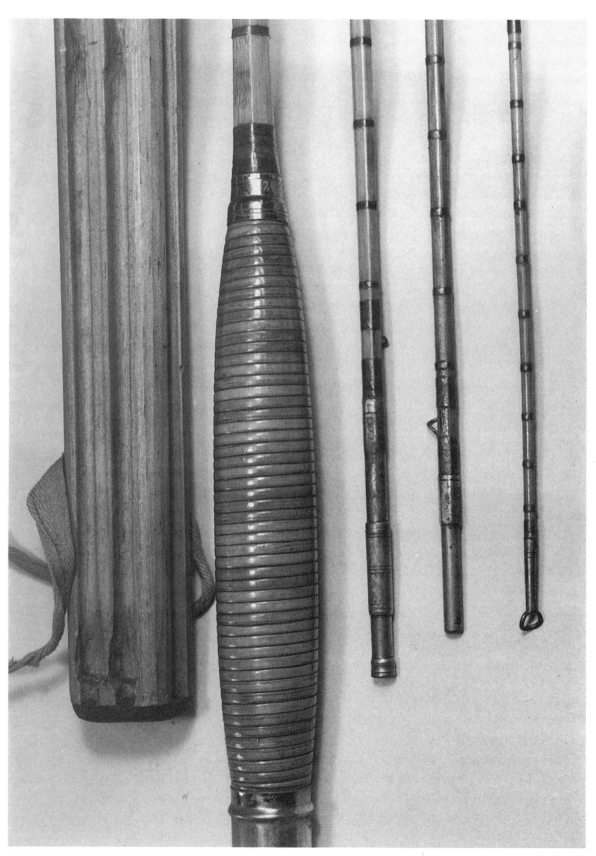

Details associated with J. B. Daniels rods include (left to right): a fitted pine form with leather end pieces; a Chubb grip check and flat rattan handle; hexagonal ferrules; and stiff "streamer" tips. The Chubb hardware was purchased from fellow Auburnite Charles Nason.

sportsmen in the Lewiston area were fishing with Daniels rods. Senators William P. Frye of Maine and John Foraker of Ohio fished rods made by Jim Daniels. Henry O. Stanley of Dixfield, at one time the Maine fisheries commissioner and a pioneering angler of the Rangeley and Kennebago region, owned 25 of them.

Most of the early Daniels fly rods were probably built to the length of the times—for models of the 1878-to-1880s era, 10 to 12 feet. Of the two examples I've examined, the late-1890s rattan-gripped Auburn model was a 9-footer, and the Scarborough fly rod was 8½ feet long. The tips of both three-section models were rather stiff, indicating that they'd been designed for streamer fishing. Jim never owned a pair of eyeglasses, and he once inferred that perhaps he needed a pair. Since Daniels didn't build a rod until he was 38 years old, when a lot of people start wearing glasses, he may never have built a delicate tip.

For upcountry landlocked salmon and togue angling, his rods must have excelled, and, delicate or not, Jim was proud of his products and glowed when he mentioned that he'd made $15,000 in a lifetime's hobby. In the 1920s, Daniels managed to reacquire the first rod he'd ever sold, to Mrs. Charles Cushman, wife of a well-known shoe magnate. Jim would wag it for admirers who were amazed at how long it had lasted. Shoemaker puns aside, Jim Daniels made his rods till the end. In November 1930 a Portland reporter interviewed Daniels for an article headlined: "Scarboro Man, 90, Still Fashioning Famous Rods." "I do all my work here," Daniels explained from his kitchen. "I have at present seven rods on the rack, drying out, ready to make this winter."

J. B. Daniels's greatest desire was "living until I'm 102," but he didn't quite make it. He did, however, continue fashioning rods in a rocking chair until his death at 95 years of age. Jim Daniels never came close to producing the number of rods built by his Maine contemporaries, Hiram Leonard, Charlie Wheeler, and Fred Thomas, but he outfished and outlived them all. The important thing to remember is that Daniels, like so many small-time makers,

bought his bamboo, reel seats, and rod trimmings from a manufacturer named Chubb.

S. W. Goodridge of Grafton, Vermont

Another obscure New England pioneer, S. W. Goodridge, was an active rodmaker during the 1870s and '80s. Born in Vermont in 1824, Goodridge entered the China and East India trade in 1843, making some money then, like so many, losing it. Returning to the small Vermont town of Grafton in

S. W. GOODRIDGE,

Grafton, Windom Co., Vermont,

MANUFACTURER OF ALL KINDS OF

Fine Fishing Rods,

For Trout and Black Bass Fishing. All kinds of Repairing neatly done at low prices.

PRICE LIST.

(The American Museum of Fly Fishing Collection)

Windham County, he began the "Manufacture of All Kinds of Fine Fishing Rods" in 1872.[3]

Goodridge's output can be corroborated only by his 1879 price list and sales sheet. At this time his rod customer "references" included William W. Warner, editor of the *Lone Star State* (Wills Point, Texas), and Fred Mather, the fishing department editor of *Chicago Field*. Twenty-six other satisfied patrons were listed from such various regions as Valparaiso, Illinois; Fremont, Nebraska; Boston, Massachusetts; and two customers from the province of Quebec. The sheet listed a number of wooden three-piece trout fly rods, a five-piece "trunk" rod with "silver plated" mountings, and two black bass fly rods.

It's interesting to note that, although Goodridge did offer a six-strip split-bamboo fly rod, he questioned its serviceability. "This is as good a split bamboo rod as is made, but I do not recommend split bamboo rods for many reasons, among which is its great liability to damage from moisture, and the impossibility of repairing in case of accidents, without making a new joint." Hmmm! Mr. Goodridge seems to have been his own worst enemy in the bamboo sales department.

The S. W. Goodridge rod line was built by arduous bench-planing methods, using white ash for the butt sections with the rest of the shafts fashioned from lancewood. Upon request, Goodridge would use greenheart for upper sections, but he believed that this heavier wood also tended to warp with time. All the grips were wound with rattan in the style of the times, and various fly models were available from 10 to 11½ feet in length. Like so many other makers, S. W. Goodridge made a "general rod," his version named after a well-known sports clothing manufacturer, Mr. W. H. Holabird. The Holabird General Rod could be fitted as a trolling rod, a bass bait rod, and a trout fly rod.

The bulk of Goodridge's output seems to have been built for the modest angler, since his prices ranged from $8 to a maximum of $15. Goodridge's operation obviously used Chubb's capped ferrules and nickel-plated furnishings, and is an example of the emerging "rod manufacturer." I wonder just how many rods this conservative rodmaker built by his old bench-planing process. Goodridge died in 1892. Any collector who happens to stumble upon Grafton, a little northwest of Bellows Falls in the southern part of Vermont, can view a genuine Goodridge rod at the Grafton Historical Society building.

Boston's One and Only Major Rodmaker

As a budding major tackle manufacturer, Boston's B. F. Nichols relied on machine-men to make key components for his products. Like so many others, his capped ferrules were from Chubb's plant, and other features, like the rattan cane handle material, appear to have been shipped from Post Mills. Some reel seats on Nichols products appear to have originated in Bangor, at the Philbrook & Payne facility.

Benjamin F. Nichols started his original rod shop at 36 Beach Street in 1880, with the pretentious title "Boston Split Bamboo Fishing Rod Co." Quite a tag, but by the end of that year the business name was shortened to "B. F. Nichols & Co." His first efforts in split bamboo featured rods built with a six-strip butt section, but four-strip mids and tips. On the one B. F. Nichols & Company fly rod I've examined, the butt was built in a true hex shape, and the upper sections were rounded in the manner of the 1870s. The original ring guides and red wraps of this 10-footer were in tough shape.

Four-strip construction was discontinued within a year or two, and later rods were built in a complete six-strip configuration and, possibly, a nine-strip style. Rod furnishings were made from a good grade of German silver, and the ferrules were capped jobbies that probably came from Chubb, as did the swaged reel seats. The sliding bands and grip checks were rolled and soldered. Early Beach Street rods were stamped "B. F. Nichols & Co., Makers, Boston, Mass."

The most distinctive feature of the best Nichols reel seats was an absence of the two raised rails usually seen on rods of the era. Instead, the barrel had a recess so that the reel's foot would fit below the contours of the seat. On fly rods, the butt cap held the lower end of the reel foot, or the angler had the option of using the upper fixed band. These are

Forest and Stream, *January 20, 1881*

The butt cap at left shows B. F. Nichols's early die mark, used in a partnership with unknown parties. The stamping at right is the more common one. A very rare third marking, used from late 1888 on, read "The B. F. Nichols, H. C. Litchfield & Co., Boston, Mass."

unusual seats, also found on McHarg's and Leonard's early products, and presumably designed by Frank Philbrook of Bangor, Maine.

In 1882 Nichols moved down the street to 28 Beach, finally settling at 153 Milk Street, where he remained until 1888. During this period, and certainly after 1883, Nichols started to wholesale his rods to Dame, Stoddard & Kendall, the 1884 successor to Bradford & Anthony. During that same year Nichols changed the structure of his company, perhaps buying out a partner's interest, and subsequent rods were marked "B. F. Nichols, Maker, Boston, Mass." They exhibited a rather flat rattan grip, and the wraps were dark red as was the custom of the time. One model, attributed to Nichols but marked "D.S.& K. HUB, Trade Mark," had a black celluloid-wrapped handle.

By 1886, while living in nearby Brookline, B. F.

Nichols was granted two U.S. patents for rod construction. The first of these, on February 16, was given for laminating nine strips of wood and bamboo into a singular rod shaft. The arrangement, should a collector ever run across such an oddity, was glued from three different woods (perhaps including bamboo) so that the rod was round in shape. The second patent, granted on April 6, was for a female ferrule that incorporated a bushing and bearing plate. This was probably an attempt to circumvent the patent for the common waterproof female given to Leonard some years earlier. In any event, neither of these two patents set the rodmaking world on fire, although the nine-strip idea must have made a pretty fly rod.

Two years later B. F. Nichols again moved his rod shop, this time to 302 Washington Street. The *Boston Business Directory* listed him at this address in 1889,

In less than 10 years, Nichols had moved from a little shop on Beach Street to the big time and become Boston's most famous and only expansion-age rodmaker.

The Charles F. Orvis Company

After 20 isolated years of making wooden rods, Charles F. Orvis designed the first "modern" reel in 1874, and in 1876 added split bamboo to his product line. The reel became a classic, and the rods incorporated a unique spring-loaded reel seat developed by his master rodmaker, Hiram Eggleston. Although introduced prior to 1872, the now famous Orvis catalog became the company's main sales vehicle from that date on. Bringing quality yet affordable products to the American angler, the small Manchester tackle business grew steadily with each ensuing season.

Of the many collectible Orvis gadgets built during the period, none are more dear than the reels, especially if found in the original walnut case. The original reels lacked a click and were made in the narrow trout size only. Charles Hallock, editor of *Forest and Stream*, suggested two changes: a click, and a larger model suitable for salmon anglers. Hallock got the click mechanism, which was added in the summer of 1875, but Orvis never built a large model for the salar, which may have been a marketing mistake. A wider version, the bass reel, joined the narrow model, and the production of both models commenced at the Manhattan Brass & Manufacturing Company in New York City.[4]

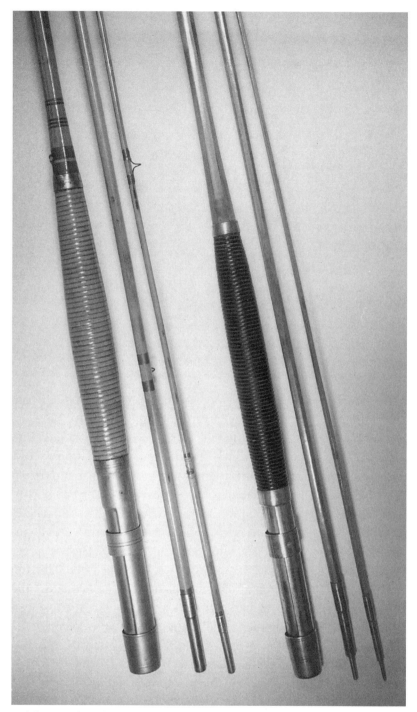

At left, an early Nichols rod marked "& Co." and having a six-strip butt with four-strip mid and tip sections. Years ago, the piece was refinished incorrectly and a hook keeper added. The second rod is marked "The Hub, DS&K, Boston" (Dame, Stoddard & Kendall). This Nichols rod has been stripped and is ready for a proper restoration.

and Washington Street was the place to be if you wanted to increase your business. Here, along Tackle Row, a prospective customer could shop at Dame, Stoddard & Kendall, at William Read & Sons, or at J. S. Trowbridge & Company.

Forest and Stream, *January 27, 1881*

For most of the production years, the two models were built with a riveted construction from nickel-plated brass and were marked "C. F. Orvis, Maker, Manchester, VT, Patented May 12th, 1874." By September 1875 Orvis was offering the reels in German silver, and around the turn of the century the company added an aluminum version. It's difficult to pinpoint the years of production of the many models, but around 1900 the riveted construction was replaced by screws. Reels so built had new markings: "C. F. Orvis, Maker, Patented, Manchester, VT.,"

and displayed a deviation in the circular plates at the reel-foot area.

Other oddities may show up as collectors root out early Orvis reels. Great finds would include the "Heavy Gold Plate" model advertised in *Forest and Stream* in 1876. Another would concern the original Orvis patent application, which mentioned hard rubber as the material for the face and rear plates; yet there is no indication that the material was used in any Orvis reels, including prototypes. Perhaps hard rubber was nothing more than a bad idea.

Other variations of the Vermont reel included the Orvis-Pennell model and the subsequent Orvis Protective Rim version. The basic Orvis reel, exported to Great Britain, was modified by Farlow's at the suggestion of writer H. Cholmondeley-Pennell. Farlow retrofitted a double handle, which extended down over the lip of the front side plate of the standard Orvis model; this made the fly line less apt to foul under the crank.5 After 1895, Orvis produced a version with a protective rim that accomplished the same thing. As it was, the trout and bass models were produced for 40 years, making them available to collectors today at modest to high prices, depending on how much plating has worn away and whether or not the reels come with their original natty little wooden box.

The rods that matched these great reels are also very appealing items. Built with six-strip construction and capped ferrules, they sported some of the handsomest fly-rod grips ever designed. Models built from 1882 until 1905 or even later featured Hiram Eggleston's reel seat, patented on June 6, 1882. It had a spring plate mounted in a wooden barrel; this enabled the reel bands to hold any form of oddball reel foot. Remember, in the 1880s as well as today, all reel feet are not created equal.

Nineteenth-century Orvis rods, typified by the cuts in Orvis's 1884 catalog, sported four styles of grip. The least appealing was solid cork—an early use of that material, although it was also employed by Charles E. Wheeler in Maine. Other grip styles included the

The C. F. Orvis fly reel was one of the earliest "contracted spool" models, which allowed the angler to gain a maximum of line with each revolution. This made it a popular item, and all those little holes helped dry the line after use.

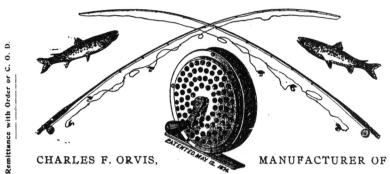

somewhat scarce. With nickeled German silver furnishings, capped ferrules, red wraps, and loose-ring guides, the rods were put up in a bag and a cylindrical wooden tube. In the mid-1880s, standard rods were available in 10- and 11-foot lengths, plus a special 9-footer billed as a "ladies' model." They all sold for $20. Even today they're a good buy, although they cost a bit more. I believe that Charles Orvis made all his rod fittings, which were unique in design and continued to be so until well after his death in 1915.

Wound Hand Piece (black celluloid), the Fluted Hand Piece, which enabled an angler to grip the rod so that it didn't "slip from the hand" (probably while eating a can of sardines and casting at the same time), and, finally, the Sumac Hand Piece.

Orvis "Hexagonal Split-Bamboo Fly Rods," which carried these obtusely exotic grips, are now

By 1905 the heavy wooden hand pieces were dropped from the Orvis line, and the only grip available was made from sheet cork. The price of the top-of-the-line rod dropped to $15—an 8-foot, 3-inch model described as "the perfection in rod building." A series of fly rods was offered for sale at $10, in lengths of 9 to 11 feet. All models continued to carry

A fine example of Hiram Eggleston's patent reel seat and the Orvis capped ferrules. Eggleston's seat was fitted to the company's rods as late as 1920. The circa-1882 ferrules were technically sound for the time.

the Eggleston-patented reel seat, and they were also serial-numbered. The original Orvis factory was not a large one, and rod production didn't seem to be a high priority during Charles's lifetime. The company remained in business and grew, but the original Charles Orvis period ended as the last vestiges of his personal catalog commentary were spirited away. The founder remained an active angler to an advanced age, and on January 16, 1908, he penned, "I am getting old [76] yet I have not lost my love for fly fishing in the least." Seven years later the industry lost this pioneer tacklemaker, but the Charles F. Orvis Company would survive to its preeminent status of today, a subject we'll delve further into in chapter 9.

The Obscure and the Odd

Orvis, Chubb, and Nichols were all major rodmakers, but there were other New Englanders in the profession—some extremely obscure, others just plain odd. Many were akin to Jim Daniels, although for lack of documentation they almost slipped through the cracks of history.

E. H. GERRISH, BANGOR, MAINE

In an attempt to fill the void in locally made rods, Gerrish began producing his work directly after Hiram Leonard left Bangor for New York. The original E. H. Gerrish shop was located at 24 Broad Street, and the only listing I found for it was in the 1882 *Maine Business Register*. Although Gerrish lasted only a short while in the fly-rod business, he would go on to make his mark in canoes. Today, the E. H. Gerrish canvas-covered canoe is a classic, eagerly sought by collectors. A signed Gerrish fly rod would be an even greater find.

H. A. MERRILL, BANGOR, MAINE

Another maker who tried to take up the Leonard slack was H. A. Merrill, a contemporary of Gerrish, judging from the features of his fly rods: lengths to 11½ feet, nonserrated ferrules, red wraps and intermediates, plus six-strip Calcutta construction. His rods are also quite rare and were signed in two lines on the bottom of the butt cap: "H. A. Merrill, Bangor." The

Merrill ferrule was unique. The male was made from three separate pieces of German silver and then capped; it looked a little like a cut-off metal-covered dowel. Merrill reel seats were made in the Bangor style, with a recess for the reel foot. Late rods sported cork grips and snake guides, indicating that Merrill built right into the 20th century.

ASA GILE & SON, READFIELD, MAINE

Asa Gile & Son appeared in the same directory that recorded Gerrish. The town of Readfield, located on the northern end of Maranacook Lake, was not exactly a metropolis. As "fishing rod manufacturers," Gile & Son were evidently active in the 1880s, and possibly before or after that decade.

CLARENCE SMITH, NORWAY, MAINE

For the well-versed geography buff, Norway is just across the river from South Paris. There, from the 1880s until just after the turn of the century, Clarence Smith fashioned a small number of fly rods thought to be made of split bamboo. In an article published in the February 23, 1936, edition of the *Portland Telegram*, Smith was "remembered by older residents of Norway in particular and by older fishermen in Maine as a rod maker par excellence."

WILLIAM C. GAREY, BETHEL, MAINE

William Garey was obscure *and* odd. He apprenticed under Clarence Smith and built split-bamboo rods from 1900 until at least 1936. In a particular quirk in an already obtuse trade, Garey became Maine's leading exponent of the "split wild pear rod." (His claim to the ultimate in odd fishing implements was exceeded only by Southport's Cecil Pierce, who in the 1970s built hollow split-bamboo fly rods with six strips of graphite glued inside.) The definitive split-pear model, built into a hexagonal six-stripper, was not easy to make—probably because not one in 30 million people could recognize the wood for what it was.

Said Garey, "Now wild pear isn't the easiest thing in the world to find unless you know what you're looking for. I've had woodsmen tell me they never heard of it. Did you ever notice when you were driving through the woods in the Spring, a tree with pinkish blossoms like the cherry? That's wild pear."

Maine Sportsman, May 1898

Funny, but I've never noticed that. Even William Garey admitted that "by the time a man had to hunt over miles of forest for half a dozen good pieces he'd want something for his time." Ayuh.

FRANK VARNEY, LONE INHABITANT OF SMITH'S MILLS, MAINE

Around 1920 Frank Varney moved to Smith's Mills from Lynn, Massachusetts, and built a number of fishing cottages on the shores of Sebago Lake. He also rented rowboats and operated a popular campground until 1930, when the Portland Water Company condemned the village as a possible contaminant to the water supply. The hotel, movie house, general store, and town hall were all torn down and the village died.

Only Frank Varney remained behind, master of a bleak and deteriorating landscape. In an attempt at revenge, he amassed a flock of 600 chickens that "shat" their way into history as an excellent source of nonpoint pollution. On a constructive note, Varney "received a large supply of wonderful bamboo from Tonkin, China," and began making rods in 1931. He split his culms by hand, devised his own cork-grip turning lathe, and designed a silk-wrapping machine. Interviewed for a November 23, 1932, article in the *Lewiston Journal*, Frank admitted, "I could turn them out fairly fast, but it is every bit hand work, and one must wait for glue, shellac, and varnish to dry. So, the process slows down. However, I believe I

am producing the artistic and serviceable bamboo rods that the sportsmen want. And the work helps me forget all I had hoped to develop here at Smith's Mills."

GEORGE H. BURTIS, WORCESTER, MASSACHUSETTS

George Burtis built a number of quality fly rods for two decades or more. The earliest advertisement I've seen dated from May 1895; the latest appeared in 1916. Burtis manufactured a complete line of flies as well as signed rods, one of which received a gold medal at some posh pavilion. Regular rods intended for regular anglers looked a lot like a Thomas & Edwards model, having midgrade nickle-silver hardware and delicate shafts. One example, a 9½-footer marked "Geo. H. Burtis, Worcester, Mass.," had a cherry reel seat with a mortised German silver foot plate and full German silver furnishings. This three-piece fly rod was made from Calcutta cane, possibly around 1900. We need to know more about this mid-Massachusetts maker and his "celebrated *Gold-Medal* handmade *Split-Bamboo Rods.*" For instance, who gave him the Gold Medal?

T. J. MANLEY, BROOKFIELD, MASSACHUSETTS

Another obscure maker, T. J. Manley appears to have built wooden rods during the 1890s-to-post-1900 era. One of his models sported a bird's-eye maple handle and reel seat plus rather Montague-esque nickel-plated fittings, including ferrule plugs. Judging from the examples I've seen, Mr. Manley was catering to a lower-income clientele. Not even 100 years of aging can make a silk purse of a sow's ear. Once a cheap rod, always a cheap rod.

W. HUNTINGTON, WILTON, CONNECTICUT

I bumped into rodmaker W. Huntington in the 1884 edition of James A. Henshall's *Book of the Black Bass*. Huntington's claim to fame, or infamy, depending on how you view it, was his great affinity for a native American wood. He advertised, "Hornbeam A Specialty." Now hornbeam, or ironwood, as it's also called, was pretty much a thing of the past with anglers by 1884, so it's unlikely that Huntington had a horde of anglers beating down his door—especially the fly-flipping types.

HENRY ANDRUS, HARTFORD, CONNECTICUT

Active in the teens and '20s, Andrus built a number of rather conservative rods for the period. On one model, a four-piece trunk rod, the three lower sections were fashioned from greenheart and the tips from split bamboo. Another Andrus, an 8-footer inscription-dated to 1918, was a standard 3/2 model with a wooden reel seat offset by a German silver cap and ring; it had close-spaced intermediates along the cane shafts. Henry Andrus was patronized by many Hartford-area businessmen and built a number of

---

W. HUNTINGTON,
WILTON, CONN.
MANUFACTURER OF
FINE HAND-MADE FISHING RODS.
HORNBEAM A SPECIALTY.
SEND FOR CIRCULAR.

---

rods. Although not common, his production isn't rare, and Andrus-built rods appear often enough that antique tackle collectors may hope to acquire one.

H. R. SEDGWICK, HARTFORD, CONNECTICUT

A maker of excellent rods, Sedgwick remains obscure due to his limited production. His delicate 1920s models have been compared to those of Jim Payne, which is as high a compliment as any maker could hope for. The last time I saw a Sedgwick was in 1992, at Lang's Spring Auction. A charming little 7-footer, the rod was built in three pieces with two tips and had a German silver sliding band over a cork reel seat. Of the several practically unknown makers listed in this section, H. R. Sedgwick carried on the highest form of workmanship. His rods, built for some of Connecticut's wealthiest fly fishermen, are very collectible.

WILLIAM R. WHEELER, HARTFORD, CONNECTICUT

Wheeler is a fine representative of a rod craftsman with extremely odd ideas. An active Connecti-

cut maker from 1900 until around 1915, Wheeler held a couple of somewhat obtuse United States patents. The first, granted on March 28, 1905, was for a criss-cross metal winding, or "ribbon," that extended the entire length of each section of the rod (according to Mary Kefover Kelly's *U.S. Fishing Rod Patents*). This idea was a reinvention of the same dorky arrangement used by the Foster Brothers of Ashbourne, England. The Foster ribbed rod had been built prior to 1895. The only difference between Wheeler's criss-cross and the Fosters' arrangement was the flat copper wire. The Foster Brothers used steel.

William Wheeler produced his rods in split bamboo and in several other woods, the species being hard to distinguish underneath all those Xs. I cannot imagine a more time-consuming and fruitless occupation than covering beautiful wood and bamboo with heavy metal. In eccentric rod construction this even beats split wild pear, although not by much.

William's second patent was obtained three years later, on February 11, 1908 (according to Mary Kelly). Here we had a new kind of oddball reel seat sliding band that "travels on an interior screw," a feature possibly installed on some of his rods. With these two patents to his credit, Wheeler became one of a slew of American rodmakers to hold fast to arcane ideas. The ribbing patent was assigned to the Anglers Company of Hartford, and possibly some rods were built with this marking. Known rods with the feature were stamped "W. R. Wheeler, Hartford, Ct,. Maker," and had full metal reel seats. They were put up in a compartmented canvas bag, the tips protected within a brass-appointed bamboo case.

At least one W. R. Wheeler rod has come to auction; it sold for $300. It was an 8¼ foot, three-piece fly model with a very old-style Chubbesque seat. The butt cap was reinforced with two small screws. The general consensus was that, under the copper criss-cross pattern, the round rod might have been built from greenheart. In the long run, Wheeler's ribbing and dopey reel seat went the way of all gimmicks unacceptable to knowledgeable anglers.

WILSON J. HUBBARD, ANSONIA, CONNECTICUT

We end the minor New England maker section on a complimentary note, addressing the rare Hubbard rods, built by two totally different makers, and then taking a look at a minor invention by Fred Richardson. The earliest craftsman of the group was Wilson J. Hubbard of Ansonia, Connecticut. On March 15, 1870, he patented a ferrule that locked on the screw principle, an idea also born in Britain at an earlier date. As with most screw ferrules, the lower section held the male, tapped for a machine screw. The screw itself was in the female at the end of the upper joint—and so on, depending on the total number of sections.

As far as I know, few rods attributed to Wilson Hubbard exist. The unique ferrule system should be a tip-off if you find a rod that you can identify as being American, circa the 1870s. I have hopes that such an item might exist, since the idea for this juncture was rather sound for its time.

JOHN H. HUBBARD, WESTFIELD, CONNECTICUT

Our second Connecticut Hubbard was John H., of Westfield. He was a little late to be a New England pioneer, so I'll just say that he helped usher in the classic era. Active from the very late 1920s into the mid-'30s, John Hubbard built some fine but scarce fly rods. As a young man he was a jock—an All American first-team football player in college (Amherst, 1905). He learned the rodmaking trade while working for his father-in-law, Eugene Prentis Bartlett, the sagamore of Montague City. Later in life John Hubbard was an avid sportsman and fly fisherman, building a few rods when he was in his 50s. Evidently, the rods were signed à la Edwards, on the lower end of the split-bamboo shafts.

John Hubbard's cane work was exceptional; known fly models had a slight swell at the shaft's emergence from the grip area, and were of medium-colored Tonkin cane. Supposedly there are only three known John Hubbard rods; two were recently sold at auction. The first model was a sweetheart, accented with brown wraps tipped cream, nickel-silver fittings, and a reel seat made from cork with a lower wooden spacer. It was a 9-footer built in three

pieces with two tips, all contained in a bag and aluminum case. The second rod, a two-piece, two-tip model, had lesser hardware but the same beautiful cane. This fly rod was an 8½-footer and came in a similar case and bag. They sold for $275 and $130, respectively. The third rod is in a private collection.

Without a doubt the John Hubbard rods were low-production, bench-planed items of the home-hobbyist genre. It's unlikely that many were built, but collectors certainly hope that more of these classics turn up on the market.

FREDERICK RICHARDSON, PROVIDENCE, RHODE ISLAND

Our last New Englander of minor importance is Frederick Richardson, who may or may not have built a few rods. He is "famous" for a very small item, a guide, but this little fixture showed up on many antiques and early classics built between 1880 and 1900.

Patented on November 23, 1880, the Richardson guide was ideally suited for the production methods of the time, being stamped in quantity from sheet German silver and then bent to its final shape. Frederick assigned the rights for this item to William Mills & Son, but I don't believe T. B. Mills ever manufactured it. Most likely he reneged and, in turn, sold the rights to Chubb. From there it was carried into the Montague line. The guide has been found on many rods with no relation to the Mills operation, including rugged Chubbs, Conroy & Bissett salmon rods, and, in much later years, to a Brewer-built Thomas.

Quite strong for its weight, Richardson's guide was ideal for salmon sticks and actually may have been inspired by the old soldered and labor-intensive standing guides that originated in Great Britain. Fred Divine produced a similar model, the figure-8 guide, made from oval German silver stock that was heated and then twisted into shape. Quality rodmakers, however, considered the Richardson guide the best of the lot.

I have no idea if Fred Richardson ever made a rod. It would seem logical that a man who built at least a small amount of tackle would come up with the idea. Then again, maybe Fred's hobby was beekeeping.

LEONARD ATWOOD'S VARIABLE TENSION REEL

We close this chapter on pioneering New Englanders and their tackle with a reel made during the post-1900 classic period. Leonard Atwood, its designer, was a gifted 19th-century engineer, a contemporary of others in the "mechanic" trade, such as Chubb and Philbrook. Like theirs, his active years began in the mid-1870s.

The son of Isaac and Miriam Atwood, Atwood was born in Farmington Falls, Maine, in 1845. He proved to be mechanically gifted at an early age, building a working model locomotive when he was 13. After a Civil War stint in the navy Atwood went to Titusville, Pennsylvania, and drilled a 500-foot-deep oil well in the backyard of a local hotel. Although it proved to be a gusher, Atwood received little financial compensation. During the next three decades, the Maine-born engineer lived at various addresses in Philadelphia, New York City, and Boston, yet he always returned to the Farmington area to summer at the old homestead and fly fish the Rangeleys.

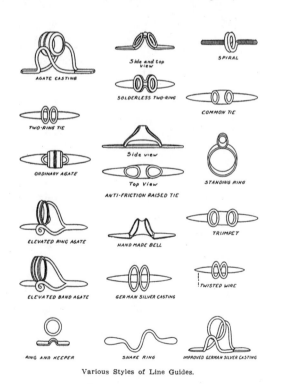

Various Styles of Line Guides.

Richardson's invention is shown in two perspectives, top center and just below, where it's called a "Solderless Two-Ring" guide. We also deviate from the 1912 nomenclature by calling A. R. Harding's "Ring And Keeper" a loose-ring guide.

In the mid-1880s Atwood entered his inventive period, along the way receiving patents for an internal combustion engine for submarines, an "aeroplane," and a racy new style of locomotive. On June 24, 1902, he patented a reel with a split frame that hinged back at the rear pillar, allowing the spool to be removed, oiled, or changed. It never went into production. In 1905 he designed the "Variable Tension Reel" and applied for his initial patents in that year and in January 1907.

During June and July 1906 Atwood purchased tools and dies for reel construction from a number of Boston machine companies. The Atlantic Machine Screw Company made the majority of his key dies, including a press shoe and punches and dies for his reel handles, pawls, spring holders, and spool sides.

On September 6, 1906, Atwood conveyed a quarter interest in the reel to John M. O. Hewitt of Philadelphia for $2,500. Additional startup funds were provided by Hewitt's brother, W. D., and A. H. Gillard, also of Philadelphia. Starting in January 1907 Atwood and J. M. O. Hewitt were busy trying to find machinery to produce the finished article. Some of the machines were built by Waltham Clock, which made a No. 3 lathe capable of holding four tool holders.

U.S. Patents 869,406 and 869,474, were issued to Leonard Atwood on October 29, 1907; later he acquired British and German patents as well. The new reel featured a foot that could be adjusted sideways to hang the fly reel in "perfect balance" and a unique drag tension system an angler could change while fighting a fish. "When the sudden strain of the fish is greater than the tension, line will be drawn off the spool even though the angler be operating the crank." The drag was adjusted from the center of the fly reel's left side plate by a machine-threaded rod that traversed *through* the spool arbor and connected to a disc brake on the opposite (handle) side. When the knurled drag wheel was tightened, the threaded rod pulled the disk against the right spool plate.

Atwood's Variable Tension Reel made its debut at the 1908 New York Motor Boat & Sportsman's Show at Madison Square Garden. It was offered at the New York Sporting Goods Company booth, and sold for $5 retail or around $36 per dozen wholesale.

The reel was also touted in a magazine advertising campaign.

In New York, A. H. Gillard approached a number of other tackle-house owners about carrying the new product, including E. S. Horssman, A. G. Spalding, and Ezra Fitch (formerly A&F). Conroy was impressed with the novelty and said he could probably use 200 of them a year. Neither Fitch nor Conroy liked the knob controlling the tension "device," how-

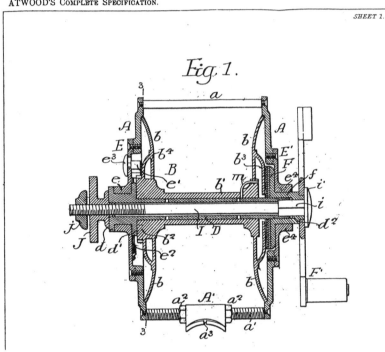

Leonard Atwood's reel, as pictured in the 1907 British patent drawing. Note the transverse drag chamber bored through the spool shaft.

ever, thinking that it would tangle the fly line. At least three of Gillard's letters mentioned the design deficiency, but Atwood wouldn't change it, and the reel became a turkey.

There were problems in manufacturing as well. The Holbrook Manufacturing Company, a small machine firm just south of Boston, went bust shortly after the first few hundred reels were produced. Atwood and Gillard looked to a larger machine company they referred to as the "Worcester people," but I don't think production ever got off the ground.

Less than 500 of the Atwood reels were ever made over a period of less than one year. On May 6,

1908, Gillard inventoried 103 in New York, leftovers from the trade show at Madison Square Garden. The last mention of Atwood's reel came in a July 6 letter from W. D. Hewitt: "I see the reels advertised in *Field and Stream* and suppose you are doing some business, but if they are to be made to go on the market they should be gotten out in quantities and ready for this fall or another year will be lost." That never happened.

The Atwood Variable Tension Reel was made in both single-action and multiplying styles, and in at least three known sizes. Examples were clearly marked "THE ATWOOD REEL" and had hard-rubber side plates pressed with a checkered gunstock texture.

The two "first model" Atwood reels looked very similar to the patent illustration, including the raised handle post that tangled fly line. The smallest Atwood model had 2¾-inch side plates and was made from nickel-plated metal parts. A midsized "improved" variation had a shortened post and a protective rim. This "second model" had 3¼-inch plates, and its nonrubber parts may have been built from either plated brass or German silver.

A larger version of Atwood's "second model," the salmon winch, had 4⅛-inch side plates and featured a German silver balanced handle, spool ends, pillars, and rims. All models had a left-hand click button and side plates made from stamped rubber. The U.S. patent application mentioned glass side plates, and recently collector Steve Vernon informed me that a "glass" Atwood reel has been found. Of course, any Atwood reel is very collectible, and the "second model" is a very appealing fly reel.

On September 23, 1913, Atwood again patented a reel design, half being assigned to Charles F. Pope of New York City. The inventor received a fourth reel patent on November 5, 1918, and a fifth on November 9, 1920, both with no assignees and granted while

The culprit! Leonard Atwood's prominent tension knob, topped with a smashing acorn nut, could have used some refinement.

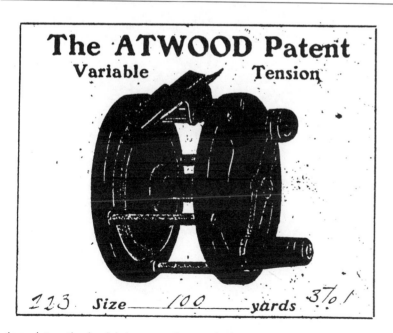

A very interesting box label, apparently unused, gives the size as "100" yards and the style as "3 to 1." It also includes the number "223," which could be a production number, indicating that at least 223 Atwood reels were made.

he was living in Boston. A prototype of the 1918 model showed up at Lang's 1994 Summer Auction and sold for $440—a very fair price.

Atwood died of a stroke in 1930 at his home in Farmington Falls. He was a true pioneer in reelmaking. His use of specialized lathes and pressing machines exemplified the tackle industry's move from the age of the gunsmith to the age of the mechanical engineer. And had he changed the tension screw knob on his 1907 model to a lever or a flush-mounted wheel, he would have been the father of the modern lever-controlled disk-drag "gold" saltwater reel.

The Honorable Bill Frye's salmon reel. Senator Frye lived in Lewiston, Maine, and was (in 1868) a founding member of the Oquossoc Angling Association. As a trustee, he helped construct Camp Kennebago, the most famous fly-fishing camp of its time. (The American Museum of Fly Fishing Collection)

Notes

[1]Bigelow, Paul. *Wrights and Privileges*. Athol, Mass.: Haley's, 1903.

[2]Kelly, Mary Kefover. From correspondence in 1977.

[3]Mathewson, Jon C. From an interview in 1994.

[4]Hogan, Austin and Paul Schullery. *The Orvis Story*. Manchester, Vt.: The Orvis Company, Inc., 1980.

[5]Brown, Jim. *A Treasury of Reels*. Manchester, Vt.: The American Museum of Fly Fishing, 1990.

5 New York State Makers

Although the core of tacklemaking was New York City, the trade gradually spread throughout America's rural East into New England and New York State. As we've seen, the original craftsmen were usually gunsmiths, such as Phillippe, Leonard, Krider, Philbrook, and Wheeler. Central and upstate New York held the same breed of individualistic mechanics who were likewise often engaged to produce fishing implements for local sportsmen. At first the rods and reels were hand fashioned, but as decades passed and the industrial revolution spread throughout the country, other types of makers joined the gunmakers in what became the expansion age of fishing-tackle manufacturing.

High-quality German silver rod fittings were now readily available from Chubb's Vermont facility, E. P. Bartlett's Massachusetts factory, Staples & Philbrook's in Bangor, and yet-to-be-discovered machine shops. For the angler with a modest wallet, Chubb and Bartlett produced lower-quality rod trimmings of thin-walled, nickel-plated brass.

At the same time, tradesmen could buy Calcutta culms and seasoned rodmaking woods from Conroy, Krider, Chubb, and other large establishments. Upstate rodmakers, such as John McHarg and Fred Divine, entered the trade on a yearly basis, developing their skills in relative isolation. Nowhere was the provincial expansion of rod- and reelmaking more evident than in the Mohawk Valley, along the upper Hudson's western drainage. Here, in the towns of Rome, Ilion, and Utica, a group of tacklemakers started to produce their own unique wares.

Even though many had patents to their credit, most New York State makers remained obscure. The majority of their reels were built on fragile designs and of inferior materials, such as nickel-plated brass; this, combined with low production numbers, means few have survived the ravages of time. Of course, this also means that those few remaining upstate reels in very good or better condition now bring premium prices.

The same is true of upstate New York rods. Few were made and those that were were odd, with the exception of the products of Fred Divine. Upper New York State also held a number of craftsmen who built eight-strip split-bamboo rods, a local eccentricity. But the really eccentric upstate kicker is the unique design of regional reels. With few exceptions, the upstate New York makers belonged to the Cult of the Side-Mounted Reel.

THE TRAGIC TALE OF MORGAN JAMES

Born around 1815 in Oneida County, New York, Morgan James began his career as an apprentice to Utica gunsmith Riley Rogers. In 1842 he started his own shop, first at Bleeker Street (1842–50); he moved next to the Fayette "corner of Seneca" (1851–66) and finally to Liberty (1867–68). By the time he was 30, James was acknowledged as one of the best gunmakers in the country and had constructed some of history's earliest telescopic sights. His products were more or less the province of gun collectors until Richard Oliver's second tackle auction in 1986.

With little fanfare, the leadoff reel moved to the auction block. It was a most curious revolving-plate, side-mounted click model inscribed "Greene Smith, Presented by, M. James." At the time, nothing was known about "M. James," although an accompany-

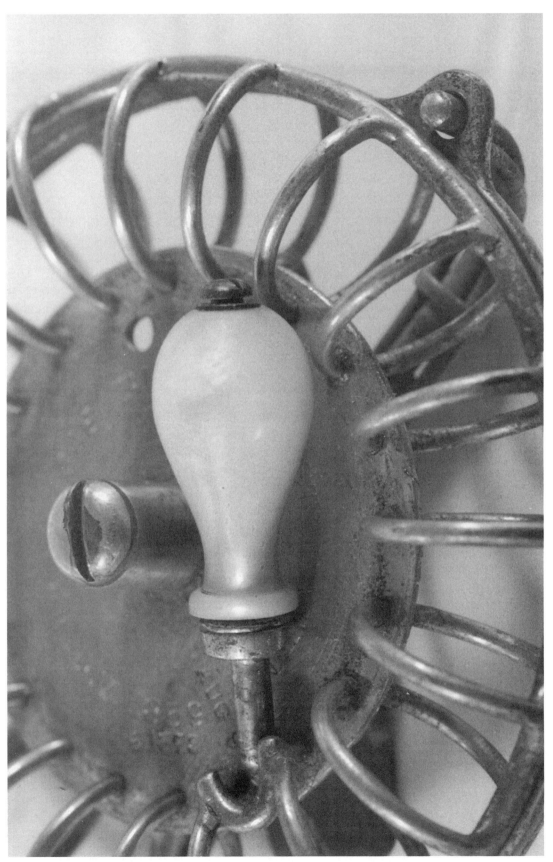

The classic styling of Billinghurst's "optional" mother-of-pearl reel handle. Most known examples sport the common wooden knob. Comparison, in William Billinghurst products, is like pointing out the difference between a 9 and a 10. (The McGrath Collection)

One of two known Morgan James fly reels, this superb representative of the gun-maker's skill features a narrow spool profile designed at least a decade before C. F. Orvis had the same idea. From a photo in the 1986 Richard Oliver auction catalog.

James's "production" was perhaps the smallest of any known smith-age reelmaker. The man had one of the finest gunmaking businesses in the United States, at one time producing 175 hand-fitted rifles annually. He owned a substantial brick home valued at $6,000 in 1865. Then, in a turn of tragic events, Morgan James lost his business, his property, and even his family, finally moving to Ilion and working for Philo Remington. In the end, James was living "in a poor district" of the village of Mohawk. A presumed alcoholic, he died on November 28, 1878. According to his brief obituary in the *Ilion Citizen*, Morgan James was "burried Sunday last by charity . . . one of the many victims of that class who say I can stop when I please."

Today, the Morgan James reels are considered to be among the finest early American fly models. Built over a span of 20 years, they were also rugged and hardy items, so there's a good chance that other James reels will show up in the future. The James product typified the output of all smith-age makers; built in limited numbers as fine sporting items for a select clientele. The rugged Morgan James reel was a far cry from the fragile Mohawk Valley models that followed his pioneering lead, as seen in the work of yet another New York State gunmaker.

ing letter, dated 1864, placed Smith as a Civil War officer. Oliver's established that Smith was also the son of a congressman from New York State and the grandson of one of John Jacob Astor's partners in the early fur trade. The reel itself was machined from brass, 2⅝ inches in diameter, and only ½ inch wide. It had an internal click held by a large-diameter yoke, finely trimmed rims, and an ivory handle. Despite its obscure origin, the little side-mounted specimen sold for $2,700.

A second reel, identical in construction to the first but with a dark hardwood handle, was purchased by Brian McGrath at a 1989 Bourne auction. The back of the side-mounted foot was stamped "M. James, Utica, N.Y." With this scant regional information, Brian finally located a little paper entitled "The Gunmakers of Utica," by H. J. Swinney, and at last the mystery of "M. James" was unraveled. To date, these are the only known Morgan James reels, although others probably exist. They were made between 1841 and 1863, at the height of James's career; both are considered the earliest American versions, if not the originals, of the side-mounted genre.

BILLINGHURST AND AMERICA'S FIRST PATENT FLY REEL

The exotic Billinghurst reel, made in a flourishing community where the Genesee River dumped into Lake Ontario, was also the product of the imagination of a well-known gunsmith, the coinventor of an early form of machine gun. Just before the Civil War, William Billinghurst (1807–80) patented a wire-framed gadget designed to cut down on weight at the rod's end as well as let a fly line dry without removing it from the spool. The earliest models were often plain brass bird-cage reels, but Billinghurst soon began to nickel-plate them for sales appeal. Like those of Morgan James, the reels were side mounted, and one wonders if Billinghurst was influenced by the earlier examples of the ill-fated James.

A very good example of William Billinghurst's reel, with nickel plating, showing the line guard and style of closure. (The McGrath Collection)

In any event, the Billinghurst reels were hand soldered on a jig and bore no resemblance to the lathe-turned James reels. The "first model" had a central stamping, "Billinghursts Patent Aug.9 1859 Rochester, N.Y." Most sported a black rubber folding handle, but a few were made with mother-of-pearl. Originally, each example was packed in a circular, dark green cardboard box with a light green covering on the lid that showed a line drawing of the reel. Needless to say, almost all of these boxes helped start some sport's wilderness campfire on a damp chilly morning.

From the outset, Billinghurst's fabulous idea became fairly popular, even though the reels were delicate and prone to dings. Eventually, others would copy the original Rochester invention, but as the first and foremost Billinghurst's model has a special niche on collectors' shelves. It has become one of the highest-priced antique American fly reels, and is largely responsible for kindling an interest in the side-mounted genre that includes a number of later New York State reels, some of which were automatics.

A "second model" was introduced after 1873, when Billinghurst renewed his patent. It had a cast foot with raised bosses to retain the frame rods, and appears to now be less common than the original

1859 version. At least one later reel had a mother-of-pearl handle. The newer reel was marked "Billinghursts Patent, Aug. 9, 1859 & 1873, Rochester, N.Y." The major problem with the Billinghurst was its nickel plating, which usually wore off at key points. Condition, then, is very important in determining any Billinghurst reel's ultimate value. A few extraordinarily rare examples of William's reels were made in German silver; naturally, these would bring a premium price.

THE JAMES ROSS FLY REEL

James J. Ross was a Buffalo, New York, carpenter who patented an unusual and somewhat mysterious reel on October 12, 1869. It's unknown whether the reel was produced in any great number or perhaps built for only a year or two; furthermore, the production model looked nothing like the reel pictured in Ross's patent drawings. The only feature that did not deviate from the original patent illustrations was the spool, which had an arbor constructed from a series of rods or pillars. This was a very effective method of allowing a fly line to dry once it was back on the reel, and the basic idea was used later by others, such as A. F. and William Meisselbach and the British firm of Allcock.

The James Ross reel also had a few other odd features, including a couple perforations on the spool plates and a unique skeleton frame shaped like a cross. This frame was similar in principle to the raised-pillar design so often attributed to Frank Philbrook of Brewer, Maine. The Buffalo-designed reel, however, was unique and can't be compared with the work of any other maker. Only one example is known, identified by reel historian Jim Brown. It was built in brass with a single crank tipped with an oval wooden handle. In my opinion, the reel looked like crapola.

I mention the reel and its inventor in passing, since additional James Ross reels may be discovered. All signs, however, indicate that Mr. Ross bowed out of the reelmaking trade shortly after his patent was granted. He returned to carpentry, and "only in his later years did he attain some modest degree of pros-

The plain brass version of the Billinghurst reel, mounted on McHarg's patent reel-seat handle. The seat is raw brass, and the grip appears to be fancy burled sumac or lilac. (The McGrath Collection)

McHarg or McHarg-style rods were usually made entirely from wooden sections. Several woods, including light greenheart and ash with lancewood tips, seem typical of his style. Due to the unique design of his patent reel seat, most rods were detachable, hand-grasp models. A few were built in the very old reel-above-the-hand style. Considering prevailing fashions, these latter rods were throwbacks to earlier decades. McHarg's operation was small, but obviously had wood-turning machines. His rod furnishings appear to have been the products of Chubb and, possibly, Staples & Philbrook.

perity as the superintendent of the Ross-patent refrigerator company," to quote Jim Brown in *A Treasury of Reels.*

JOHN B. MCHARG, ROME, NEW YORK

The earliest small-time rodmaker to take advantage of production rod fittings, John McHarg was an inventive type who fished southwestern Adirondack waters and lived near the banks of the Mohawk River. Among the most obscure of rodmakers, McHarg patented a two-piece, ferruled reel seat that was so sound in theory it was copied by several others, including Fred Thomas. This seat design, patent assigned on March 18, 1873, was the prototype of the "detachable butt," which in a modified form is still used on modern saltwater sticks.

Not all of McHarg's rods carried the detachable-butt reel seat, and a good many examples were unsigned. They had a chamfered reel-foot area that was lower than the round of the rest of the seat, a feature also seen on B. F. Nichols's rods and some of Hiram Leonard's Bangor models. John McHarg used a very distinctive butt cap, shaped much like the rubber crutch or Daisy caps of today, attached at the bottom by a round-head screw in traditional English style. This single item is the easiest way to now identify a possible specimen from the Rome maker.

The one example illustrated, and attributed to McHarg, is a 10-foot fly rod with the reel mounted above a two-tone rattan grip. All the furnishings are heavily nickel-plated, drawn tubes; the ferrules are swaged to a two-piece style and spiked. The shafts are light greenheart and the tie guides are Chubb's, as patented in 1882. A pleasing rod of late McHarg vintage, it was obviously built for the kind of modest angler being wooed by production makers. Yet, perhaps due to his antiquated design, I doubt that J. B. McHarg built a large quantity of rods.

During a production span of a quarter century, the John B. McHarg Company also made a line of flies and early trolling spinners. In 1901 Horrocks-Ibbotson bought out McHarg, and the machinery was transferred to Utica.[1] Unsigned McHarg products usually sell for modest prices.

A. H. "DOC" FOWLER'S RODS AND REELS

The "Doc" Fowler tale actually starts at the very beginning of the industrial revolution, when Charles Goodyear discovered the process of vulcanization. First applied to the stabilization of rubber, it was also used to create a workable gutta-percha, a material employed in the casting of dentures. And this, in turn, brought about several other innovations, including the Gem reel and Poli-Grip.

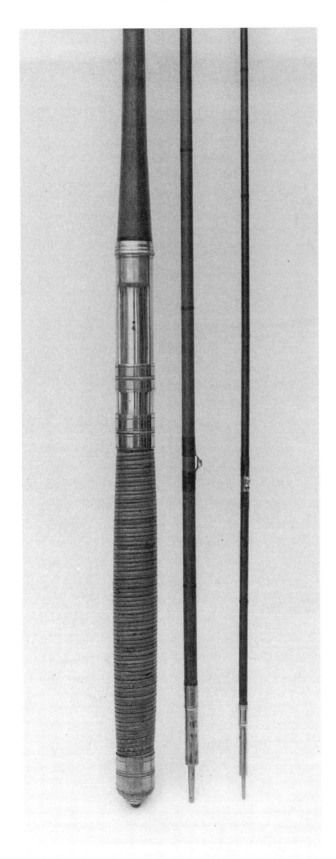

Built in the old style, this McHarg greenheart fly rod has his special touch. Note the seat design, old-fashioned spiked ferrules, and English-style mounting of his distinctive butt cap.

Forest and Stream, July 8, 1893

Alonzo H. Fowler (1825–1903) was a Batavia dentist who seized the moment, on June 18, 1872, to patent a perforated-spool reel cast from gutta-percha. A somewhat fragile version of Billinghurst's side-mounted model, the Gem had the same basic shape, an external click spring, and round perforation holes to aid in drying the line. The reel would become one of the earliest of the hard-rubber genre, but since it was about as hardy as an eggshell, few are left.

At least two little Fowler reels have come to auction, one in excellent condition in 1988 ($14,000), and another with both ends of the foot broken in 1989 ($4,000). Since then several other Gems have been discovered, meaning that these high auction prices will probably never be seen again. The reels are

An incredible Gem reel that's survived the years in amazing condition. The fragility of this dental-casting wonder accounts for the current scarcity of Doc Fowler's reels. (The McGrath Collection)

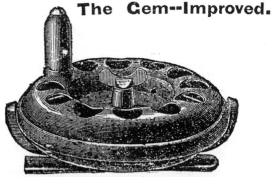

DR. FOWLER'S
PATENT
HARD RUBBER REEL.
The Gem--Improved.

This beautiful Reel has now been before the public two years, and all who have used it speak loud in its praise.

IT HAS BEEN IMPROVED,

and all the Reels made this year will pass through the hands of the Inventor, and none allowed to go out, except those that are perfect.

For sale by all first-class jobbers and the trade generally.

A. H. FOWLER, Inventor,

Feb. 4   ITHACA, N. Y.

Forest and Stream, *April 1, 1875*

marked "Fowler's Pat.—June 18, 1872," or "Fowler's Patent—June 1872." At least two sizes were made: a small 40-yard model and a slightly larger 60-yard version.

Like Humpty Dumpty, one good fall would bring a Gem to ruin, which resulted in the destruction of many a Fowler reel. Like so many products of the Side-Mounted Cult, the Gem is one rare bird. A most interesting feature of its construction is the fitting of the external click spring and its accompanying pawl. Both items were fixed to the reel with a gunmaker's cheese-head

screw. The use of cheese heads on American reels was limited to Alonzo Fowler, Frank Philbrook, and a very few other makers. Such screws should not be confused with a multitude of slightly crowned and similar fastenings, such as those of Edward vom Hofe, which were not truly flat headed.

The talented dentist also produced a modest number of split-bamboo rods, at first rounded but later either in a hex shape (on his six-strip versions) or an octagonal (on his eight-strippers). The most notable design feature of these rods was the shape of the ferrules near the shaft. On an eight-sided rod, for instance, the ferrule end conformed to the shaft as an eight-sided fitting. These ferrules were similar to those used by Jim Daniels, and their origin can be traced to Thomas Chubb's Vermont plant. The Fowler ferrules were spiked until about 1885, when these were superseded by an early drawn capped version.

Writing in 1883, William Mitchell stated that Fowler was one of the first makers to offer six-strip split-bamboo rods. We also know that he was one of the first to build rods with a classic hex shape, vying with Leonard, Wheeler, and Conroy, Bissett & Malleson as the originator. In 1873–74 Fowler moved to Ithaca and began making his rounded fly rods to accompany the Gem. These products were slowly

A close-up view of the Fowler reel, showing the patent marks, click spring mechanism, and cheese-head gunsmith screws. (The McGrath Collection)

refined, and by May 1877 Doc was leaving the entire enameled surface of the bamboo intact: a hexagonal rod.

By April 1878 the Ithaca maker hooked up with a chap named Tisdel. The Fowler & Tisdel models were advertised as "The Original Hexagonal Rods"; the ad stated that "the superiority of the Fowler Rod, AS IT IS NOW MADE, is beyond comparison with any of the imitations offered by parties in the trade." I assume the reference was to Conroy, Bissett & Malleson, who were running an extensive contemporary ad campaign touting Thomas Bissett's hex-shaped models. At this time Fowler & Tisdel were producing enough rods that they could market them in the "City," where CB&M and Leonard wares dominated. Fowler & Tisdel's New York dealer, John W. Hutchison, ran a tackle house at 81 Chambers Street. This may have been a branch store, since Genio C. Scott mentioned a "Mr. Hutchinson" of Utica, an early designer of fly wallets.

The rods made by this partnership, and by Fowler alone, appear to be as rare today as the Gem. The most notable A. H. Fowler model, the D. F. van Fleet presentation rod, was built in 1891. This fine example of Doc's craftsmanship had engraved German silver fittings and ornate reel-seat furnishings, and was an example of Fowler's later work (according to Martin Keane.[2]) I don't believe the Ithaca maker's career continued far beyond this date, and, judging from the relative scarcity of Fowler rods, they were not production items. Today, the remaining Doc Fowler rods and reels are coveted by those who collect early expansion-age tackle.

FRED D. DIVINE, UTICA, NEW YORK

Born in Utica in 1856, Fred Divine became a major American rodmaker, saved from relative obscurity by an overactive and slightly twisted imagination. Divine is best remembered as the first to make a short bait-casting model, a 6-foot, 3-inch lancewood stick constructed in 1885 for J. M. Clark

---

CARD.
Dr. Fowler's Celebrated Six Strip Bamboo Fly-Rods.

Improved and finished without cutting away the enamel. The strongest and handsomest rods in the world. Private orders solicited for all kinds of fine tackle. Rods sent per express on approval. Send for circular. A. H. FOWLER, Ithaca. N. Y.

Forest and Stream, *May 17, 1877*

---

of Chicago. In his lifetime, Divine built some absolutely smashing fly rods, including models trimmed to the teeth. His successors, George F. McDuffee with head maker Frank Becraft, carried the Divine line into the classic period and designed the wonderful Fairy and rugged Gloriwest fly rods. As one of the giants in the early production biz, the Fred D. Divine Rod Company also built rods aimed directly at the fly fisherman with a skinny wallet.

Although he had made his first fly rods a few years earlier, Divine was initially listed as a "fish rod manufacturer" in the 1879 *Utica City Directory*. His early work was bench made, usually from lancewood, and built in a barn in back of the Divine residence on State Street.[3] In the early 1880s Fred was an active tournament caster—a diversion that probably helped him design some reasonably good fly-rod tapers. At this time, Divine extended his rod materials to include bethabara, greenheart, and even exotic Maltese and Dagama woods. The first bamboo rods were offered around 1883, and in less than two years, Fred's split-Calcutta production almost equaled that of his wooden-shafted line.

Through the years Divine acquired a couple U.S. patents. The first, issued on December 1, 1885, was for a semilocking reel seat, which had a ridge running along the back of the barrel and a matching groove in the sliding band. This patented reel seat appeared on many of his products, good and bad, and was often unmarked as to its pedigree. It showed up, as a full metal, nickel-plated brass affair, on heavier models, including fly rods designed for salmon and grilse. On quality trout rods, the nickel-silver version was marked "F. D. Divine, Maker, Utica, N.Y.,"

1882 Utica City Directory

along with the patent date. Any lucky collector who finds this marking can be assured that the rod was actually built by Fred himself. This earliest of Divine stampings was applied until 1888, when Frank Wolcott and then young George Becraft, Frank Beecraft's brother, joined him at the bench. Rod markings were then changed to "The Divine Rod, Utica, N.Y.," and used for just two years, according to Divine historian Mike Sinclair.

In 1890 the business was restructured as Fred D. Divine & Company. Once again rod markings were altered, to the last and the most common Divine reel seat stamping: "The Divine Rod." In 1892, Fred patented the process used to make his twisted Spiral Rod, a method that made the blank stronger for a given length. He also produced a silk-wrapped Silken shaft, a patent actually issued to John Kenyon in 1897 and reportedly assigned to Thomas Chubb. These off-the-wall rods were surprisingly popular with fly fishermen at the time, and enough of them were made that collectors still see an occasional model.

Fred Divine finally built a new rod factory in 1898, a substantial brick building constructed at the corner of State and Roberts Streets, equipped with the newest lathes and

presses, all powered by an overhead belt system.4 The benefits of the new mechanized factory would elude Fred Divine, however. On the afternoon of March 16, 1900, his coat caught in the giant belt when he was inspecting the overhead after a fire, and he was hauled through the feed six or seven times at the belt speed of 500 revolutions per minute. He lived until 8 the next morning, fully conscious the whole time. His obituary in the *Utica Daily Press* read: "He realized his condition, and his only concern was not for himself, but for his wife and mother, and the anguish his death would cause them."

While the majority of original Divine products were not top-quality items, some fly rods were built with care and a great number of man-hours, as seen by their close-spaced intermediates, German silver drawn and welted Chubb-style ferrules, and superb imitation-cherry reel seats. Since many of Divine's rods were sold unmarked through eastern and midwestern hardware stores, the sliding-band "cherry" seat is practically a Divine trademark for identification. Early ferrules carried spikes, omitted on the following two designs. The low-end mid-

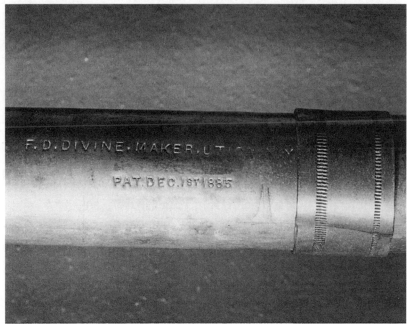

Fred Divine's 1885-patent reel seat allowed the sliding ring to slide lengthwise on a raised rail. Not exactly an earth-shaking development; it looked better than it worked.

1880s ferrule style, seen only on lesser Divine products, had a fake welt at the top of the female made by bulging the ferrule to look like a more expensive fitting. In contrast, Fred also advertised a "double shoulder waterproof ferrule," an actual predecessor of the famous Super-Z ferrule and obviously his best effort. Some split-bamboo models, in true Mohawk Valley tradition, were built in the eight-strip configuration. The Divine rod would continue into the classic period under the able direction of George McDuffee (see chapter 9).

THE BUCKINGHAM AND WARING EIGHT-STRIPPERS

A few other New York State makers, such as H. P. Buckingham and N. K. Waring, would follow Fowler's and Divine's lead and market eight-strip split-bamboo rods. Unlike Doc and Fred, however, they never settled back to the superior design of the six-stripper. Sometime after Divine became active, Utica's H. P. Buckingham and a man named Perrie began offering their own rods for sale. Perrie had vanished by the time Buckingham advertised in the June 11, 1881, issue of *Chicago Field* that he had available "Split Bamboo Rods—Eight Strip from Butt to Tip—The Finest Made For Trout, Bass, Grilse, or Salmon."

With a little push from the Erie Canal, Utica and Rome were linked to the Great Lakes and Chicago—a direct pipeline to the sportsmen of the Midwest. Buckingham, like McHarg, Fowler, and Divine, took advantage of the midwestern market, and I believe that many of the makers' products went in that direction. The Buckingham eight-strippers were definitely obtuse, and very few have surfaced. Old H. P. "Eight Strip from Butt to Tip" Buckingham summed it up in his own motto—just one reason to be banished from rodmaking's mainstream into eternal obscurity.

At this same time another New York State maker, N. K. Waring, produced a very modest number of eight-strippers. Living in Newark Valley, just above eastern Pennsylvania, Waring built his rods into the 1890s or perhaps later. The estimate of his production is based on the offerings of antique- and classic-rod dealer Len Codella, and the listing dates back to the 1970s. At the time Len offered three rods marked "N. K. Waring, Newark Valley, N.Y." One was a 9-foot, 6-

Forest and Stream, *April 24, 1879*

inch trunk rod, built in a six-section design with two extra tips, a ring cork grip, and a German silver sliding-band reel seat. It also had German silver ferrules, loose-ring guides, and red wraps and intermediates. The rod, marked with serial number 264, was contained in a unique tapered, canvas-covered wooden case, with each section inserted from the end.

The second and third Waring rods were 10-foot, 6-inch salmon models, circa 1890, also made from eight-strip Calcutta with a classic swelled butt and a rattan grip. These three-piece rods had two mids and four tips, and German silver fittings; the seats were stamped by the maker, with the serial numbers 234 and 277 added. Probably grilse models, the two rods had loose-ring guides and were put up in cases similar to those used for the shorter rod. What we know

Chicago Field, *June 11, 1881*

Obscure Upstate Rodmakers

Among the more obscure upstate New Yorkers making fly rods in the late 19th century was W. L. Hoskins of Oswego, who in 1879 was guaranteeing his eight-sided rods "to be the best made rod in the world." Two years later we see ads from Cazenovia's Will Cruttenden, a wood turner who used McHarg fittings on his ash and lancewood rods. In 1887–89, the Syracuse Split Bamboo Fish Rod Company was making hexagonal models, available packed in fiber tubes, wooden forms, or tips cases, as the price increased.

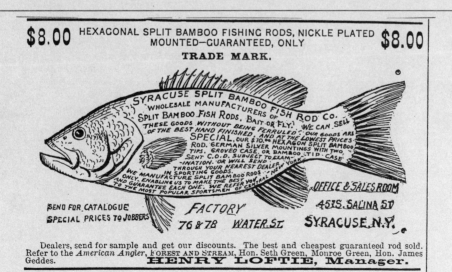

about N. K. Waring comes from these three rare examples, but it's obvious that he built at least 277 rods—or more. In terms of *known* production, however, N. K. Waring is still a most obscure upstate maker.

PETTENGILL'S PATENT REEL

As the expansion age blossomed, more Mohawk Valley reelmakers entered the trade, carrying the perforated-spool genre through the late 1880s. One such inventor-maker, Albert N. Pettengill, built a new reel in Ilion, a town just below Utica. Patented on April 26, 1887, the reel came in two styles: side-mounted on the rod, or attached in an upright position. A notable feature of the side-mounted Pettengill was its one-piece tubular frame, which encompassed the rotating spool.

An examination of the few Pettengills available shows that these reels were built in narrow-spooled, odd-numbered models and wide-spooled, even-numbered counterparts. The No. 1 contracted model looked like the No. 3 but was side mounted. The No. 2 reel—marked only with a large "2" on the face or foot, along with the usual stamping, "A. Pettengill Pat. P'D'G"—was the wide side-mounted version. The No. 3 and No. 4 reels, similarly marked and with the Pettengill stamp on the front of the spool, had the upright foot for mounting on a reel seat. Most of these models seem to carry an average diameter of 2¾ inches.

Both styles of Albert Pettengill reels are scarce today and often have more wear than collectors would like to see. Built from nickel-plated brass with a wooden handle, they were vulnerable to abuse. Thus these models, in good condition, sell for around $300 or more—a lot of shamolas for a light stamped-metal reel. As far as production years go, the facts are still a mystery. Was Pettengill so poor that he couldn't afford a new stamping die after the reel was accepted by the patent office? Maybe he just didn't care. Considering the relative scarcity of all New York State products, the Pettengills sell for a reasonable price. And the old saw—"They just don't make 'em like that anymore"—fits well here.

THE ABRAHAM COATES REEL

A unique example of the Side-Mounted Cult, the Coates reel was built according to an odd con-

A pair of Pettengill fly reels: a standard wide-spooled model (left), and a narrow-spooled side-mounter (right). (The McGrath Collection)

cept. It was conceived and purportedly built by Abraham Coates of Watertown. On March 20, 1888, he was granted two U.S. patents—one for the old rod-and-reel combo idea and one for a really unique fly reel that actually went into production. The heavily nickel-plated brass reel had two speeds—sort of, provided the user possessed tremendous agility and digital dexterity. This rather small reel functioned as a single-action model with its handle located in the center of the face plate. For the adventurous, however, a screw nut that rose from a post near the edge of the plate could be removed, and the handle transferred to this new location. Now the angler had an impressive 1½-to-1 retrieve ratio. Talk about gainin' line, fella!

The major fault of the design, other than the silly gear ratio, was the pea-sized retaining screw. After moving the handle into its new position, an angler had to screw the little retainer into the center hole. Oops! It just dropped into the stream to nestle among a zillion pea-sized pebbles. This absolutely dopey design gives the Coates reel a very special charm, and when it's found totally intact, with the little nut still on one of the posts, it's a great treasure.

Although I'm not positive, the reel may well have been the first American two-speed model. Judging from the current supply of Abraham's antique, which was well marked as to its maker and origin, it had a short production life. The Coates reel was one of the most eccentric of all American fly reels. I love models that fall into the wicked-odd category, and this obtuse fly fisherman's gadget reigns near the top of the heap. Collectors usually *pay* for this small chunk of brass.

FOLLETT AND CLINTON "KNOCKOFF" REELS

The patented reels of Billinghurst and Fowler brought imitations out of the woodwork. Somehow these knock-off reels made it through the U.S. Patent Office—probably on a slow day, perhaps a Monday following a hellacious weekend. Follett's "patent reel" was a stamped, plated-brass version of Billinghurst's 1859 original. It even looked like a Billinghurst, but without the expense of arduously soldering brass rods into a doughnut shape. This was the brainchild of Edward P. Follett, an itinerant lanternmaker and reel inventor from Rochester who would later (1889) actually patent a rod–auto-reel combo that was manufactured for a very short time in Olean, New York.

Very little is known of Follett or his years of reel production. In the late 1870s he was in the tin-knocking trade, which certainly was handy for producing a stamped sheet-metal gadget like his reel. Follet was recorded in the *Rochester City Directory* as a "reel-maker" from 1884 to 1889, which may indicate his major productive period in the tackle business. There seems to be

One of the most charming of wicked-odd American angling inventions, the Abraham Coates reel was a (kinda) two-speed model. (The McGrath Collection)

no corroboration of any actual patent date for the "Follett reel." Neither Steve Vernon nor Mary Kelly included "Follett's patent" in their excellent antique-tackle books. Jim Brown, who researched every available reel patent for his now expensive little tome, *Fishing Reel Patents of the United States*, didn't find it, either. So I believe that Edward Follett's patent lived inside his head, where it rattled around long enough to seem real, as was the case with old William Mills's "patent" salmon reel seat. In the time of Follett and Mills, phony patent claims were easy to foist upon rubes living in the sticks.

The Follett trout reel has shown up at several national auctions, usually described in such glowing terms as "Scarce Early All Original Follett Side Mount Trout Fly Reel. Difficult to find with original wooden handle, this 3⅜″ dia. reel has two tiny patches of brass showing through the nickel plating otherwise excellent condition." So stated the blurb in Oliver's First High Roller's Auction catalog. The Oliver reel, by the way, sold for $500. At the next High Roller's Auction, in 1989, a similar Follett sold for $220.

Another obvious knockoff, and an almost spittin' image of Doc Fowler's dental accomplishment, was the famous Clinton reel, noted by collector Steve Vernon, who discovered the remarkable similarity. As Yogi Berra said, "It's déjà vu all over again!" Ithaca's Charles M. Clinton (1834– 1909) patented the reel on October 29, 1889. The only design changes between Doc's version and the new one were Clinton's oil reserve in the hollow shaft and the use of metal instead of hard rubber.

The Clinton reel made a small name for itself, being used as a cut in Henshall's 1904 edition of the *Book of the Black Bass*. It was also sold through Abercrombie & Fitch until 1910 or later, according to Steve. Today, a Clinton reel, intact and in good condition, is worth a tidy sum, though less than Fowler's original. Two ver-

sions can be found—but not very often. The German silver Clinton, the hardier of the two, was the model with the highest chance of surviving intact. The other reel, made from aluminum, was a fragile affair. In view of the overall scarcity of all New York State side-mounted reels, the Billinghurst, Fowler, and Clinton reels sell at high market prices—and they should.

Copying another product would become commonplace in the tackle world, and it was evidently okay to make an item that was the same, only different. The reels of Clinton and Follett were blatant imitations, but sometimes pinning down the origins of the egg and chicken is not all that easy. Such is also the case with the "double-built" fly rod.

THE EUGENE M. EDWARDS DOUBLE-BUILT ROD

Eugene M. Edwards of Hancock, New York, was an active rodmaker from roughly 1882 until the mid-1890s; there may be a connection between this Mr. Edwards and the family of Eustis Edwards, who came from Maine. In 1895–98, a Eugene M. Edwards, listed as a "traveling salesman," lived at 7 Newbury Street in Bangor. This may just be a case of "the same name, only a different guy."

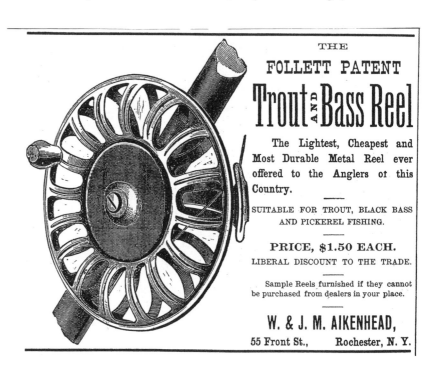

Forest and Stream, *December 21, 1882*

A rear-view shot of Follett's "patent" reel, featuring the combination foot-and-spool-post screw. Note how easily the spool sides could be stamped from thin sheet brass. (The McGrath Collection)

John Harrington Keene[5] extolled Eugene's rods, mentioning that they were built with double layers of bamboo, one inside the other. He quoted Edwards:

I have, for about ten years, made the single enamel split bamboo rods. During this time

An immaculate aluminum version of Clinton's oil-reserve, ventilated-spool reel. By paging back to Fowler's Gem, you can see the uncanny similarity.

I have used nearly all kinds of ferrules to prevent their breaking off at the ferrules. Finding this was a failure, generally, as the bamboo was strong only on the outside—the inside being at best very poor—the idea occurred to me to plane away as much of the inside as was poor, and glue another piece of enamel in its place, thereby making the whole rod out of enamel.

Eugene M. Edwards received a patent for his idea and method on November 17, 1885—years before the Cross Rod & Tackle Company would use the same configuration. The notion, however, was not strictly American. Hardy Brothers not only made a "built up" model contemporaneously with Edwards, but added a steel core down the center. This, of course, made for a good heavy item, one that could be used to beat back the Hun should one of Her Majesty's soldiers accidentally lose his Martini.

THE AUTOMATIC REEL

As an angling curiosity, the automatic reel also stemmed from the New York State Side-Mounted Cult. The first of these wicked-odd thingies was patented on December 7, 1880, and again on July 5,

1881, by Francis A. Loomis of Onondaga. On the second patent, Loomis assigned 50 percent of the rights to a James S. Plumb, who lived nearby in Syracuse and had the wherewithal to manufacture the thing.[6] The twangy arrangement was soon in production by the new Loomis, Plumb & Company. It was a big hit, available in several differing metals, including plain and nickeled brass, plus three finishes in bronze. Not exactly a lightweight, the Fisherman's Automatic Reel probably helped balance the standard 10- to 12-foot rod of the period.

By the mid-1880s Loomis and Plumb sold the business to Phillip H. Yawman and Gustave Erbe of Rochester. Original Yawman & Erbe models were built from the older company's stamping dies, complete with L&P's patent markings; the only difference was the new maker's name. In ensuing years, Phil Yawman made design changes that culminated in two new patents, assigned on June 16, 1891, and later on August 1, 1899. The latter patent involved the

Y&E Automatic Reel, the H-I gadget was marketed until 1923 or later.

Due to the immediate success of the original automatic reel, others got on the bandwagon. In Syracuse, sporting goods retailers Willis S. Barnum and Frederick A. Morehouse began the manufacture of a second auto-reel under the patents of Franklin R. Smith, which were secured on July 26, 1881, and June 20, 1882. This reel had a very short life. Our next participant, jeweler Herman W. Martin of Ilion, would fare much better.

Martin received three patents for his much-improved reels on July 26, 1892; December 17, 1895;

MARTIN AUTOMATIC FISHING REELS

and November 30, 1897. The original models came in four widths, ranging from a narrow 5/16 inch to 13/16

The Loomis, Plumb & Company automatic fly reel, later Y&E edition, from the Maine Sportsman, *August 1895.*

unique key winding device, and the new models were called Improved Automatic Reels.

The "newer" versions were constructed in brass, nickel-plated brass, bronze, hard rubber, and aluminum, but they appear to have been made in only two sizes: the No. 1, which carried 90 feet of fly line, and the No. 2, with a 150-foot capacity. Sometime after 1910 Yawman & Erbe sold the company to Horrocks-Ibbotson, and the patterns and machinery were transported from Rochester to Utica. Now called the

inch, and appropriately named by Thomas Bate Mills the Creek Whipper, Fly Casting Reel, Bass Reel, and Salmon Reel. Early reels by the Martin Novelty Works were actually quite handsome, with a compact design and an etched filigree that encircled the face plate. Around 1907–08, the company's name was changed to the Martin Automatic Fish Reel Company and, in 1921, the business moved to Mohawk. The Mohawk reels had cosmetic changes to the face, and the handsome filigree was replaced by a couple trim bands. Through the years the Martin automatic reels underwent further changes, but these later models have less collectibility. The Martin Reel Company, as it is now called, is still in business.

As an aside, the automatic reel showed up again, in New Jersey, where a fancy aluminum model was made by August Meisselbach. The side-mounted auto-reel has a certain following among collectors, and early Loomis & Plumb, Yawman & Erbe, and Martin filigrees are now prized.

AN OLD IDEAL-GEM-KELSO MYSTERY

Certain tackle products can't be pigeonholed, and the Ideal, Gem, and Kelso reels all fall into this category. The Ideal was built to the patent of Andrew Wollensak of Rochester, New York, the rights being assigned to his optical company. The patent, for a three-position, lever-operated drag spring, click, and "free spool," was obtained on September 6, 1910—almost two years after its so-called maker, the "Rochester Reel Company," supposedly went out of business.[7] This upright Perforated Click Reel came in two sizes, the No. 1 or trout model (80 yards), and the No. 2 or bass size (120 yards). Both were stamped from German silver; they were priced at $1.25 and $1.75, respectively.

The second model, the fragile Gem, was also built from German silver or from nickel-plated brass. A traditional side mounter, it had an adjustable click and a line-guiding arm and sold for a mere 50 cents. The No. 1 Gem held 60 yards of fly line, and the larger No. 2 could accommodate 80 yards. The Gem was similar to but even less rugged than the Ideal, having no frame to speak of and a central spool shaft attached to the foot. Few Gems in good condition have survived, and collectors value them highly.

So far, only the Ideal models have been found in sufficient numbers to yield much information, but what we find is no help at all. The company, Rochester Reel, was a one-year successor to Carlton Manufacturing, maker of the famous nine-speed multiplier. The address usually found is "90 Chambers St. New York," which was occupied by Herbert J. Frost's manufactory and salesrooms. I believe that Frost only wholesaled the reels, though he could have produced them.

So who built the Ideal and Gem? I'm not sure, but a likely candidate is Wollensak or perhaps a small fly-fish-by-night manufacturer in the Rochester area. It's apparent that Herb Frost was the promoter and wholesaler. The reels sold through some prestigious houses of the 1907–11 era—not only H. J. Frost & Company, but also Orvis and Abercrombie & Fitch. Most of Frost's products originated in upstate New York, and I believe the region also produced these somewhat rare items. In the future, an itinerant collector will probably crack this old mystery—and meanwhile we can just enjoy the "Rochester" reels for what they are without getting constipated about it.

The New York State makers offer collectors an array of very unusual tackle. Upstate craftsmen left us their unique eight-sided fly rods, and we can also see an amazing design continuity in the products of the Side-Mounted Cult, extending from the Morgan James reel to the Gem. In the same vein, there's a relationship between the Loomis & Plumb, Martin, and Kelso automatic versions, be it coincidence, fate, or karma. Just plain wicked-odd!

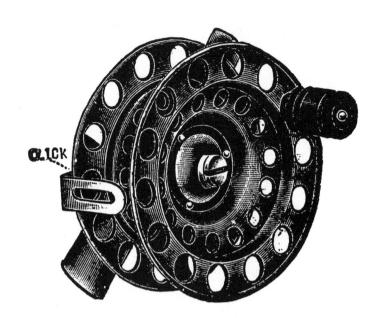

The Gem perforated fly reel. I haven't seen many of these lately; if they exist, they're sleepers.

Notes

[1]Sinclair, Michael. *Bamboo Rod Restoration Handbook.* Grand Junction, Colo.: Centennial, 1994.

[2]Keane, Martin J. *Classic Rods and Rodmakers.* New York: Winchester Press, 1976.

[3, 4]Sinclair, Michael. *Fishing Rods by Divine.* Grand Junction, Colo.: Centennial, 1993.

[5]Shields, G. O. (ed.) *American Game Fishes.* Chicago: Rand McNally, 1892.

[6]Brown, Jim. *Fishing Reel Patents of the United States, 1838–1940.* Stamford, Conn.: Trico Press, 1985.

[7]Brown, Jim. A *Treasury of Reels.* Manchester, Vt.: The American Museum of Fly Fishing, 1990.

6 Brooklyn Tacklemakers

In the history of American tackle production, no location can compare in antiquity with Brooklyn Ferry and the Fulton region of lower Manhattan. This area spawned the earliest of our fly-fishing implements, going back to the early 1800s and the original shops of Conroy and Welch. These were followed by rod- and reelmakers J. B. Crook, William Mitchell, the Pritchard Brothers, and the vom Hofe family. The majority of these artisans lived in Brooklyn, which then included the boroughs of Kings and Queens, and several individuals had shop fronts across the East River on the isle of Manhattan.

Until 1883, when Roebling's Bridge was completed, many tacklemakers went to work by ferry. By the end of the century this core of lower Manhattan tackle shops and retail houses had spread from Fulton Street to Pearl, Vandam, Chambers, Reade, and Beaver Streets, as well as Maiden Lane. Only in London could a tackle buyer have such a diversified array of rods, reels, flies, and paraphernalia from which to choose.

When the great tackle expansion age arrived, several rod and reel "factories" were actually established on the east side of the river, in Brooklyn proper. This was where Julius vom Hofe's reel operation was located, as stamped on his earliest products. Here, too, the "Kosmic" boys would make the majority of their rods—and this was where Otto Zwarg started his short but exemplary career. No other location in the United States would produce as much consistently high-quality fly tackle as would the Brooklyn-Manhattan area.

One of the least-known Brooklyn manufactories, the U.S. Net & Twine Company operation, pro-

duced a lot of cheap tackle yet made a number of extremely collectible reels marked with the Kosmic label. Another obscure manufacturer, John G. Landman, built a number of high-quality rods—some of the most beautiful ever to enter the market—yet he never got credit as their maker; most were marked with the retailer's name.

To lend a full and uncensored view of the metropolitan tackle trade, I should mention that at least one of these late-1800s shops appears to have employed child labor for piecework—the sort of operation condemned by Beatrice Webb and Upton Sinclair. I shrink at the thought of small hands soldering ferrules and assembling reels for 12 hours a day, but these operations were not typical in the tackle biz.

THE HOUSE OF CONROY

As the oldest tackle establishment on Fulton Street, Conroy's had enjoyed a sterling reputation in the hand fitting of fly reels and wooden rods for almost half a century. As a wholesaler it sold everything that eastern retailers could possibly need, including fly lines, landing nets, creels, wet flies, and rubber bugs. In 1875 the Conroy company was restructured, and Thomas J. Conroy, son of J. C. Conroy, took on Thomas M. Bissett and Frederick Malleson as partners. Up until this time there is every indication that Conroy's had maintained an extremely conservative attitude in the making of rods, fashioning monster sticks from ash and lancewood.

Under the new ownership, and at least by 1877, Conroy, Bissett & Malleson was producing "Six Strip

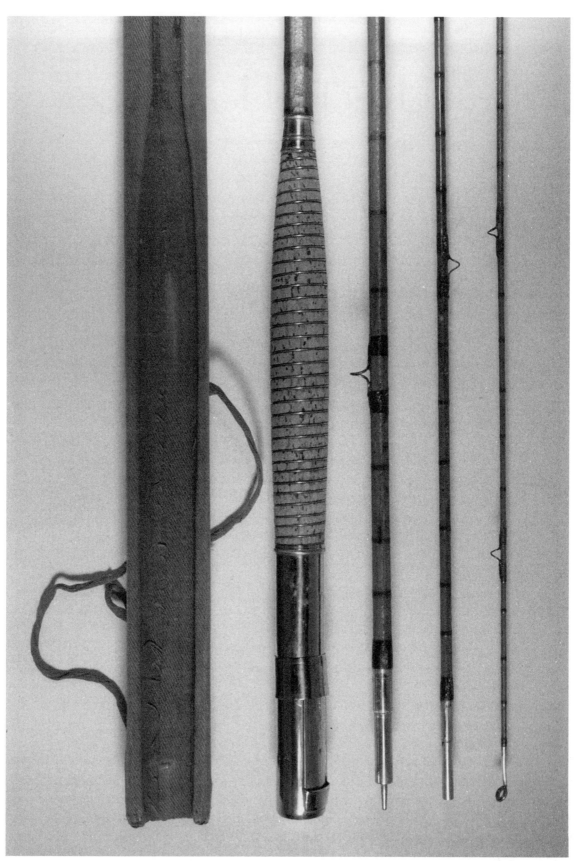

An early-1890s-vintage fly rod showing the infrastructure of the tackle trade. The piece is marked "Thos. J. Conroy, Maker, N.Y.," has the same reel seat that's found on most Thomas & Edwards rods, and was actually built by the John Landman factory in Brooklyn.

CONROY, BISSETT & MALLESON,

MANUFACTURERS OF

FINE FISHING TACKLE,

65 Fulton Street, N. Y.   Factory, Brooklyn, E. D.

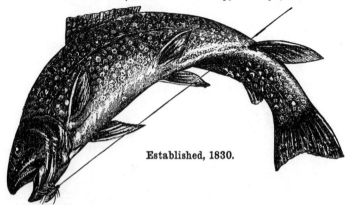

Established, 1830.

SEND FOR NEW ILLUSTRATED CATALOGUE. Price 15c.

Book of the Black Bass, J. A. Henshall, 1881

Split Bamboo Rods—either round or hexagonal." This giant leap into the competitive world of split bamboo was accomplished by the tedium of long hours per product, and by high price. "These rods are, in all their parts, WHOLLY MADE BY HAND. Positively no machinery is used in any part of their construction," stated its advertisement in *The Country*. I doubt that more than a few anglers ran into Conroy's and slapped down $80 for a salmon rod—a lot of money in the days when a worker might make 80 cents a day—or even $40 for the trout model.

By December 8 Conroy, Bissett & Malleson dropped prices in an attempt to hit the Christmas trade. The trout rod sold for $25, the salmon stick went for $50, and a bass model's price fell from $50 to $30.

Judging from the relative scarcity of all CB&M rods, they were made in limited numbers, the old-fashioned way. I have examined the bass model, and outwardly it was a beautiful rod, with strips of cedar running between the six-strip bamboo handle. The ferrules, hand seamed and rolled, of course, were built in two pieces (both male and female) and had a spike. The shafts were formed in a very crisp hex shape, considering that they were made while Hiram Leonard was emerging from his "round period."

Conroy, Bissett & Malleson rods were not built at the 65 Fulton showrooms, but were produced in the factory at the foot of South Eighth Street in Brooklyn. In 1881 Frederick Malleson, who had operated Conroy's rod and reelmaking department, left the partnership to start his own company. The remaining partners became "Conroy & Bissett." Malleson replaced the old seamed Conroy fittings with new swaged reelseat tubes and butt caps. The nicely rounded Malleson butt caps appeared around 1882, ushering later Conroy & Bissett rods into the machine age. It is likely that Fred Malleson rented the Eighth Street factory from his former partners, as neither Thomas Conroy nor Bissett had an interest in manufacturing. The Malleson rod shafts continued the same mortised swell that distinguished the CB&M models.

Of all the various markings of Conroy products, "Conroy & Bissett, NY, Quality Guaranteed," is the toughest to find today—not that CB&M rods and reels are all that plentiful. Recently a metal reel tin was found, marked not only with the Conroy & Bissett logo but also with the address "Soho Square, London, England," indicating a distinct possibility that the company may have had a British branch store. The C&B stamping was used for only two years; in 1883 the old tackle firm became the Thomas J. Conroy Company.

One of the few Conroy & Bissett items I've examined, a 12-foot fly rod, had a construction similar to the Conroy, Bissett & Malleson bass model. The C&B rod had a cedar mortised grip, the female ferrule rolled in one piece, and a two-piece male with a waterproof spike. Extreme care was needed to make a mortised rod, allowing for a large swell where the six strips of cedar tapered to meet the split-bamboo shaft. Often built to this style, CB&M and C&B rods exhibited top craftsmanship, although fine contemporaneous mortised examples were also made by Chubb and Leonard.

Conroy & Bissett's 12-footer was one rugged fly rod. I didn't weigh this beast, but it was wicked-heavy—either a powerful casting instrument or an

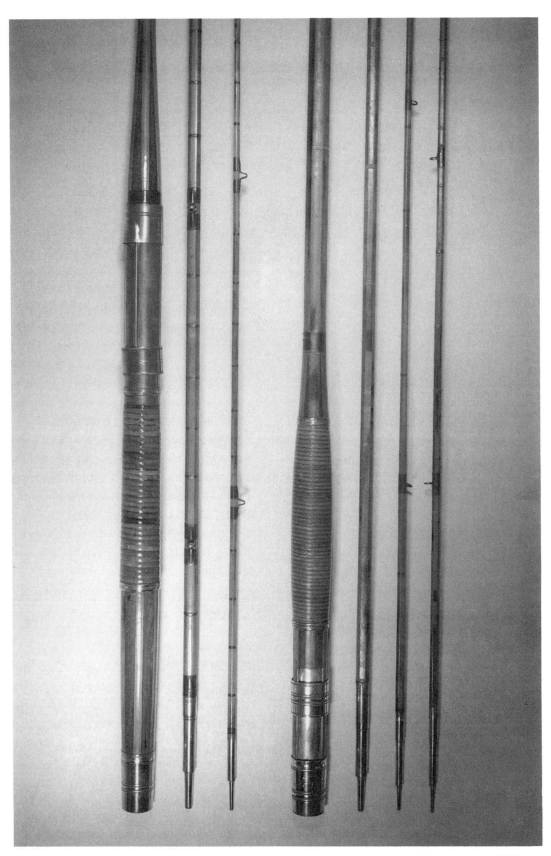

A pair of mortised landlocked salmon rods attributed to Fred Malleson. At left, an 8¾-foot bait model for sewn-smelt fishing, marked by Conroy, Bissett & Malleson. The fly rod at right, stamped "Conroy & Bissett," is a 12-footer designed for a heavy line and for casting into early-season winds.

Details of the markings and butt-cap style of the two previous Malleson rods. The trim bands are identical, with much more wear showing on the CB&M cap. The bottom disks were stamped from the same die.

excellent pry bar. Considering the delicacy of an average Leonard or Orvis of the period, it's no wonder that the C&B clubs were slow sellers. Unlike the British, the American sportsmen of the 1880s weren't into weight lifting, and "T. J." was certainly aware of the antiquated state of the company's products.

In debuting a new line of tackle, T. J. Conroy

wasn't always candid with his valued customers. Conroy was a pitch man. He really didn't fib; he just left out a few details. A September 27, 1884, advertisement in *The American Angler* touted "New Pattern Steel Pivot Multiplying Reels" and "New Styles" of rods. Within a long-standing Fulton Street trade noted for integrity, T. J. was now selling reels obvi-

J. C. 1930.	J. & J. C. C. 1840.	J. C. C. & CO. 1864.	C. B. & M. 1875.	C. & B 1881.

CONROY & BISSETT,

Importers and Manufacturers of

FISH HOOKS, FISHING TACKLE, &c.

LAWN TENNIS,

ARCHERY & CAMPING OUTFITS, &C.

T. M. BISSETT }
T. J. CONROY. {

65 FULTON STREET.

New York, _____ *July 2* _____ *1883*

The American Angler, September 27, 1884

ously built by Julius vom Hofe and marking them "Thos. J. Conroy, Maker, N.Y." He used the identical stamping on rods built by an independent Fred Malleson, or by John Landman, and on several models carrying Gilbert Bailey's new patent reel seat. While many of these products were still made in Brooklyn, they were contract items, and the famous old house of Conroy had turned into just another lower Manhattan retail tackle house.

In 1892 Conroy moved to Broadway, where his theatrics were more appreciated. Thomas Conroy always liked a gimmick, or "novelty," as he termed it, and he was constantly looking for the unusual. By 1897 the company again relocated, to 28 John Street, and continued selling tackle until 1911 or perhaps later. Although the 80-year-old business finally died with a whimper, it had outfitted three generations of American anglers, catering to the needs of the saltwater and fly-fishing brotherhoods. Today, collectors revere the Conroy products, the wonderful old click reels and split-bamboo rods that are veritable museum pieces. We may joke about the crudeness of the earliest Conroy items from the smith age, but deep inside we respect them for the tradition they spawned.

FREDERICK MALLESON'S REEL

There are indications that, even before Frederick Malleson left Conroy and Bissett, he'd applied for the

A John Landman–built rod, as sold by Conroy. The German silver spiral grip winding was used just prior to the emergence of the famous "narrow cork rings."

patent for his new multiplying fly reel. The key features of the unique Brooklyn-built item were its ability to retrieve line quickly—helpful in playing a fish from the reel, as is often done on large trout and land-locked salmon—plus good line capacity, light weight, and a handle protected by a raised rim, as was used earlier on Frank Philbrook's classic. Some of the

German Silver Click-Reel.
(Conroy, Bissett & Malleson.)

first Mallesons were built round, without raised pillars, and were stamped "Patent Applied For" on the back gear cover. A 2¼-inch model in this configuration, with most of its nickel plating, sold at a 1991 auction for $350.

A more "common" version of the reel—for lack of a better term—was a slightly smaller raised-pillar model in a 1⅞-inch diameter. Marked "Patented by Fred'k Malleson, Sept. 4, 83," again on the rear plate, the reel was built of plated brass or aluminum and hard rubber. The similarity of this reel, in profile, to the Leonard bi-metal model is striking. But the fact remains that Malleson's product, built in his small Brooklyn plant, was one of the first multiplying fly reels to have the handle located centrally. The handle shaft extended through the spool and engaged the rear gear box; unlike most fly reels built as multipliers, the Malleson reel was free turning. Some reels, if you remove the rear cover, have Malleson's name and patent stamped on the back plate above the gears.

The reel was also made in a single-action version, which, like the multiplier, came in several sizes. Mallesons were also built with a hard-rubber side plate and back cover, probably in an attempt to reduce weight. There's a striking similarity between Malleson and CB&M raised-pillar mounts, so it's possible that Malleson used some of the old CB&M dies. When Fred left the partnership, he did so on good terms, a fact punctuated by a C&B logo found stamped on a signed Malleson—a single-action model built from aluminum and hard rubber.

All known Mallesons also have an arrow-shaped click button on the recess of the front plate; when engaged, it twitches to each click. For its time, Frederick Malleson's reel was a breath of fresh air. For whatever reason, Malleson closed his facility sometime between 1886 and 1889—as the story goes—moved to Trenton, New Jersey, and operated a large wholesale tackle business for another decade. Thus the well-designed Malleson multiplying fly reel entered history and became one of the most sought-after fly-fishing items for today's collector.

THE PRITCHARD BROTHERS' A&I RODS

The fifth known U.S. rod patent went to Henry Pritchard of Brooklyn, New York, on October 4, 1859. It featured tie guides soldered onto sliding sleeves of varying diameter, adjustable in case a rod should twist or warp (see figure 6-10). It's safe to assume that

A superb example of Malleson's multiplying fly reel. This small model features rubber side plates inside pillar retaining rings, and the famous "twitching" click. (The McGrath Collection)

No. 25,693.—HENRY PRITCHARD, of Brooklyn, N. Y.—*Improvement in Guide Rings for Fishing Rods.*—Patent dated October 4, 1859.—This invention consists in combining a movable thimble, (having the eye permanently attached thereto,) with a ferrule securely fastened to the fish pole, in such a manner as to allow the thimble to turn freely on the ferrule, thereby adjusting the eyes on the thimbles to one and the same line, in case the pole should get warped or twisted.

Claim.—The combination of the fixed ferrule *f* and the movable thimble *c* with its eye *d*, constructed substantially as described, for the purpose set forth.

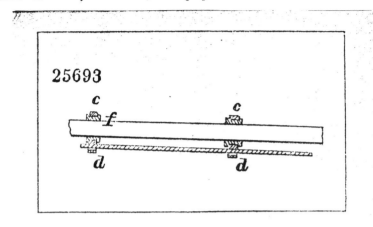

Henry Pritchard was in the biz by 1859. Just how novel an idea the sliding guide really was is open to debate; Arthur Tait's 1854 painting of Daniel Webster at Carman's Brook showed similar guides. Nonetheless, Henry Pritchard's guides wound up on a good many rods—usually the rock-slamming saltwater jobbies of marginal price.

The Pritchard Brothers, Henry and Tom, came from England and started a little tackle shop at 94 Fulton Street, Brooklyn, in the 1850s. Tom was the elder, a quiet sort who wore gray muttonchop whiskers. Although a chronic stutterer, Henry was a jovial storyteller and an accomplished tournament fly caster. In the field, Henry was a "noisy" yet successful angler, a rare combination. Fred Mather, who knew the Pritchards personally, mentioned that "the little shop upstairs was kept busy by anglers who knew their skill, and also by some of the largest fishing tackle houses . . . so the brothers found plenty of work." The store was a social hangout, and when it finally closed, Mather suggested that New York fishermen form an angling club to take its place.[1]

The brothers' September 1865 advertisement, lamenting their fly selection as "early diminished" during the season just ending, indicated just how many anglers had returned to the streams that April immediately after the end, on the 9th of that month, of the American Civil War. The survivors of this gen-

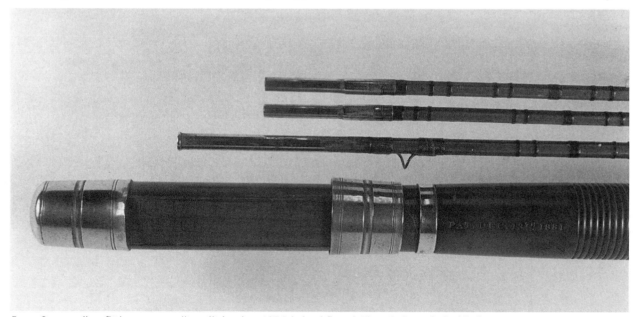

Fancy German silver fittings on a quality split-bamboo A&I Pritchard fly rod. The grip is marked with the 1881 patent.

Forest and Stream, February 18, 1875

eration of soldiers were the first to stand in line at the beginning of the next revolution: the Tackle Revolution. And of course the Pritchards were there to pass the ammunition, so to speak, in the form of popular fly patterns.

All was quiet on the Pritchards' tackle front until November 30, 1880, when Henry got his second patent award for an improved reel seat. The sliding band had internal shoulders that fetched when the band was turned, thus locking the reel into place. This is now a rather tough patent to find on Pritchard rods and, in truth, perhaps it didn't work. Oh well; maybe the next one will bring fame and fortune.

The lightbulb flickered again on December 13, 1881, when Henry (also called "Harry") received a patent for a layered handle of wood and an "elastic substance." This was a rod grip, perhaps first built with celluloid over wood, that became standardized as "Harry's rubber handle." He assigned one-half of the patent to himself and the other half to Abbey & Imbrie, just a few blocks away on Vesey Street.

After such a fine patent, Henry languished for a time, but on January 10, 1888, our inventor came up with a "double beveled wedge" in the sliding band. A reincarnation of Pritchard's 1880 version, this 1888 sliding band actually worked and can now be found on numerous rubber-handled rods.

For all Henry Pritchard's patent mania, no rods have ever been found bearing his stamp. This is odd,

and perhaps someday a signed example of the Brooklyn maker will surface. In the meantime we have a number of rods marked "Abbey & Imbrie, N.Y." that are attributed to the brothers or workmen under their tutelage. The rods are found with both the 1881 and 1888 patents on the hard-rubber reel-seat-handle area. In general they're quality items. Cheaper versions with plated brass fittings can be seen, but most have quality six-sided bamboo and German silver fittings.

Many Pritchard–A&I models were built from a very dark and stunning wood, either greenheart heartwood, called bethabara, or snakewood. The shafts were stained a rich mahogany brown, and I've never seen such a posh finish by any other maker. Most of these impressive sticks fell into the bait-rod or saltwater category, but there's always the possibility that some fly models were built. Such a unique blend of brown shafts combined with the hard-rubber handle and spiked German silver ferrules would be a healthy addition to any antique collection.

The handsome and distinctive Pritchard Brothers–style fly rods were built from the early 1880s until after the turn of the century. The number of surviving examples attests to their popularity; Abbey & Imbrie carried them as standard items for around 20 years. The only flaw I can find in the design of the rubber area of these rods is a kind of "white stuff" that emanates from the material itself. Whatever this stuff is, it can be removed easily—but it will come back again.

WILLIAM MITCHELL, PIONEER RODMAKER

William Mitchell is considered one of America's pioneers in the use of split bamboo. Many collectors fail to realize, however, that Mitchell was not only British by birth but also a very articulate man. In the May 19, 1883, issue of *The American Angler*, Mitchell wrote that "the first split bamboo rod that I made myself was in June, 1869. It was put together in four sections [strips]; made not of Calcutta bamboo but of

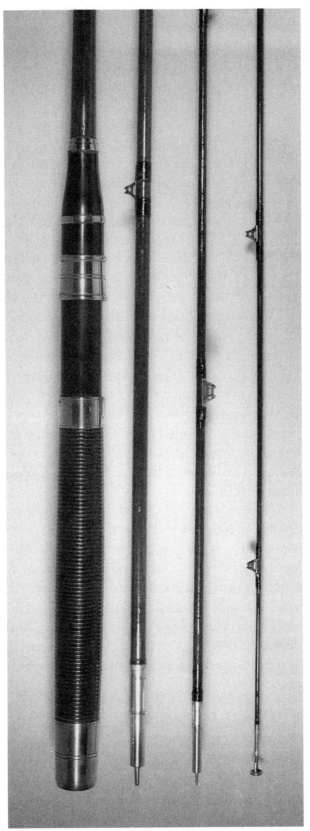

"Harry" Pritchard's patent one-piece rubber seat and grip on a bait model. The fittings on this Abbey & Imbrie rod are rolled and seamed, and the "copper-bottom" spike ferrules indicate a John Landman origin.

Chinese, which is much harder, more homogeneous and more difficult to obtain than the former."

Prior to this accomplishment Mitchell had risen to a certain expertise in angling, repaired a number of earlier rods, and was a student of the history of the sport. He never built rods in a big way; during his tenure in the business, from 1869 until about 1890, he evidently left us very few examples of his work. Although he was the first to produce a Tonkin rod and soon went to six-strip construction, Mitchell built a good number of fly rods from snakewood, lancewood, and split Calcutta. As a split-bamboo artisan, he started his career in the smith age, and all his products had that handcrafted look. Perhaps the most distinguishing feature of any Mitchell rod was its reel band (illustrated); you can immediately discern the maker at a glance.

How odd it seems that, in 1878, when Leonard and Wheeler were just beginning production and Jim Daniels was offering his first rod for sale, William Mitchell placed an advertisement in *Forest and Stream* proclaiming that he was "Still in the Field." At this time his shop was located on Vandam Street. I have no idea how large his operation was, but his rod furnishings appeared custom designed and probably originated on the premises. Many of his trout rods carried distinctive ferrules that were fluted where they met the cane or wood shafts. A thin trim wrap was located just above the six-flute area on each ferrule. His reel seats were unique, and the grips were wrapped with a fine cord, sometimes painted black. I've found one sliding reel band, looking much like Mitchell's product, on an early Bangor Leonard.

On January 9, 1883, Mitchell was awarded one of the first patents for a detachable butt section designed expressly for fly rods. The shaft of the rod's butt entered the hollow handle and slipped down to the butt-cap area, where it was screwed into position. The patent was used on many of his wooden trout models, which had a snakewood butt and mid with lancewood tips. These distinctive rods were built in lengths of from 9¾ to 10 feet, and the hollow butt contained an additional short "feeler" tip. For reasons unknown, the wooden trout rods are much more common today than the split-bamboo rods.

William Mitchell's salmon rods bear a striking resemblance to the trout models, but here you're

"The Mitchell Rods."

W. MITCHELL,

FISHING-ROD MAKER,

(Still in the Field).

jc6 3mos　　　　26 VANDAM STREET, N. Y.

Forest and Stream, *June 13, 1878*

more apt to find him using bamboo, plus adding a few typically English touches. On an impressive 14-foot, two-handed model featuring the 1883 patent, wire loops were wrapped to the ferrules to keep them from "throwing apart." Upon checking the fit, I discovered that each pair of ferrules had a good tight pop, so perhaps this was some traditional yet unnecessary carryover from Mitchell's homeland. The salmon model was built from six-strip Calcutta, and the portion that fitted into the detachable handle-reel seat had a reverse taper ending with the female screw fitting.

William Mitchell's products were built entirely with rolled and seamed German silver furnishings,

and the male ferrules sported covered-dowel construction. Although the dowel covering met the inside corners of the ferrule proper, wood or bamboo was exposed at the very end. How ironic that this pioneer in split-bamboo construction should have used antiquated fittings at a time when waterproof ferrules were the norm. The guides were also made from German silver; trout fly rods carried loose-ring versions and salmon models had double-ring standing guides.

Most Mitchell rods were marked "W. Mitchell, N-York, Patd. Jan. 9, 1883." The last of his rods were stamped "W. Mitchell & Son, N-York," with the same patent date. It's unknown when the pioneering rodmaker retired or exactly how many pieces he built, but they now remain as most unusual examples of the art. The very distinctive Mitchell wands surface now and then, and are eagerly sought by a new generation of students of rod history.

U. S. NET & TWINE

Founded in 1887, U.S. Net & Twine was a wholesale commercial-fishing supplier, making a variety of seines, stop nets, and gill nets for the active mackerel and menhaden fisheries. Shortly thereafter it entered the tackle trade, at first selling ball-handled New York–style multipliers. Apparently, its earliest reels were built by W. A. French at 210 Fulton.[2] After purchasing the rights to Frederick W. Moog's patented sheet-metal reel foot, it hired L. Levy to be

Mitchell's knurled, sliding reel band was practically a trademark. This unique seat is fitted to one of his salmon rods, and is "doubled" with corresponding double guides on the butt shaft. To avoid creating a set in the rod, the reel can be mounted on either side.

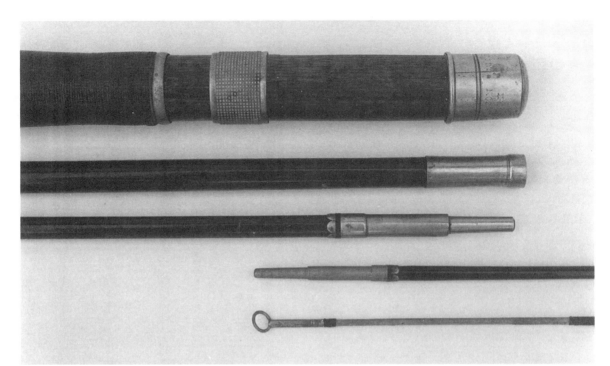

The seat details of a William Mitchell snakewood fly rod; the lower section has a metal fitting that screws into a threaded post inside the butt cap, as well as scalloped covered-dowel ferrules.

the reel's initial maker. In 1894–95, under the direction of Charles M. Pratt, the company became interested in mass manufacturing, purchasing an old reel factory at 163 Grand Avenue, Brooklyn—possibly the original Malleson reelworks—and established a storefront on Broadway.

A number of U.S. Net & Twine Company reels were produced at the Brooklyn plant, the bulk of them aimed at the angler with a modest income. One of the most often seen examples, the Climax, served as a bait-casting and double-multiplying fly reel. The Climax trademark was registered on May 21, 1894, and granted on July 3 of that year; it's usually found on nickel-plated or nickel-plate-and-hard-rubber models. U.S. Net & Twine also produced a variety of completely unmarked models designed for the budding fly fisher and bait caster. Many of these lesser reels sported Moog's patented reel foot, which was ideally suited to production methods. Some reels incorporating the stamped foot bore the Moog patent date of August 14, 1894.

The most famous U.S. Net & Twine reels were the various Kosmic-marked models sold between approximately 1894 and '98. These were higher-end products and have been mis-

taken for the work of Julius vom Hofe. A closer inspection, however, shows the production limits of the U.S. Net wares; the reels lacked the superior

The famed Kosmic fly reel, U.S. Net & Twine's most collectible winch.

workmanship found in the Kosmic rods. The reels, however, were sold as companion pieces to the work of Thomas, Edwards & Payne, and they now generate high prices on the collector's market. Many of these top-of-the-line reels had steel-pivot compensating bearings on the left side plate. It's not easy to mistake the well-known markings, and most reels were boldly stamped "The Kosmic Reel" on the front side plate.

Kosmic rods were marketed wholesale and built at the Brooklyn factory from around 1894 until '98, and then at Highland Mills (Payne), and in Brewer, Maine (Thomas & Edwards, and Thomas alone). Besides Kosmic markings, the rods carried the designation "Isaak Walton."

In 1899, U.S. Net & Twine went out of business and sold its Brooklyn reel factory to the Montague City Rod Company. Montague City continued to make and wholesale the reels, adding a multitude of newer models. The remaining rod inventory was purchased by Hermann H. Kiffe, who sold it for 50 cents on the dollar to his New York walk-in customers.

HERB FROST'S KELSO AUTOMATIC REEL

Just before U.S. Net & Twine threw in the towel its sales manager, Herbert J. Frost, left the Pratt company and formed his own business at 90 Chambers Street, just across the Brooklyn Bridge in lower Manhattan. The five-story building had limited manufacturing facilities on the top two floors, and I believe this was where the Kelso Automatic Fish Reel was built. The Kelso model arrived on the scene by 1908, supposedly designed after two patents by Edward E. Kleinshmidt of New York City, obtained on November 19, 1907.[3]

The inherent advantage of the Kelso auto was its ability to retrieve 150 feet of line without rewinding the spring, which was attached to the frame and drum

"with friction relief action at both ends." Sounds good to me, and it tickled the fancy of fly fishermen who purchased the thing into the 1920s. Made for a dozen years or more, the Kelso Automatic Fish Reel should draw some collector attention—particularly the best model, which had a sliding agate line guard. It isn't today a common item, even though it was touted in a national advertising campaign and pictured on Herb's letterhead. (For more about Frost, see chapter 13.)

Frost was an innovative man with a keenly developed marketing sense. For instance, he labeled his

Field & Stream, *May 1920*

tant, a gifted rod designer whose legacy was a rod so striking in appearance that its likes have not been seen before or since.

The Landman rod was a high-quality example of the art and came in a number of variations. The shafts, wood or split bamboo, were not built at his factory because Landman had no facilities for their construction. Instead, he purchased lots as they became available from other makers, such as Thomas Chubb and probably Thomas, Edwards & Payne. In turn, a great many of his ferrules and reel seats were sold to the trade—in particular to Thomas, Edwards & Payne, and later to Thomas & Edwards. This may sound like heresy, but certain Kosmic, Walton, Von Lengerke & Antoine, and J. G. Landman rods are so similar that it's almost impossible to tell them apart. The bamboo shafts and fittings of all these rods indicate col-

top-end goods "Frost Kelso" to play off the well-respected tacklemaking name of Scotland's Forrest of Kelso.

JOHN G. LANDMAN'S TWIST

Of the many rodmakers of 19th-century New York, the single most obscure had to be John Landman. Here was a man whose products were not even acknowledged by collectors until the 1980s, a full century after he entered the trade. Today, we know a little more about his operation, pieced together from my own research and that of Oregon collector Dave Holloman.

Active in the 1880–1900 period, J. G. Landman entered the trade upon purchasing a ferrule and reel-seat rolling machine that had been constructed in France. He was not a fisherman, nor did he personally make any of the furnishings that were shipped to a number of eastern rodmakers. The fittings and rods built in his Brooklyn facility were, for the most part, constructed by women and children. John Landman was a shrewd businessman and, most impor-

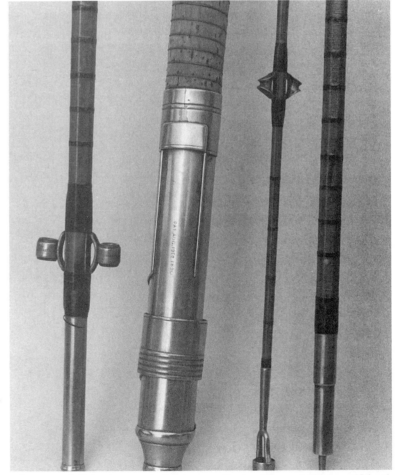

A very rare J. G. Landman seat, shown on a saltwater striper rod. This beautiful article may also appear on Landman salmon models. A heavy-duty version of his patent, the locking band rides on a spiral German silver rail.

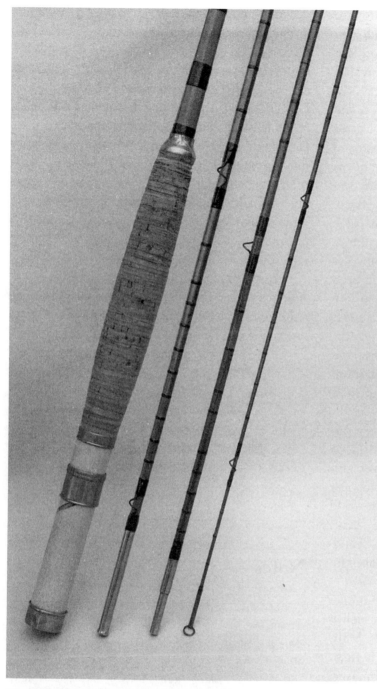

The exquisite VL&A Landman fly rod, shades of Kosmic and Thomas & Edwards.
The shafts appear to be of Kosmic origin, and the narrow 1/8-inch cork rings may
have been Landman's idea and product.

blanks or trying to give the lowest quotes on a gross of ferrule sets. As a rod designer, John Landman's trademarks included a spiral of German silver wire over a cork grip, and also a distinctive, spiral twist-locking reel seat of his own invention. He established an outlet with the house of Conroy sometime around 1884, after the heavy Eighth Street–built T. J. Conroy rods were discontinued. At this time, Conroy started to sell lighter and much more pleasing rods, stamped "THOS. J. CONROY, MAKER, N.Y." The reel seats common to these fly rods were somewhat plain and very light, with a soldered pocket and sliding band. And the grips carried the spiral Landman German silver wrap.

Many T. J. Conroy-marked rods were fashioned with rolled-and-soldered fittings that originated from Landman's French machine. The eccentric quirk of wrapping a spiral band along the grip was probably done because Landman knew that sheet cork tended to delaminate from the grip's softwood underlayment. The spiral wrap was also present over the sheet cork grips on VL&A rods during the next decade.

On August 19, 1890, John Landman was granted his patent for the spiral-locking reel seat. As far as I know, this was the first American reel seat to employ the "screw"-locking principle. Other ultimately more practical screw-locking bands, such as Heddon's and Jordan's, were yet to come, but for its time the Landman seat was workable and aesthetically pleasing.

Throughout the 1890s Landman made a limited number of rods under his own stamping—"J. G. Landman, Maker, Brooklyn, N.Y."—but most of his products were sold through the firm of Von Lengerke & Antoine in Chicago (the most common stamping) and possibly Von Lengerke & Detmold in New York. The rods were built in fly, bait, bait-casting, and saltwater versions. Most known

lusion between the five men, and, while the real story may never be unearthed at this late date, Landman appears to have been a direct supplier and purchaser.

As a businessman, Landman spent countless hours on the telephone, one of the first tackle producers to use Bell's invention to advantage—finding the cheapest prices on a lot of Chubb lancewood

examples were constructed of wood—the first two sections of snakewood, the tips of lancewood. The highest-grade models were built with standard six-strip, hex-shaped bamboo shafts, some of which appear to be Tonkin.

All models, with the exception of salmon and marine versions, were fitted with a beautiful ivoroid reel seat with the 1890 patent, giving the bamboo sticks a look much like that of the contemporaneous Kosmic rods. Indeed, dealers and even knowledge-able collectors have told me that a particular Land-man rod was a "Thomas, Edwards & Payne." One very distinctive feature will aid in correct identifica-tion, however: At the intermediate wraps found the first 6 inches above and below the ferrules of each section, John Landman instructed his makers to dou-ble the wraps so that the space between wrappings was cut in half. In the same vein, the wraps near the tiptops were also close wrapped. This practice was singular to Landman, and probably done to beef up the areas of the shaft that absorbed the most stress during casting and fighting a fish.

In a trade noted for obtuse thinking, it's not sur-prising that Landman had a few twisted ideas, as exemplified by his seats and German silver grip wraps. He also capped his male ferrules with copper (or perhaps brass with a deep patina). This odd method of waterproofing a male ferrule seems unique to Landman. If we add up all of J. G. Land-man's twisted quirks, his rods cannot be mistaken for the work of any other maker.

The confusion between Landmans and Kosmics is a direct result of several design similarities, such as red wraps tipped black and similar reel seats. Many of Landman's shafts appear to have been built by Thomas, Edwards & Payne, or by Thomas & Edwards. These could very well have been a trade-off for ferrules (used in the Waltons and T&Es) and the celluloid (ivoroid) Kosmic reel seat. There's every indication that this most famous of reel seats was built at Landman's Brooklyn plant, which had the machinery to work celluloid into any given shape. To further confuse modern collectors, Landman also discontinued his sheet-cork grips in the mid- to late 1890s, replacing them with grips built from narrow, ⅛-inch cork rings.

At this point Landman was marketing a product through VL&A in direct competition with the Kos-mic rods sold by the Spaldings, Wilkinson, and Whit-temore. Yet, for whatever reason, his rods disappeared from the market by 1898, when the Von Lengerkes replaced Landman's products with those of Thomas & Edwards. Perhaps the times had finally passed by rolled-and-seamed ferrules, Landman's major product. High-quality, one-piece capped fer-rules—in particular the Varney ferrule—were avail-able from Montague, and most rodmakers were moving to the new style. In the same vein, "modern" one-piece, hot-swaged reel seats, the best by Varney, replaced the seamed versions.

For his time, John Landman produced some of America's finest rods. He left the business as quietly as he entered it, but he also left behind a number of products that exhibited a new twist in an old trade.

Artificial Crawfish.
(Conroy, Bissett & Malleson.)

Notes

[1]Mather, Fred. *My Angling Friends.* New York: Forest and Stream, 1901.

[2]Munger, Albert. *Old Fishing Tackle and Tales.* Philadelphia: Munger, 1987.

[3]Brown, Jim. *Fishing Reel Patents of the United States, 1838–1940.* Stamford, Conn.: Trico Press, 1985.

7 The Legacy of the Rubber Reel

Few pieces of fly tackle have been more drooled over by collectors than the rods and reels—especially the hard-rubber reels—of Brooklyn's vom Hofe family. Under four separate businesses and for three generations, the vom Hofes built the finest fly reels of their time, a tradition later maintained by their old foreman, Otto Zwarg, from the sunny climes of Florida. All told, vom Hofe fly reels were produced for 80 years. Like Leonard rods, vom Hofe reels are far from scarce, but collector demand keeps prices high.

Certainly these aren't the only great reels—a wedding-cake Fin-Nor or a Seamaster, for example, is excellent by any measure. But with vom Hofe reels we're talking about a tradition, a relationship, and, most important, a legacy. They were the classics from which all other vom Hofe types were born, including reels made by A. L. Walker, Thomas & Thomas, and even Stan Bogdan. If imitation is the highest form of flattery, the vom Hofe reel, especially as built by Edward and his sons, has been flattered like no other.

Collectors' reverence for vom Hofe reels makes them the most popular reel on the market. Fortunately, many are still showing up every year, and the prices are high but stable. This is less true of Zwargs, simply because so few were made; consequently, their current prices are a tad higher. Years ago, when vom Hofes and Zwargs brought a little less attention, I used an Otto Zwarg 3/0 multiplier for general saltwater fly casting. I took fish upward of 130 pounds, and there was nothing wrong with the reel's 25-year-old drag. The vom Hofes, Zwargs, and Walkers not only perform well for anglers, but they also look great doing it. And that's why they'll never go out of vogue.

FREDERICK "VONHOF" & SON

The son of a silversmith, 39-year-old Frederick Vom Hofe came to America in 1847, probably with his brother, Louis. They settled in Brooklyn, in "Little Germany," and worked for others within the trade they had apprenticed in. The following year the rest of Fritz's family came over, including Frederick Jr. (born March 2, 1833), Julius (born August 6, 1836), Ludwig (born January 28, 1839), and baby Edward (born in 1846). For the next decade the "Vonhof" family, as their name was commonly misspelled, remained obscure.

Then, in 1857, the *New York Directory* listed Frederick "Vonhof" near Fulton, at 545 Pearl Street, as a "reelmaker." The neighboring shop was occupied by Louis "Vonhof," a "smith." Thus we have the beginning of the firms "F. Vom Hofe" and "F. Vom Hofe & Son." The Son, Julius, always claimed to have built reels "Since 1857." The early Pearl Street products appear to have been knockoffs of the reels built by Conroy and Crook, although they were cruder than Conroy's. The interior screw ends, for example, were not filed but simply snipped with a pair of end cutters.

Many early winches were marked either "F. Vom Hofe, Maker" or "Hawks & Ogilvy, N.Y." Most reels, however, were not marked at all. The single feature that distinguished them from the products of other New York makers was a slash mark at the bottom of a multiplier's face plate and the adjoining cover plate. Unfortunately, this does not help the collector of fly reels. The single-action winch, however, usually had the standardized rear bearing cap found on

One of the earliest hard-rubber salmon reels found to date, this 1855–65 curiosity appears to be an English Birmingham product originally fitted with a 2-to-1 ratio, New York–style gear box. Wicked-odd!

all very early Vom Hofe reels. Unlike the permanent "domes" found on Conroy specimens, the F. Vom Hofe "cup" was threaded so that it could be removed to oil the end of the spool shaft. It was the earliest American-built removable bearing cap.

Any small Fritz Vom Hofe fly reel is a scarce little item today. While examining the wares at Lang's 1992 Spring Auction, I noticed a cheap wooden rod and very small brass reel combo with a catalog value of "$100–$125." A close inspection revealed a genuine Fritz bearing cap; I pointed this out to noted collector "Doc" Herr, who smiled knowingly. Unfortunately, someone else also spied this obscure telltale, and the rod-reel combo was bid higher than I cared to go—perhaps a mistake on my part. Someone has a rare and charming Fritz Vom Hofe fly reel in his collection—whether he knows it or not.

Using that fly reel as a guide to features, the single-action Frederick Vom Hofe winch looked much like any other nonclick New York product, which in turn looked a lot like the earlier British versions they copied. The handle was a straight single crank with a carved knob of rosewood or, possibly, ivory. Perhaps at some point a real signed example will surface.

When it does, I hope not to make the same error twice.

In 1860, Fritz became partners with his son Julius, a relationship that lasted until the elder vom Hofe retired in 1882. Most marked reels of this period don't have a "N.Y." stamping; they read "F. Vom Hofe & Son," sometimes with the word "Maker" added. It was Julius, apparently, who changed the "Vom" to lower-case. Aside from Hawks & Ogilvy, the reels were probably sold through Bradford's in Boston (but so were Conroy's products). Doubtless other retailers also carried Fritz reels, but the various models and production details remain obscure. One dated example has been found with the inscription "Wm. R. Renwick, 1868."

During this period the first rubber reels appeared on the scene. Whether or not they were originally made by Fritz & Son, or even first built in New York, is a point of contention. The material itself was native to the Amazon delta, coming from a tree in the castor bean family. Connecticut Yankee Charles Goodyear stabilized the natural polymer in 1839 and patented his vulcanization process five years later. Rubber made under Goodyear's process was first pro-

The simple idea of producing a removable bearing cap to facilitate lubrication appears to have been Fritz Vom Hofe's brainchild. Contemporaneous American reels made by Conroy lacked this option.

duced in Europe in 1851, but the actual origin of "hard rubber" is difficult to pinpoint. The British called the stuff "ebonite" and, at least by 1877, some of S. Allcock's revolving-plate reels were incorporating the material.

American reelsmiths had first used hard rubber at least a year earlier, when an obscure gunmaker's machinist in Bangor, Maine, employed the material in Leonard's prototype reel. Hard rubber probably appeared in the New York tackle trade just after the Civil War. The earliest rubber fly reel I've found—unfortunately undated and unmarked—is seen in this chapter's opening photograph. It appears to be of Birmingham, England, origin, with an oil hole on the rear bearing cover, sculptured pillars, and a cowhorn crank knob. The handmade screws are steel, indicating pre-1865 methods. The atypical crank almost conforms to the standard New York ball handle, and the reel has a 2-to-1 gear ratio. The face and rear plates are constructed of machine-turned hard rubber. The design is flawed; the small gear used to turn the large spool causes excessive torque—to the point that regular use slightly bends the handle. Nonetheless, it's one of the earliest multiplying salmon reels to employ hard rubber.

The existence of this British reel points out that neither Fritz Vom Hofe & Son nor Edward vom Hofe nor Frank Philbrook was the originator of the rubber reel. These reelmakers were, however, among the first in America to use the substance. Very few known pre-1880 hard-rubber reels marked by any maker have been found, however; and of those made by Vom Hofe & Son, most were saltwater models.

Of course the Fritz & Son fly reels are extremely collectible—superb smith-age specimens. They may have been crude, but then so were most fishing implements of this era. The majority were made of either brass (the most common material) or German silver. Originally, the reels were finished to a fine polish and then coated with a protective lacquer. This

coating, of course, wore away during the first streamside outing. The screws were hand cut and threaded, and the plates were rather rugged. This made for a substantial winch, one that would survive the rigors of use for many years—perhaps until today.

JULIUS VOM HOFE, MAKER, BROOKLYN, NEW YORK

In 1882, Julius vom Hofe continued making reels with his own stampings. Long before this, however, he'd been an accomplished reel designer and possibly a rodmaker. On May 20, 1862, he received a patent for a "pulley tip" for fishing rods; he must have had some rodmaking experience to follow through on the idea. Julius vom Hofe also patented a new style of gear bridge in 1867, probably the first noteworthy design change in the traditional New York reel. Julius had by now revealed himself to be a talented, perhaps even gifted, tacklemaker.

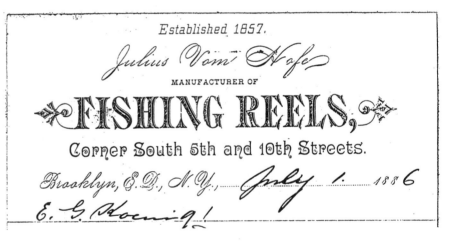

Very few fly reels remain from Julius's early career; most of his energy went into the design and construction of his various multiplying models, particularly those built for salt water. Some of his earliest fly reels lack the hole in the middle of the foot and are marked with a cloverleaf logo, used perhaps from 1882 until 1885. The turning point in Julius vom Hofe's production came when the William Mills interests contracted him to produce the H. L. Leonard reel, formerly built by Philbrook & Payne. Thus, in 1885–86, Julius began the manufacture of the Leonard-Mills reel and also incorporated key fea-

Julius vom Hofe's "little round hole" die-punched into the center of his reel foot. It was a common feature and now helps identify unmarked or tackle-house-stamped vom Hofe fly reels.

The classic lines of a fancy Julius vom Hofe fly reel. On this model the handle can be removed by turning it counterclockwise, and the pillars appear to be somewhat narrow. (Bob Corsetti photo)

tures of the Leonard model into his own line.

From this time we see the early square-tipped, raised-pillar model, which was built until the early 1890s. The reel is not found with the Mills stampings but was marked by the maker, or to Dame, Stoddard & Kendall, or to Abbey & Imbrie. The reel was typified by a recessed hard-rubber front plate, heavy German silver frames, a small hole in the middle of the foot (between pillars), a solid frame connection between the two foot-pillar screws, a nonadjustable click, and a single-crank handle. The knurled rear bearing cap was removable for oiling, and the reel was made in several sizes.

A similar square-ended, raised-pillar Julius vom Hofe trout reel, built without the dished area in the rubber front plate, carried the January 19, 1892, patent stamping of William King, also of Brooklyn. The click-drag patent, which only worked when the spool gave line, was assigned to Charles Imbrie. The reel was marked "Abbey & Imbrie, N.Y." This single-crank model, like many small Julius-built trout reels, had a spiral-knurled bearing cap that may or may not have been removable. Any Julius vom Hofe fly reel with the square-topped raised pillars is a great find.

Throughout the same period Julius built a selection of "Leonard-Mills"–marked raised-pillar fly reels, including some larger salmon models. These carried the double-pillar design for the foot screws, essentially as built originally by Frank Philbrook. The reels had rounded pillar ends on their German silver plates. The smaller sizes had a single crank; the larger models had a balance

At left, a Julius vom Hofe square-ended raised-pillar fly reel, marked by Abbey & Imbrie. The model at right, a protective-rim version, has a "Dame, Stoddard & Kendall, Boston" stamping.

handle. Small reels often had the spiral-knurled rear end cap, and some large models had a slotted, screw-adjustable tension cap. This was a modification of Julius vom Hofe's first adjustable-bearing patent of 1882.

Later Leonard-Mills models, large and small, had aluminum spool ends and increasingly larger holes along the center of the foot—features added to lighten the fly reels as their corresponding rods grew lighter. Many models also had a back-sliding click button. Perhaps one of the most charming of the Julius vom Hofe Mills reels, the little 2-inch so-called midge reel, incorporated most of the above-mentioned design features.

A very unusual style found on at least one Julius model used the 1889 patent of Peter Giroud of New York City. The reel profile shown in his patent application looks remarkably like the solid-framed foot-pillar area of vom Hofe's square-ended models, previously described. The patent, assigned to Thomas B. Mills, was for a button-operated spring drag. It was built into the rare raised-pillar Dry Fly Multiplying Salmon Reel, which was marked "William Mills & Son, N.Y., Pat. June 25, 89." The balance handle of this reel was made in the classic Julius vom Hofe style that was affixed to many of his smaller saltwater models.

Standard Leonard-Mills raised-pillar click reels, as built by Julius vom Hofe, have either a dished hard-rubber inset on the face plate or a smooth version. The midge size was 2 inches, and the trout models went up from there, at 2⅛ inches and 2¾ inches. Large trout or grilse sizes included the 3-inch model and a 3¼-inch size. Perhaps the rarest of the J. vom Hofe Leonards was the 4/0-sized raised-pillar salmon reel. It carried the 1877 Frank Philbrook patent number and had a dished rubber plate, a German silver frame, and a balance handle.

The raised-pillar reel, as carried from Philbrook & Payne to Julius vom Hofe, was also made entirely of nickel-plated brass. A small 2-inch model sold through Abbey & Imbrie had the seller's name plus "Pat. Oct. 8, 1889" stamped on it. This patent was commonly featured on saltwater reels, but not often on tiny trout models. It was granted for a pivot bearing and "spider" washer under the rear cap. The great variation among Julius vom Hofe raised-pillar models gives reel buffs a wide choice; an entire collection could be built around this style of frame construction.

Julius also designed a number of other fly-reel styles, including a copy of the traditional English-Scottish revolving plate and a unique perforated model. One of the most pleasing models featured protective rims or bands wrapped around turned hard-rubber plates. The 1877 raised-pillar Philbrook-style models had a protective rim to keep the fly line from jamming under the handle, but they were set inside a heavy frame. Julius borrowed the banded-reel idea from his brother Edward, who had developed the model just prior to 1880. The resulting reel was substantially lighter than the earlier raised-pillar model, yet could accommodate the few bells and whistles (clicks and drags) deemed necessary by the angler of the mid- to late 1880s.

Typical rimmed Julius vom Hofe trout reels ranged from a size 2, at 2⅜ inches in diameter, to a size 1, which measured 2⅞ inches across the plates. In all, Julius made fly reels from a size 5 (the smallest) to the huge 6/0 salmon reel for two-handed rods. Most of the rimmed models had hard-rubber side plates and German silver bands, feet, and spools. Some of the modestly priced versions were built from nickel-plated brass, and a few delicate trout models were cut entirely from hard rubber, aside from the feet, pillars, handle, and spools.

The big salmon reels are sometimes found marked with the 1889 and 1903 patent dates, and a few fancy models have an adjustable drag lever as well as a sliding click button. Handles are always counterbalanced and straight; I have yet to see an S handle on one of these old workhorses. Small reels

Highest Award and Medal at the World's Columbian Exposition, Chicago, U. S. A., on Fishing Reels, awarded to

JULIUS VOM HOFE,

WHOLESALE MANUFACTURER

FISHING REELS ONLY,

No. 351 South 5th St., - Brooklyn, N. Y.

Fine Rubber and German Silver Reels. Made in sizes 40, 60, 80, 100, 150 and 200 yards.
All genuine Reels bear my name. For sale at all retail stores. No branch store in any city. Established 1857.

Send stamp for Catalogue.

Forest and Stream, *July 28, 1900*

The "banded" reels were marketed through vom Hofe's established network, mainly New York's William Mills & Son, Thomas J. Conroy, and Abbey & Imbrie. In Boston, they were sold by Dame, Stoddard & Kendall, which was the successor to Bradford & Anthony. This latter association actually went back to the days when the firm was known as Bradford's and retailed the original F. Vom Hofe & Son products. Many Julius vom Hofe reels were also marked "Dame, Stoddard, Boston" after Kendall left the firm. Another Boston outfit, the large sporting goods store of Iver Johnson's, sold the rimmed reel as late as 1917.

and trout models often had single or balanced handles. Many banded reels, even smaller ones, such as the little 2⅜-inch, size 4 model, had the patent spool journal and spider washer marked "Abercrombie & Fitch." This reel also was marked "Julius vom Hofe, Pat. Oct. 8, 89."

Julius vom Hofe's banded or rimmed reels were some of the earliest of that rather famous style. Collectors find that these lightweight models have a charm all their own. They were quality production items, made on the piecework principle and under vom Hofe's direct supervision. The Julius vom Hofe trout reels, raised-pillar or rimmed, were built with attention to detail at his factory in Brooklyn until

Julius's death in 1907. He was succeeded by his son, Julius vom Hofe Jr. (1871–1939), and reel manufacture continued well into the 1920s or later. It was a family tradition. The vast variation in Julius vom Hofe fly reels enables collectors to acquire a selection of totally different models from just one Brooklyn maker.

THE CRAFTSMANSHIP OF EDWARD VOM HOFE

In 1867, Frederick Vom Hofe's youngest son, Edward, opened his original shop at 102 Fulton. At the time he was a mere 21 years old and a confirmed tackle tinkerer. Like Julius, he had an early fascination with rod design but, unlike his brother, Edward would continue to make rods throughout his life. It is interesting to note that, from the outset, Edward vom Hofe claimed no relation to the business of his father or brother and went his own independent way.

This staunch individualism came to the fore in his early career. During the 1870s, when the majority of rodmakers were bringing the six-strip fly rod into final form, Edward was fashioning *seven*-strippers. It's unknown how many seven-strip rods Edward built; at least one or two have come to light. These are very rare rods, and just seeing one is more than most collectors could hope for.

By 1880 Edward had settled back to conventional six-strip, hexagonal split-bamboo rods. Of course, he also built a great number of rods from greenheart, lancewood, and hickory. Although many of his ocean rods survive, fly-casting models are far less common. Perhaps it's because Edward vom Hofe's rod designs got trapped in a time warp, with such antiquated features as the old rolled-and-seamed reel seats and rattan grips, which appeared on rods he built into the 1930s. Today an Edward vom Hofe fly rod is a major acquisition. The many models were handsome affairs, usually wrapped red and built to the popular lengths of each decade, from 10 feet down to 8 feet. The "Bonbright Salt Water" fly rod, one of the first models built for the marine fly fisher, was named after the originator of the Bonbright tarpon fly.

Built in two grades, Edward's fly rods were well described in his many catalogs and the available reproductions of them. I won't run through a blow-by-blow hit list of which Edward vom Hofe fly models are good or bad. To an Edward vom Hofe fancier, any fly rod built by this high-quality company will do—until a shorter one is found. The rodmaking end of Edward vom Hofe & Company was sold to M&H Sporting Goods of Philadelphia in 1939. Owned by the same partners who operated Ocean City, M&H retained a few of the top vom Hofe rodmakers, but no further fly rods were built at M&H. The firm concentrated on hickory sticks for tuna and marlin pumpers.

Throughout his career, Edward vom Hofe maintained a high level of craftsmanship, not only for himself but for every person who worked under him. At some point, a little after 1883, he started to oversee a core group of employees, yet standards were so rigid that his tackle was never mass-produced. This fanatical adherence to quality was passed on to his shop foremen and to his sons, Clarence and Edwin.

Evidently, Edward was also a very frugal man. Collector Bruce Wilson discovered that, after Edward's move to 97 Fulton, around 1880, the tacklemaker continued to use old 102 Fulton stationery, scratching out the old address and writing in the new one. The entire vom Hofe family, which still lives in Brooklyn, was extremely conservative.

As a reelmaker, Edward was the champion of graceful rubber. He did things with it that no other maker—with the exception of Zwarg and Walker—could. Edward's earliest reels were patterned after the classic New York saltwater models; they were usually

Edward vom Hofe,

Manufacturer of and Dealer in

FISHING TACKLE,

97 FULTON STREET

New York, _Sept 20_ 188 2

A post-1910 Edward vom Hofe salmon reel fitted, on the rear plate, with his 1896 adjustable drag. This is an "arrow-indicator" model. (The American Museum of Fly Fishing Collection)

made from German silver and had his distinctive early reverse-S handle. At some point prior to 1875 he started using hard-rubber side plates, and the material was then incorporated into his fly reels.

Very early fly models—built between 1875 and 1878—are extremely scarce, have no patent markings, and are stamped "Edward vom Hofe, Maker." At least one of these first models had a permanent click and a nonadjustable drag that was always on. The reel was built with hard-rubber side plates encased in thin annular bands of German silver, a material that Edward called "white metal." I believe that this was the first "rimmed" fly-reel design, later copied by his brother Julius. The pre-1879 model had a sliding rear oil cap surrounded by the knurled wheel that operated the drag. The front-plate area was finely turned and sported a reverse-S handle, with the S curved the "wrong" way.

On September 2, 1879, Edward received a patent for a variable "tension device" that used cams and a large spring to press leather pads against the rear of the fly reel's spool.[1] This was possibly the first effec-

tive overrun system incorporated in an American salmon reel.

No known trout models date from this first patent period; known specimens are single-action salmon reels in sizes 3/0 (a grilse size) to 6/0. The larger sizes had a bigger foot to fit the rugged two-handed salmon rods of the day. I once had a very nice 1879 model inscribed "Wm. Mitchell," before I knew William Mitchell was a famous rodmaker. Another case of collector error. The 1879 Edward vom Hofe reels, by the way, were the maker's first standardized fly-fishing models. Previous to that date, his reels are believed to have been built to custom orders only.

This first model established Edward as the leading maker of American salmon reels, and his business grew in the ensuing years. The model, eventually termed the Restigouche, was typified by the drag wheel under a standard S handle (not reversed). Early versions had holes in the wheel that accepted a cammed stopper for adjustment; the very first reels were marked "Edward vom Hofe Maker & Patentee, Fulton St., N.Y., Patd. Sept. 2, 79" on the wheel. Later models had a marked solid drag wheel with an arrow indicator.

Edward's next patent, registered on January 23, 1883, was also incorporated into his fly reels. It covered a click mechanism in which a flopping arrow-

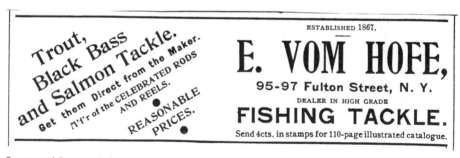

Trout, Black Bass and Salmon Tackle. Get them Direct from the Maker. M'f'r of the CELEBRATED RODS AND REELS. REASONABLE PRICES.

ESTABLISHED 1867.

E. VOM HOFE,
95-97 Fulton Street, N. Y.
DEALER IN HIGH GRADE
FISHING TACKLE.
Send 4 cts. in stamps for 110-page illustrated catalogue.

Forest and Stream, July 28, 1900

shaped pawl moved in and out of a circular spring. Usually mounted on the back plate and operated by a button, the click became the standard of the industry. Reels containing the patent included single-action salmon and trout models and an offset handle multiplier, which later would be called the Pasque. Although a few 1879-marked reels have been found with the click, most new models, combined with the

Although the beautiful Edward vom Hofe Perfection trout reel was far less expensive than his salmon models, it now commands the highest prices.

original tension device, have the 1883 patent date and maker's name on the sliding oil-port covers.

At approximately the same time, Edward vom Hofe entered into the retail trade, a venture that grew into a substantial business that sold not only his own growing line of products but also tackle made by other companies. This was also the period that saw Edward's entry into the trout-reel market. With the reel that would become known as the Perfection, vom Hofe had designed a true classic—a reel so beautiful that it would be copied by others, the first being his brother. The model had an arrow-indicator 1879 tension drag under the handle; through the years, aluminum came to be substituted for German silver on the spool ends and, sometimes, on foot.

As late as 1907, the Perfection was still simply called the "celebrated fly reel with automatic silent tension drag." A second trout model, the Peerless, had the click mechanism but commonly wasn't fitted with the drag; it was made in sizes down to 2¼ inches in diameter.

In 1896, Edward vom Hofe improved the original tension device with a patented adjustable drag that worked only when line was pulled from the spool. Awarded its patent on July 14, the "automatic drag" had brake shoes that pushed against the outside of the spool.[2] Models with this very effective drag

included the standard single-action Restigouche and the later Tobique. The latter reel was a center-handled triple-multiplier, and both models were built with little change throughout the tenure of Edward vom Hofe & Company and into the Zwarg years.

A slew of Edward's salmon reels also have the May 20, 1902, patent date stamped on the oil-part covers. These were originally conceived of as saltwater models and were adjusted with a "key wrench," which often fell overboard in an angler's haste to adjust a screaming reel. This last of Edward's patents, without the silly wrench, was used on many salmon models.

Around 1910, the front-adjusting drags on many Edward vom Hofe fly reels were moved to the rear plate—a concession to fly fishers who complained about bruised fingers while trying to adjust drags while the handle was spinning wildly backward during the run of a salmon.[3] In the 1920s, the company introduced the Col. Thompson multiplier, which was a throwback to the Pasque-style offset-handle model. This version was named after Colonel Lewis S. Thompson, the angler who pioneered the use of hair wing salmon flies. It was a fizzle, as was a similar reel, the Griswold. The famous Tobique was introduced in the late 1920s, and remained a popular item.

After Edward's death in 1920 he was succeeded by sons Clarence J. and Edwin C., who were already active in the business. Clarence was eminent in reel production, while Edwin is best remembered as an avid tarpon and big-game angler. Around this time, the plain old salmon reel, then known as the Cascapedia, was discontinued. It was copied by the Hardy Brothers—who even used the same name. Of course, almost anything vom Hofe was copied by other makers, and fly fishermen are grateful for the works of at least five modern reelmakers who, as Bruce Wilson deftly stated, have produced near misses with "lesser degrees of success."

In 1939, the old Brooklyn firm was purchased by the Ocean City Reel Company. The Edward vom Hofe reel inventory was moved to the Philadelphia plant, and Ocean City started to assemble the reels with bearing covers stamped with its new address. When it ran out of vital parts, such as handles and the sliding bearing covers, Ocean City made cheap

National Sportsman, *April 1923*

cast pot-metal reproductions and attached them to the remaining Edward vom Hofe reels. Blasphemy! Luckily, in the beginning years of World War II, Ocean City ceased to assemble this junk.

THE OTTO ZWARG STORY

Otto Zwarg was a successful dentist in his native Berlin who emigrated to the United States around 1923. Upon arriving, he took a temporary job at Edward vom Hofe's manufactory, hoping to eventually bring the rest of his family to this country and, in due time, get a license to practice his craft. Fortunately for anglers, this never happened.

Zwarg evidently found reelmaking to his taste. By the 1930s he had risen to foreman at Edward vom Hofe & Company, and controlled the fashioning of its superb reels. When the company folded at the onset of World War II, Otto acquired the greater portion of the Edward vom Hofe tools, dies, and turning machines, saving the vom Hofe reel from oblivion. To assist in earning a living, Zwarg helped develop the Norden Bomb Sight and spent his spare hours repairing the products that he knew so well. Oddly

enough, he also made a few split-bamboo rods during this period. He was living in an apartment over his Edward vom Hofe reel-repair and rodmaking shop in Brooklyn.[4]

After the war, Zwarg took on as silent partners concert pianist Dorsey Whittington, Gene Fitzgerald, and Wilbur Prorr—all ardent fans of Edward vom Hofe tackle.[5] In 1946, Otto started to produce his reproductions of three saltwater models (4/o, 6/o, and 9/o) and two of Edward's salmon models, the Restigouche and the Tobique. These reels were faithful to the originals, and collectors are thankful for the talent and sensitivity of this great craftsman.

The saltwater models were actually a tad stronger than vom Hofe's versions. All the New York–built reels were marked "Otto Zwarg, Maker, Brooklyn, N.Y.," and have an "A" prefix in the serial number. In 1947 Otto moved his family and business to St. Petersburg, Florida. The partners hoped to build a sporting lodge and cater to itinerant saltwater anglers, but evidently the lodge was never constructed. Zwarg occupied a house on Maximo Point, and the saltwater reel models became known as Maximos.[6]

Otto Zwarg modified Edward vom Hofe's indicator to form a small lever, enabling an angler with wet fingers to adjust the drag while playing a large fish.

The Zwarg factions rented a structure at 635 First Avenue South from Oscar Steinert, a contractor and fellow fisherman who also bought into the partnership. From this time on, the Zwarg reels would carry the famous "St. Petersburg" address and perform many new duties in the growing fly fishery. They became the first viable saltwater reels on the market, and fly fishermen like the Albrights and Joe Brooks sang their praise. John Alden Knight used one for steelhead angling. My own 3/o Laurentian helped me catch some memorable marine fish for two decades. In the end, even with some pitting on its aluminum spool flanges and handle, I sold the great old reel for three times what I'd paid for it.

Fly reels were built from a delicate 1/o size up to a 4/o; the single-action model was dubbed the Saguenay. The triple-multipliers, like my Laurentian, were available in the same sizes. St. Petersburg models were built with a "B" serial number prefix starting in 1947, a "C" in 1948, and so forth. A few odd Zwargs built in 1947 have a simple "Z" stamped on the oil-part cover plates—obviously an interim marking until Otto received a new "St. Petersburg" stamp from his diemaker.

As far as I know, Otto Zwarg published only one catalog, called simply "Custom Reels." Undated, the little catalog had an introductory letter written by Edwin C. vom Hofe, who sanctioned the Zwarg reel and praised its maker. The pamplet also mentioned that the names and addresses of all Zwarg reel owners were recorded in a "register." Most likely the reels were assigned a serial number when they were sold, not when they were built. Otto's reel production ceased in 1955, when Oscar Steinert convinced Zwarg that sagging sales weren't tolling a profit. From this point on Otto continued his electronics business only. Zwarg sold the remaining reel inventory piece by piece until 1958, when he dropped dead of a heart attack on the streets of St. Petersburg.7

From that time on, the few reels left were sold and numbered by Steinert. Zwarg collector Norm Pinardi purchased three reels from Oscar Steinert's niece, two of which had an "N" prefix. Further, a couple "O"-prefixed reels are known, and there were still a few reels in Steinert's inventory as late as 1963—one of which was purchased by Knight.

That's the last, trickling end of the Otto Zwarg story. We're lucky that the man built as many reels as he did. Produced for only eight years, not many Zwarg reels were made, and fully half went to the saltwater sector. Now the salmon models are prized by collectors and anglers who wish to fish with the very best examples of the hard-rubber art.

In one full century, almost to the year, from the time of the first yet-to-be-found hard-rubber reel—moving on to the crude example shown at this chapter's beginning, the commendable products of Julius, the masterful models of Edward, and, finally, the zenith of Zwarg—it has been a saga. What a shame it ended. Or did it?

THE WALKER RUBBER-REEL EPILOGUE

The relationship between Edward vom Hofe and Zwarg was complete. The classic hard-rubber, alu-

Zwarg introduced his stunning salmon models at about the time that some pioneering anglers began moving into the brine. The Zwarg reel became the first trustworthy winch to tackle larger saltwater game.

minum, and German silver reel, however, remained alive even after Otto's passing. Back in New York City, a chap named Arthur L. Walker had been making similar models—along with somewhat simpler ones—for 15 years previous to Zwarg's death. And Walker continued to make his excellent reels until the early 1970s.

The Walker reel came in a variety of sizes, from the tiny midge to large salmon models. The reels built for heavy fish were often equipped with a drag system reminiscent of those of Edward vom Hofe and Otto Zwarg. The workmanship and materials were likewise similar to vom Hofe's and Zwarg's. In the final analysis, Arthur Walker would be acknowledged as one of the few master builders of the hard-rubber winch.

Walker's smallest reel, the M/3, was a mere 2 inches in diameter. It had an aluminum foot, spool ends, and crank. The counterbalance, pillars, and rims were German silver, and the side plates were classic turned hard rubber. Most of Walker's models followed the same scheme. The TR-1 had a diameter of 2½ inches, a nonadjustable drag, and a weight-reducing hole in the foot. Next up in size, Arthur's TR-2 measured 2¾ inches and sported the same drag as the TR-1. The largest of the series, the TR-3, had

the same features, a 3-inch diameter, weighed only 3¾ ounces, and had roller pillar guides. The workmanship of these reels was superb, right down to the handle retaining nuts, and established Walker as one of the two best reelmakers of his era.

Arthur Walker also produced a very few salmon reels; unlike the trout models, these are scarce—rarer than Zwargs, and far rarer than Edward vom Hofes. I last saw one back in 1991, and it sold for a fair chunk of change ($900). The reel was a 3½-inch 2/0, designated the Model 100, and had been built with a front-mounted click switch. On the rear plate was a classic adjustable drag.

Most Walker reels were stamped with either "A. L. Walker, Maker, N.Y." or "Arthur L. Walker, Maker, N.Y." on the oil-port covers. They were originally sold in a quality soft leather pouch. Since Walker's retirement, his fly reels have escalated in price, no surprise, and those models in excellent condition will continue to increase in value. Like all the truly superb products within this hard-rubber idiom, the Walkers are collectibles and excellent, fishable instruments.

Hard-rubber reels of the quality produced by Julius and Edward vom Hofe, Otto Zwarg, and Arthur Walker are unlikely to be seen again, at least

Fine craftsmanship and supreme attention to detail place Arthur Walker's TR-3 in the classic rubber-reel fold. (Bob Corsetti photo)

not until some exceptional craftsman comes along, one with a real eye for exquisite detail and willing to work with the close tolerances that hard rubber demands.

Notes

[1]Vernon, Steve. *Antique Fishing Reels.* Harrisburg, Penn.: Stackpole, 1985.

[2]Wilson, Bruce L. *Fishing Collectibles Magazine.* Fall 1993.

[3]Wilson, Bruce L. *Fishing Collectibles Magazine.* Spring 1994.

[4, 5, 6, 7]Pinardi, Norm. *Fishing Collectibles Magazine.* Fall 1993.

8 Mainers in Brooklyn and Beyond

By 1890 Brooklyn held the very heart of the American tackle industry, with a ready market nearby in the bustling shops of lower Manhattan's Tackle Row, as Fulton Street was then known. Of all the stories to come out of Brooklyn, none compares to that of a select group of Downeasters who stopped there on the way to greatness. In the world of rodmaking, three of these craftsmen — Thomas, Edwards, and Payne — remain at the pinnacle some 100 years later.

There is no better compliment than immortality, and museums and collectors have assured that status for each of these men. Collectors have long been in awe of the works of Thomas, Edwards, Payne, and Hawes. These modest Mainers didn't plan to become immortal; they just wanted to build the best fly rods possible — which was why they left the increasingly production-oriented Leonard facility. Of all the rods these men built, the most romantic examples were a collective effort, the celebrated Kosmic and Isaak Walton rods, which sport some of the most recognizable features of any 1890s collectibles.

The new line was conceived to be a total fishing package. The shafts would be built on a machine designed by Loman Hawes, an apparatus that's still tapering bamboo strips today under the watchful eye of Walt Carpenter. The metal furnishings, designed by Hawes, Edwards, and Thomas, were made by Ed Payne and Chubb/Montague. And the beautiful ivory-like reel seats were contracted out to John Landman. The reels would be built on the site of the old Fred Malleson factory in Brooklyn. The origin of Kosmic fly lines remains unknown; they were made by either Hall, Ashaway, or Gladding. One thing is certain: The total Kosmic package was a combination of the products of four or five companies. Of all the Kosmic tackle, however, it's the rod that still causes hearts to race. The Kosmic rod wasn't born in Brooklyn, though. Brooklyn is just part of the story.

THE KOSMIC EXPERIENCE

To tell the Kosmic-Waltonian tale, we must go to Newburg Village, Penobscot County, Maine, where Fred E. Thomas was born on September 9, 1854. Young Thomas, the son of a bookkeeper, grew to become a river-driving boss, but it was clear he had considerably more talent. At the age of 26 Thomas left the logging industry and, urged by his friend Eustis W. Edwards, moved to Central Valley, New York, to work as an apprentice rodmaker for Hiram Leonard, with fellow Mainers Edwards, the Hawes brothers, and Edward Payne.

Thomas soon became disenchanted with an atmosphere that stressed production over quality. After seven years with H. L. Leonard, in late 1889 Thomas convinced Edwards and Loman Hawes to leave the Leonard factory and form their own operation. The company, "Thomas, Edwards & Hawes," started production in 1890, using a new beveling machine designed by Loman, at Highland Mills, New York.

Around 1891 Loman Hawes left the group (some sources say he died) to be replaced by Edward Payne. Now titled "Thomas, Edwards & Payne," the manufactory produced some of the most endearing rods in history. These very distinctive instruments were built in two grades, the highest-quality rod marked as the Kosmic, and a slightly lesser model titled the Isaak

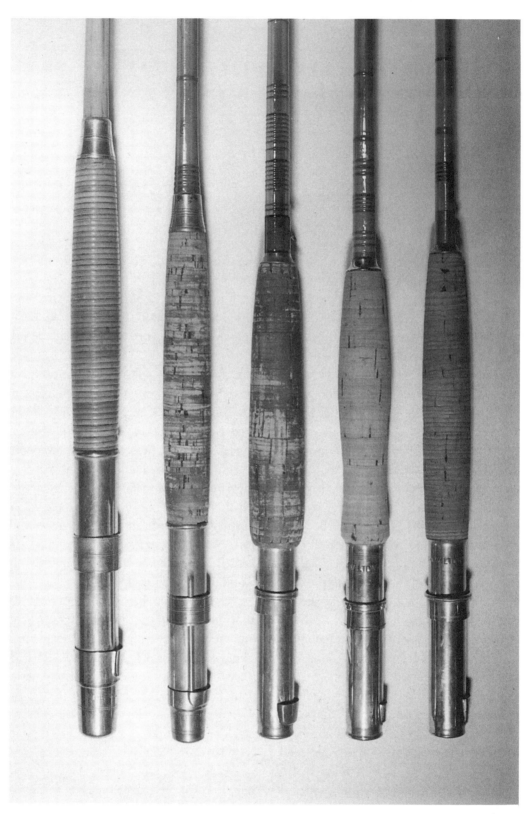

More than just a companion piece to the Kosmic rod, the Isaak Walton is the first model whose progression can be documented from the antique period to the classic years. Built from 1890 until 1901, these five examples show the changes in rod design (left to right): a 10-footer marked "A. G. Spalding & Bros., THE ISAAK WALTON"; another 10-footer marked only by Appleton & Bassett, Boston; a 9½-footer marked as the William Read & Sons "Premier"; a 9-footer marked "THE WALTON"; and a delicate 7½-footer marked "THE ISAAK WALTON."

Patented May 6, 1890, and May 27, 1890. Registered March 18, 1890.

Walton. Both brands were stamped in upper-case letters and introduced through their first major outlet, the A. G. Spalding Company, which had stores in Chicago and New York.

Of the two grades of rod, the Waltons seem a bit easier to find; the specimens I've examined were trout fly rods. Waltons were usually wrapped in Chinese red with no tipping. Kosmics had black tipping on red wraps above the winding check and on the guides. The Kosmic-grade came in several styles, with three- and four-piece trout models, light salmon or bait rods, and two-handed salmon versions. Even "valise" rods were made, such as a five-piece example. Collectors will encounter Kosmics in sizes from 7 feet for a 4-weight line to double-handed 15-footers that probably pushed an 8- or 9-weight. The company also marketed a range of reels and lines under the Kosmic name. The reels, from a little raised-pillar trout model to a multiplying nickel-plated jobbie and even a large salmon model, were built at the U.S. Net & Twine factory in Brooklyn.

The Kosmic rods were fitted with unique drawn ferrules designed by Hawes and Edwards. Loman Hawes patented a waterproofing cap for the female joint on May 27, 1890, which complemented the May 6, 1890, Eustis Edwards–patented celluloid collar that jammed under the ends of both male and female ferrules, where the metal met the bamboo shaft. Most of the high-end trout models were fitted with a celluloid reel seat with two German silver raised rails and a German silver sliding ring and cap.

Indeed, this seat was made from the same materials used by Landman—without the twist, of course—and may have been manufactured for Thomas and associates at the former's Brooklyn sweatshop. Conversely, some Von Lengerke & Antoine rods made by Landman seem to have

Thomas, Edwards & Payne shafts. The shafts may have been traded to Landman for reel seats. This is purely speculative, but the Kosmic boys had the ability to produce cane shafts and Landman did not. Landman, however, had the facility to make the finest reel seats and ferrules of the period.

Larger Kosmics, such as light and heavy salmon models, were fitted with all-German silver seats. The two-handed salmon rod seat had serrated locking rails in conjunction with a locking band that looked remarkably similar to the seat that Mills had surreptitiously tried to patent in 1881, and that Fred Thomas would successfully patent in 1913. This handsome, yet-unpatented seat was built with the utmost care. The Walton rods carried drawn waterproof ferrules, probably made in the rebuilt Chubb factory in Vermont. Although modern looking, the early Walton ferrules were not serrated. Walton reel seats were plain, almost austere, but the total look of the rods, especially the later versions, was surprisingly modern.

The Highland Mills operation lasted until about 1894–95, when Thomas and his two partners sold the business to the U.S. Net & Twine Company. All the equipment was moved to Brooklyn—many knowledgeable collectors say to the old Self Winding Clock building owned by the Pratts, probably the same address once used by Fred Malleson. With the financial boost supplied by U.S. Net & Twine, production increased, a full-time salesman was hired, and additional retail outlets were established. Kosmic's traveling drummer, a regular huckster named Herb Frost, also sold his own line of Frost's Kelso products.

Through Frost's smooth tongue, Kosmic-Walton rods were well marketed and sold through Spalding, The Wilkinson Company of Chicago, H. A. Whittemore & Company of Boston, and the E. G. Koenig Company of New York and Newark. Shortly after the move to Brooklyn, perhaps in 1895, Eustis Edwards left Thomas and Payne to their own devices and followed the belated advice of Horace Greeley. The Kosmic rod was still alive and well in 1896 when it was advertised by The Wilkinson Company, but at some point Fred Thomas also left the production

Two classic Kosmic salmon models. At left is a big two-handed rod, with the interrupted-thread reel seat, marked by H. A. Whittemore, Boston. The second rod is a smaller 10-foot version, stamped by Spalding & Brothers, Chicago.

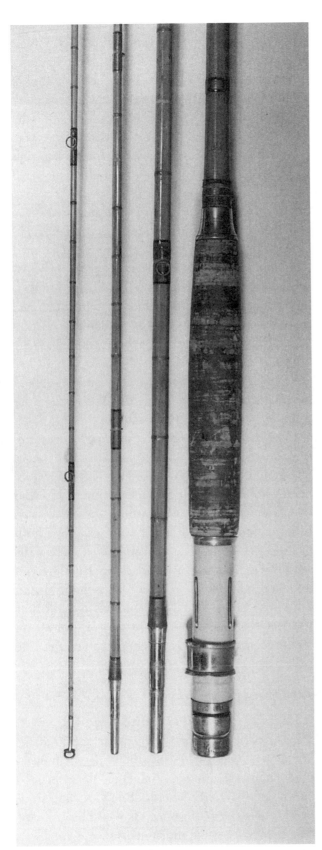

The Kosmic fly rod—a 10-foot model showing the famous ivoroid reel seat, Hawes-Edwards–patent ferrules, and thin ⅛-inch cork rings. It's marked by U.S. Net & Twine.

line. At this time only Ed Payne continued manufacturing the Kosmic models.[1]

In late 1896 or early '97 Thomas arrived back in Brewer and continued the Walton line. He also produced a few rods under his own mark, the extremely short-lived "F. E. Thomas Co., Brewer, Me." This rare rod-marking die was used prior to and possibly after his partnership with Eustis William Edwards. Thomas rods dated to this era include Brewer-marked Kosmic-style models and Isaak Waltons. An early rod in the Maine State Museum, stamped "F. E. Thomas Co.," is a rattan-handled model with a Kosmic reel seat and grip check, and has the Brewer address. Other than the short rattan grip and Montague ferrules, this piece could be mistaken for a Brooklyn Kosmic.

During the same period, Ed Payne fashioned the last of the marked Kosmics. Recently I restored a typical Payne-Kosmic fitted with an agate stripper and tops plus large English snakes. A wonderful old 10½-foot light salmon rod, it's marked "Spalding & Bros., Chicago," and has the Hawes- and Edwards-patented ferrules, Kosmic colors, and Ed Payne's signature wraps (five narrow wraps; a wider red wrap tipped black; three narrow wraps; another wider, tipped wrap; three more narrow wraps; and a final wide, black-tipped wrap). This classic beauty was built around 1898, when Payne started the E. F. Payne Rod Company back in Highland Mills.

The late Kosmic product was one of the finest rods of its day and came in a wide choice of models. Like the Thomas & Edwards fly rods, these last Kosmics often had Tonkin shafts and English tungsten snake guides. Both products were the earliest examples of the true "classic" fly rod. The only others that compared were contemporaneous Leonard sticks, plus a few Landman models built on Tonkin blanks originating from the Hawes beveler.

The Kosmic and Walton models spanned the decade that progressed from antique to classic. Thus we now find Thomas, Edwards & Payne models that have rattan grips, narrow cork grips, and wide cork grips; standard loose-ring guides and the newer English snakes. Given the variety of the rods, an entire collection could be created of them. And the good news is that Kosmics are rare but not exceedingly so, appearing on the market and at auction with relative

An early Thomas rod, stamped with the Brewer, Maine, address and featuring a Kosmic grip check and reel seat.

frequency. Waltons are a bit easier to find and not as pricey.

THOMAS & EDWARDS RODS

Back in Maine in 1898, Fred Thomas acquired the original Thomas, Edwards & Hawes beveling machine from Payne. Joined by Eustis William Edwards, Thomas formed the fragile association known as the Thomas & Edwards Company, Fishing Tackle Manufacturers. Among the most exotic of semimodern rods, the Thomas & Edwards creations were never marked by the makers. The distinctive grips were built from narrow ¼-inch (or less) cork rings—the same little rings found on at least 50 percent of the Kosmics. The bamboo shafts looked a lot like Waltons, except that the Calcutta cane was upgraded to Tonkin on later models. The rods carried only one seat style, a full-metal, railed version with a soldered pocket—not a Thomas & Edwards exclusive—that looked remarkably similar to, and perhaps was, Landman's work. These seats were rolled and soldered, as were the ferrules, which were waterproofed and "capped" and probably of Landman origin.

Thomas's original Maine rodmaking shop was located at 52 Center Street; this was most likely the address where the Brewer-

marked Thomas rods, as well as the T&Es, were built. Thomas & Edwards fly rods were made for less than two years, which places these truly modern rods in the truly scarce category. I've seen a number of examples, however—even one sold by Abbey & Imbrie. Early rods had loose-ring guides, while the later models sported English snakes. Wraps were fairly standardized and simple, and some late rods had a nice additional touch—an early and functional fly-keeper ring.

Thomas & Edwards rods were available in 8-, 8¼, and 9-foot models. The grips were nearly identical in all the rod lengths. Wraps varied from plain dark red ("old red") to dark red with black tipping; the intermediates were usually red as well. In many respects, one Thomas & Edwards rod looks pretty much like another—a harbinger of the standardization that would typify production classics to be built by many others.

Known T&E products were usually marked "Von Lengerke & Detmold, N.Y.," "Von Lengerke & Antoine," or "V.L.A." (Chicago); also found is an "Empire City, Special Grade" stamping, an Abbey & Imbrie trademark. It's entirely possible that Thomas & Edwards rods were marked with other retailers' names; time may provide a better picture of just who sold these rare items. Meanwhile, any Thomas or Edwards collector will consider the examples built by this great partnership a valuable find. A few years ago I was at Ben Clark's Northeast Tackle Show and discovered a very nice 9-foot T&E lying on an exhibitor's table. Literally thousands of collectors had walked by it, including dozens of dealers, yet there it was, priced at $75. I bought it, and this fine old classic is still in my collection.

F. E. THOMAS IN BANGOR

In 1899–1900, Eustis William Edwards again left Thomas, setting up a photographic studio at his home at 3 State Street, near the bridge in Brewer. Fred Thomas continued production alone. In 1900 he formed the "F. E. Thomas Rod Co.," and since that humble beginning a legion of fans have discovered the great casting instruments he designed. His early work mirrored the lines of the previous Kosmics, Waltons, and T&Es. Heavy-duty rods sometimes had trumpet guides, and there was the subtle

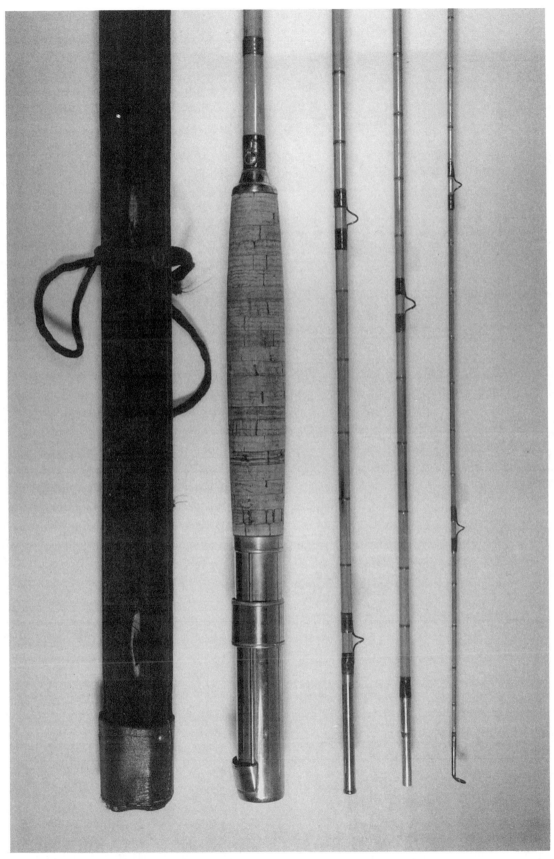

The no-nonsense Thomas & Edwards production fly rod helped usher in America's "classic" tackle age. This Von Lengerke & Antoine–marked model came with a Landmanesque reel seat, "new"-style snake guides, and a leather-tipped velvet case.

One of the earliest classic "hook keepers," original to an 8½-foot Thomas & Edwards fly rod.

swell in the rich straw-colored cane where it met the grip. Some models even sported the Kosmic grip check, and many of the classic Waltons and newly introduced Specials carried the original T&E hook-keeper ring.

The last of the Waltons were built into the early 1900s. These fishable classics were for the adventurous traditionalist, since they had Calcutta bamboo shafts and often handled light lines. Currently, the price of an average Thomas-built Walton is low, bringing it into the user category. These wonderful rods ranged from 7½ to 10 feet and had drawn Varney serrated ferrules topped with a rather large welt. The 7½-foot Walton was an ideal tool for light stream work, carrying a 4-weight line with ease. Even the 9-foot Isaak Walton was built for 4- or 5-weight casting.

Maine Sportsman, *March 1901*

It appears that these last Waltons were designed for the delicate presentations needed by midseason anglers. Rigged with rather small English snake guides for silk lines, wrapped in Chinese red, and with a swaged, soldered-pocket, German silver reel seat, the rods sported handles built from "modern" ½-inch cork rings and wide, classic grip checks.

Thomas introduced the Dirigo model in March 1901, in an advertisement that accompanied a glowing article on the rod by Herbert W. Rowe, publisher of the *Maine Sportsman*. The Special first appeared in Thomas's January 1904 ad in the same periodical. While the ad claimed that only one grade of rod (Dirigo) was built, the printing on the new aluminum rod tube read "F. E. Thomas Special." These two new models became the backbone of the Thomas business.

In 1902 Fred moved his business across the bridge to Bangor, to 117 Exchange, the Stetson building. Subsequent rods marked by Thomas carried the "Bangor, Me." stamp. A good many products, like the Walton, were stamped with a seller's brand name, and the period from 1900 to the mid-1920s produced the famous Thomas tackle-house Special. With a name that was perhaps a continuation of the original VL&D–Empire City key phrase, the Special or Special Grade showed up in other retailers' offerings as Thomas expanded his sales base. Boston became a stronghold. Bob Smith, Sporting Goods was an early Thomas dealer, and the William Read & Sons Special can be attributed to Fred Thomas. His top retailer, however, was Dame, Stoddard & Kendall, also on Washington Street.

Samuel Clark, a salesman at Stoddard's in 1927, stated, "Our top fly rod was the Thomas Special . . . selling for slightly less than $50. We often stocked rods made by Leonard and others but the one we pushed was the Thomas Special. The Dame Stoddard Company manufactured no fishing rods . . . but

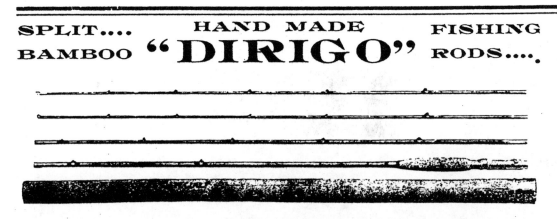

Maine Sportsman, *January 1904*

its name appeared on some so the rod . . . was a Thomas Special marketed by Stoddard."

Thomas strengthened his New York market as well, selling through his old associates at Madison Avenue's VL&D as late as 1923 (*National Sportsman*, April 1923). Thomas rods were also sold by Abercrombie & Fitch, often as Yellowstone Specials, a series of robust versions for big western fish. Many rods built during these first 25 years have rolled-welt Montague female ferrules. The early F. E. Thomas Rod Company fly rods were fitted with a full German silver seat, much like the last of the Waltons. From around 1910 on shorter and light-line rods had a delicate cedar cap-and-ring seat. One of the most diminutive models of this period weighed 1⅛ ounces and had a cedar barrel. Another charming classic, Fred's 9-foot Pack Rod, came in four sections. An example of this model, built in 1926 for big-game hunter Matt Lahti, took a 6- or 7-weight line. Appointed throughout in German silver, the rod had a full Wells grip and remains a fine example of the western style.

Standardized Special Grade fly models were furnished with Fred's unique tanged, pear-shaped tiptops and a keeper ring that was soldered to the German silver winding check. Tonkin blanks with "water marks" or blemishes, not quite good enough to be called Special, were used on Fred's second-grade models, the Dirigo line. Dirigo rods were fit-

ted with less fancy furnishings and had no tangs on the tiptops. These were excellent rods, and many collectors who don't dare use their Specials have found the Dirigos to be wonderfully castable rods, especially the crisp 8-footers designed for a 4- or 5-weight line.

Fred Thomas was an early champion of English snake guides, incorporating them into the Kosmic-Walton line. (Viewed from below, English guides snake to the left, while American versions twist to the right.) At Bangor he continued their use, and many early fly rods were completely snaked, including an English snake stripper. At one time Fred had the largest stock of English tungsten snakes in the industry. An old story goes that, when the E. F. Payne Rod Company ran out of this commodity, it turned to Thomas.

LEON THOMAS AT BAPTIST HILL

Around 1916 a few rare Mahogany models were built, precursors of the famous Browntone series. In 1917 Fred Thomas began making Browntone rods, and two years later these dark-shafted models were standard catalog items. The Browntones were also marked "Special," and that they were—the company's highest-priced items. The Browntones were fitted with oxidized furnishings; this eliminates any confusion between them and other Specials. By the late teens and until the 1930s, intermediate wraps on

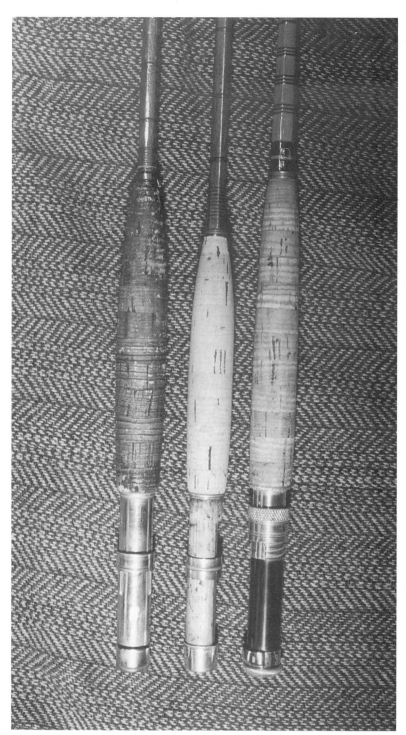

Three Thomas classics featuring reel-seat variations (left to right): a mahogany Special with the full German silver seat; a light-line Dirigo with a cedar-barreled cap-and-ring seat; and a newer Dirigo with the Hardy-style seat.

pear tops. Also in this era, the Brown-tones and Specials received the classic brown wraps that became a Thomas hallmark.

In 1923 Fred Thomas moved his business from Exchange Street to a new brick building at the top of Baptist Hill, on the corner of Center and Park Streets. By now his son Leon was one of Fred's top craftsmen, and six other people were employed at the shop. The F. E. Thomas Rod Company offered a staggering array of products. During the early Depression years the Bangor Rod was added to the line. This model was not marked by Thomas, and was built from lesser cane and furnishings than were used on the Dirigos. Often, the new locking reel seats had a black hard-rubber spacer. Today, Bangor Rods sell for robust prices if in excellent condition. They're actually less common than Specials, judging from auctions and tackle lists.

Special Grade rods designed for heavy lines carried, from the 1930s to the last rods in production (1958), a locking Hardy-style seat with a butternut spacer and a knurled locking ring. Light-line rods continued to have a cedarwood seat, although a few cork-seated rods can be found. Starting in the late 1930s the Specials received serial number stampings—a letter followed by a series of numbers. These serial numbers help place later rods as to their year of production; the records are held by collector and dealer Martin J. Keane.

Regardless of the model a walk-in customer preferred, Fred Thomas always gave every sale personal and friendly attention. Around the end of 1937 Thomas retired from an active role in the business due to failing health. He died of myocarditis on July 16 the following year at his home in Brewer. His obituary stated, in part, "Absolutely absorbed in the produc-

all rods grew farther apart, until the spacing became just a few wraps between guides. You can also "age" a rod by the stripper guide, which moved farther up the shaft as years went by. Later rods usually carried the tongued ferrules, Perfection strippers, and tanged

tion of his rods, Fred E. Thomas has given the mature years of his life to this manufacture of highest possible grade products . . . and sportsmen have visited his shop from afar and near, for a friendlier touch with the maker of their angling equipment." He had been a leader in the quality fly-rod trade for almost half a century.

Leon J. Thomas, who entered the rodmaking trade in 1910, became an able and gifted successor. The only change he made in the Bangor shop was its name—the Thomas Rod Company. Around 1939–40 Leon designed a new style of heavy-animal rod that became known as the "Atlantic and Pacific Salmon, Steelhead and Salt Water Fly Rod." It had a short, nonremovable butt, similar to those on modern salt-water fly rods. With the Salt Water Rod, Leon became one of the first makers to test the emerging marine market.

Around the same time Thomas also introduced the Streamer Rod, the great early-season upcountry workhorse designed to deliver a large hunk of feathers through sleet and snow. Thomas correctly claimed, "These beautifully balanced, powerful rods cast streamers or bucktails safely for remarkable distances, even into strong winds—imparting the darting action necessary for best results." Like his father, Leon was a rodmaker who wanted to meet his clientele in person, and in the early spring of 1942 he went down to the Boston Sportsman's Show and demonstrated the streamer rod to anyone who wanted to see it in action.

Each year thereafter, Thomas set up a booth at various northeastern sporting expositions and hobnobbed with potential customers. He also started an extensive advertising campaign in national hunting and fishing magazines. The new saltwater and streamer rods played a leading role in bringing the American fly fisher out of the staid dry-fly syndrome that had trapped anglers into a single way of fishing for almost four decades. Streamer casting proved incredibly effective for landlocked salmon, brook trout, and the newer marine species, and—in part spurred by the success of the new Thomas rods—fly angling branched out into the multifaceted sport we know today.

The Thomas Rod Company offered six different fly-rod actions. The workhorse models included Wet

and Dry Fly Salmon sticks and the Salt Water Rod. Streamer Rods were built in three sizes. The models most often encountered today are the two Streamer sticks built in a three-piece configuration, the big 9-footer at 5⅝ ounces and a slightly lighter 8¼-footer. These rods had an extralong cork grip and were designed for a 7-weight line and lengthy casts. A seldom-seen two-piece "Baby" Streamer Rod measured

only 7 feet, and came with a warning: "Recommended for use only by experienced casters."

Other unique Thomas actions include the Special Canoe Rod and the rare Special Parabolic. The Canoe Rod is another powerful three-piecer, a 9½-footer for a 7-weight line with a miniature detachable butt. Today it's a great bass or belly-boat rod. The 7-foot, 9-inch Parabolic is one of the rarest and most coveted of all Leon Thomas rods. This 3½-ounce, two-piece beauty had the ability to present accurate casts both close in and from a distance. Since only 13 Parabolics were ever made, they represent the higher-priced end of the Thomas spectrum.[2]

Thomas Dry Fly trout rods were available in lengths from 6½ to 9 feet. Wet Fly actions came in similar lengths, but, at least with Thomas, the term "wet fly" didn't mean the rod was a noodle. On the contrary, Thomas wet-fly rods could lift a sinking line from the water and were popular nymphing instruments. The Dry Fly action was also used in some unique examples, including a 6½-foot, five-piece Pack Rod. This uncommon Special tossed a 5-weight line and had a fast action.

Which Thomas rods were best from an angler's standpoint? Well, Martin J. Keane liked the 9-foot Special Streamer Rod, at 5½ ounces with an oversized grip. Len Codella espoused the 5¼-ounce, 8½-footer as "one of the finest tapers in its length." For years, my favorite for light trout work from a canoe was the three-piece 8-footer (a Dirigo, as it happened) at just under 4 ounces. Actor William Conrad sang the praises of the 7-foot, 2½-ounce Browntone, which tossed a 3-weight line.

During the 58 years of their production, the Thomas three-piece models varied from 7 to 15 feet long. Two-piecers were built from 6½ to 8½ feet. Special-order items, such as early lancewood models, fresh- and saltwater trolling rods, and diminutive 5½-foot flea rods do exist, although they're rare.

In 1947 the Thomas Rod Company employed 10 men and one woman in the shop, as well as two more

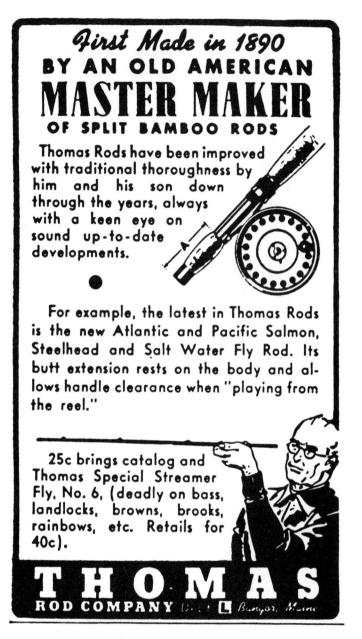

women at home as silk winders. The top Thomas craftsmen included Harold Soucy and Leslie Tenney. Through Leon's able rod designs and management, this select group managed to continue building high-end rods through the stormy early "glass" years. In 1958 Leon Thomas retired and sold the Maine company to Sam Carlson, who in turn sold the Hawes beveler to Walt Carpenter. These two extremely talented men still own the rights to the Thomas patterns and name.

Today, "well fished" three-piece Specials and Dirigos are often found on dealer lists, usually sell-

E. WILLIAM EDWARDS,
PHOTOGRAPHER
STUDIO 3 STATE ST., NEAR BRIDGE, BREWER, ME.

1916 Bangor-Brewer Business Directory

ing for no more than a comparable Granger. Thus the classic Thomas rod continues to do what it always did best: provide an eminently enjoyable fly-fishing tool for anglers of modest means. Many "previously owned" Thomas rods available today were designed for popular line sizes—4- to 6-weight—and, with a slightly short tip or mid, may sell from $350 down to $175. They are great production pieces, capable of pleasurable casting for years to come.

I make no bones in saying that Thomas rods are comparable to or better than the Leonards, both in castability and in finish. They're great rods, yet to command their true value.

E. WILLIAM EDWARDS & SONS

After leaving Thomas in 1899–1900, Eustis William Edwards remained in Brewer and returned to his occupation as a photographer; he was listed in the *Bangor-Brewer Business Directory* under that title from 1901 until 1916. Concurrently, Eustis Edwards began experimenting with cane and produced a few rods in a small shop behind his State Street home. Without question this was a humble enterprise. Throughout his unstable yet superlative rodmaking career, the thrifty Edwards usually signed his rods in ink just above the grip check, eliminating the expense of a die stamp. (This "signature" idea would be copied by a host of later part-time artisans during the classic era.) Edwards was also one of the first makers to employ black hard-rubber hardware in the form of butt caps, seat barrels, and checks.

Around 1915 Edwards, known to friends as "Billy," discovered his heat-treating process that flamed the cane to a dark brown and, more important, added amazing power to the shaft. This was the first radical change in bamboo rodmaking since the adaptation of Tonkin cane. The next year the city

directory carried a new listing: "E. William Edwards, photographer and mfr. of fishing rods." During the next four years he and his older son, William E., are said to have built approximately 1,000 rods. In reality, far fewer were made. These early "Signature" models, built in the 1916–1919 era, have not survived in numbers, and a Brewer Edwards is a great find.

As a Brewer directory entity, son William E. Edwards was first listed in 1907, as a "clerk" boarding at 7 State Street. He then became a chauffeur and, in 1910, a partner in Eddy & Edwards, Machinists, in Bangor. William E. worked with his father making rods from 1916 until his marriage in 1918.

The male and female drawn ferrules on the early rods were purchased from Montague, then cut to shorter lengths. Edwards removed the extruded end of the male and soldered in place a perfectly fitted end disk. You should note that these disks are now sometimes missing, exposing bare cane at this area. Many of these Brewer rods were sold through Abercrombie & Fitch in New York, and you can use these ferrule features to identify Edwards rods sold through that retailer. I've seen an early experimental rod (with these tommy-knocker ferrules) that had been given the heat treatment with a blowtorch after the sections were formed.

In 1919, the Winchester Repeating Arms Company of New Haven, Connecticut, purchased E. William Edward's company, and convinced E. William to move to its large factory, recently acquired from Hendryx. Bill the younger remained in Brewer as an auto mechanic, and Bill senior relocated to Connecticut in the fall of that year. High-end Winchester rods built during E. W. Edwards's tenure were graceful casting tools. Many of the better-grade rods were built for, and marked, South Bend.[3] The most memorable and collectible of the

Fishermen's Dreams Come True!

WINCHESTER tackle for each fish and wind and water. Rods and reels, lines and baits—Winchester dependability in everything for anglers, bait casters and still fishers.

Winchester rods are as stout and balanced as Winchester guns. They are built to withstand the strike and rush of hard-fighting fish. They are refined to the delicate snap of the wrist. And one and all—bamboo or steel, from trolling rods of stiff backbone to the feather-weight fly rods of expert hand construction— they are of Winchester guaranteed quality.

As to reels, spoons, plug baits, flies, lines and sinkers, Winchester suits the fish as well as the fisherman. Like the Winchester rods, they are made to satisfy completely the wants and exacting requirements of fishing. It's the Winchester way.

With Winchester Fishing Tackle, fishermen's dreams come true!

5000 Hardware Stores display this sign on their windows:

THE *WINCHESTER* STORE

They sell Winchester Roller Skates, Pocket Knives, Flashlights, Rifles, Fishing Tackle, etc.

WINCHESTER REPEATING ARMS COMPANY, NEW HAVEN, CONN.

National Sportsman, *April 1923*

Edwards Winchesters was the little 7½-foot model, which took Catskill anglers by storm and may have made that length the popular stream fixture that it is today. The low-end rods were not well-appointed products, however, and after five years as foreman over this rampant production, Edwards left Winchester to go on his own.

Thus began the short Filbert Street era—really only a few months—when "about 50" Autograph Perfection rods were produced.4 These scarce fly rods were a far cry from the Winchester models. Today, they remain examples of the best that Edwards ever made. The hard-rubber accents, almost-burnished shafts, and fine German silver seats all added to the rod's character. Each was signed "E. W. Edwards Perfection," and usually had gold wraps tipped purple and very dark cane.

In 1925, with the help of the younger son, Gene, the business was expanded to become the E. W. Edwards & Son Rod Company. The machinery and stock were moved to a larger building in New Haven's Mount Carmel district. In the space of two years the production and workforce increased by the addition of elder son William E. Edwards and Robert Van Hennick. The rod models included the top-of-the-line E. W. Edwards Deluxe, the E. W. Edwards Mount Carmel, and a lesser-appointed E. W. Edwards Special, which often sported a rubber ring at the junction of the seat and grip. The cane of the Mount Carmel–era rods was lighter than that of previous models; the windings were usually light brown tipped yellow. Early rods had a German sil-

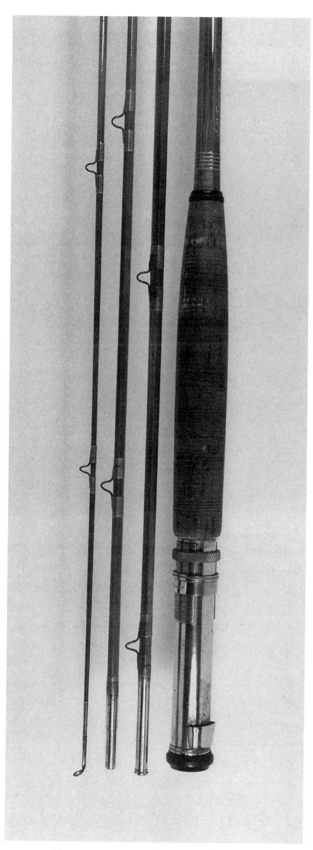

A light, 10-foot E. W. Edwards Mount Carmel salmon rod, sporting Wes Jordan's 1927-patent reel seat and clear windings from the tiptop to the first guide.

ver soldered-pocket seat. Light rods built after 1927 exhibited a classic two-piece, walnut-barreled style.

Edwards rods featured the new Perfection tiptops, patented in 1915. I believe that Eustis was one of the first makers to fit them as standard items. It is not always easy to pigeonhole an Edwards. Take the example of a 10-foot Mount Carmel salmon rod with a full German silver screw-locking seat stamped "PAT. FEB. 15, 1927." The brainchild of Wes Jordan, this 4⅛-inch seat was made by South Bend—its No. 516 Positive Thread-Lock Reel Seat. The salmon rod also exhibited standard Mount Carmel ferrules, overwrapped for the first ¼ inch. It showed a post-Winchester link to South Bend yet remained a classic Edwards.

The E. W. Edwards rods were some of the finest built during the production period. Many of his rods were private-labeled to tackle dealers across the country. Because there was a distinct similarity between the products of Eustis and of Gene Edwards, it's hard to distinguish which Edwards made the rods marked to Gray's, Hartford, Connecticut; Cook, Newton, and Smith, Special; and Abercrombie & Fitch, Yellowstone and Triton, a rod perhaps first made by E. W. and carried on by Gene. Senior Edwards rods were sold through Abbey & Imbrie, marked "E. W. Edwards Deluxe" and carrying the A&I decal.

THE HORTON "BRISTOL-BUILT" YEARS

Today, very few E. William Edwards rods show up on the market. It's a collector's choice whether to use these superb casting instruments or to keep them in state for posterity. The Bristol Edwards classics are a different story; they're readily available and sound, fishable sticks. Just before his death in 1931, Eustis Edwards sold out to the Bristol Rod Company. After the deal was consummated, the Mount Carmel factory was lost to a fire. The machinery was salvaged, however, as was most of the cane, perhaps slightly heat-treated before its time.

With the E. W. Edwards machinery transported to the Bristol plant, the second generation, Gene and William E. Edwards, began to produce their own beautiful rods. Gene supposedly designed the higher-grade Bristol-Edwards models, probably the No. 50 and No. 52 series, built with the best German

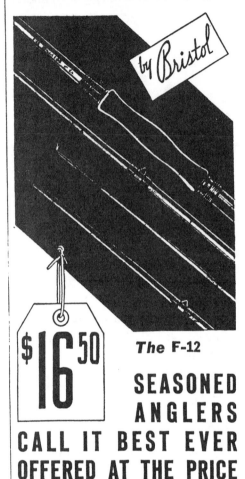
silver/aluminum furnishings and having "Edwards-Bristol" written on the shaft. These top-of-the-line Bristol models sold for around $35 in 1940–41. Mid-priced Bristol rods, the F-18 and F-12, had good-quality hardware. The penny-wise F-7 and lesser F-4 series, with nickel-plated trimmings, were often private-labeled and are still a good buy for today's anglers. The Bristol rods, like later Edwards creations, were built from "matched cane," meaning that the six strips in any given rod section came from the same bamboo culm.

For long, controlled casts, the Edwards No. 52—a 5¼-ounce 9-footer—could not be beat. The Edwards No. 50 was a handsome 8-foot model weighing in at 4¼ ounces. Each Bristol-Edwards model was available in three actions: trout, dry-fly, and bass, to cover the spectrum up to 8-weight. The No. 50 trout-series rods are superb casting tools, extremely lively and responsive, and handling popular 5- or 6-weight lines.

The Bristol years came to an end around 1941 or early 1942, when World War II finally slowed production to a crawl.

BRISTOL "MEEK" FLY REELS

During the early 1940s Bristol, a pioneer of "tuned tackle" outfits, sold the second-generation Edwards creations in conjunction with matching reels and lines, such as the Martin-made Kingfisher Deluxe or Luckie double-tapered fly lines. As was the case with Edwards-Bristol rods, the higher-grade reels were marked "Meek," while the lesser were marked "Bristol." The F-series fly rods matched the practical Bristol No. 66 reel, which looked a lot like a Meisselbach Rainbow and sold for $5. The Edwards No. 50 series just happened to complement Bristol's Meek Fly Reel No. 56, which sold for $10 and sported a Hardy-style spool retaining clip. The best of the reels, Bristol's Meek No. 55A, sold for $11–12, had a nice round, agate line guide, and is a fine collectible for Edwards fans. Judging from the few examples seen today, agate-appointed Meeks are now worth far more than their original selling price.

THE WILLIAM E. EDWARDS QUADRATES

Just before the demise of the split-bamboo wing at Bristol, William E. "Bill" Edwards and his son,

An 8-foot Gene Edwards Autograph fly rod with its original bag, owned and well used by author-angler Arnold Gingrich. (The American Museum of Fly Fishing Collection)

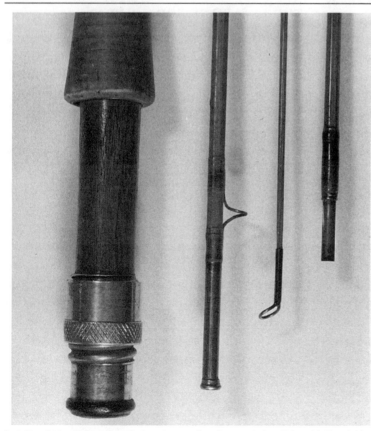

Details of the walnut-barreled reel seat, ferrules, and tip of Arnold Gingrich's Eugene Edwards fly rod. (The American Museum of Fly Fishing Collection)

Scott, moved back to Mount Carmel, set up their own shop and machinery, and built the unique Edwards Quadrate rods. These four-strip rods had square shafts; even the ferrules and grips were square. Although they may have seemed like a giant leap back to the days of Green and Murphy, these highly collectible four-strippers had punch and panache.

Today, their only problem is scarcity. Not many were made, and the few that surface at auctions are scarfed up like hot dogs at a baseball game. True, other Quads are sometimes available, like Sam Carlson's beauties, but overall, four-strip rods are rare birds.

Collectors will discover that William E. Edwards's "lesser" models (although "lesser" is a term difficult to apply to a Bill Edwards rod) were confined to the #30 series, which had plain German silver ferrules. The #34, #35, #36, and #37, built from 8 to 9½ feet in length, comprised these "common" models. Next up the line is the #40 series, built in a two-piece style from 6 to 8 feet long. These rods, as well as the models in the next two higher-numbered classes, had oxidized fittings. The ultimate Quads, the super rods in the #50 series, were three-piecers from 8 to 9½ feet in length, and the ultra-rare #60 salmon series ranged from 9 to 10½ feet.5

The Bill Edwards Quads occasionally show up at auction or in the rod listings of Len Codella, Marty Keane, and Bob Corsetti. Selling at premium prices, they're perhaps the most exotic and expensive of all the Edwards family's creations. After working with the Edwards family, Sam Carlson began making his Carlson Four series, and these jewels are today's continuation of the classic four-strip school. In recent years Sam has slowed down, and the relative value of his work has dramatically increased.

EUGENE F. EDWARDS PRODUCTION YEARS

After the Bristol factory closed Gene Edwards moved back to Maine for a while, working for Leon Thomas. Then he went back to Connecticut, and in 1945 was offered the entire machinery and inventory of the dormant Bristol factory. Although he couldn't actually meet the selling price, Gene was given permission to take the stock until such time as he could pay for it. Moving the equipment to Whitney Avenue in Mount Carmel, Gene Edwards was soon making some of the best production fly rods in the country. Within a few years the stock was paid for and the workforce increased to six, including Robert Van Hennick and Sam Carlson.

The Eugene Edwards rod was built from the end of the war until 1955, with an estimated 10,000 rods ending up in sportsmen's hands. Considering this large number, one might wonder about the quality of the product, but Gene Edwards rods were close to flawless, even those sold under various retailers' names, including L. L. Bean, Bob Smith, Stoddard's, and Skilton. From several years of personal experience tossing a 6-weight line to some large Maine brook trout, I've found the 8½-foot Skilton Supreme to be a great production rod. Gene also made the Abercrombie & Fitch two-piece Triton. The stunning Triton 7-footer, which handled a 5-weight line, was an excellent all-around rod.

Gene Edwards carried the old Bristol model numbers and designs into his own line. The pinnacle of the classic later Edwards models, two- and three-piece rods in the No. 52 and No. 53 series (the latter were three-piece), sold for $55. They ranged from 7 to 9 feet and were put up with an extra tip, bag, and aluminum case. Resolute and beautiful salmon models, the No. 75 Deluxe series rods, were three-piecers selling for $75. These were built between 9 and 10 feet in length. The least-expensive varieties, a category into which the Skilton probably fell, were still well built, handsome, and castable designs selling for $25. All models were signed on the shaft; those marked by the maker say something like "Gene Edwards Deluxe #75." Today, the Gene Edwards rods rank very high in the classic lineup.

All the Edwards six-strip rods built by these three generations of superb craftsmen are the very finest examples of the production era and are among the most enjoyable casting instruments obtainable. With proper care, they can only increase in value as time goes by.

THE E. F. PAYNE ROD COMPANY

We left Ed Payne in Highland Mills, New York, in 1898, the year he established his own company. A veteran of reel- and ferrulemaking, Payne finally began to make his own rods—and each one would be better than the last until the demand permanently outstripped the supply. In frugal Maine fashion, some of the first Ed Payne rods carried various Kosmic furnishings, especially the patented ferrules that seem to have been fitted to his early salmon rods. Reworked Kosmic ferrules were used into the 1900s, and Varney sets were used as late as 1910, when the Payne hourglass ferrule became standard.

Many early E. F. Payne fly rods had a distinctive exaggerated Wells grip, with shafts in the natural Tonkin color. When Ed died in 1914, his son Jim took over the business.6 A few years later, Jim Payne had perfected his flame-treated models, and the Payne rod became the famous dark-toned gem over which collectors now drool. Although the lighter-colored untreated models were still available, most customers were willing to pay extra for the darker versions.

During the next period, lasting into the 1940s, rods had an "E. F. Payne Rod Co.—Makers" stamping. By 1951 rods built by Jim had a stamp consisting of just the word "Payne" inside a horizontal, pointed box, with "Reg. U.S. Pat. Off." in one line under the box. Rods built before the late 1920s carried nickel-silver seats with a soldered pocket. Later models had either the Payne aluminum seat with spacer or the walnut-sliding-band-and-cap seat. With the aid of George Halstead, post-1925 rods had that ferrulemaker's oxidized fittings.

Original Ed Payne fly rods were built from 8 to 10 feet long in the one-handed versions, and up to 14 feet long as two-handed salmon rods. The earliest rods, like Leonard and Thomas models, came with an antique cane tip case, all put up in a canvas sack. By the 1920s they were packed in a modern cloth bag and an aluminum tube. All Payne examples had jewel-like metalwork, perhaps the finest in the industry. Some early models built by Ed were available with detachable handles, but most rods came in the standard 3/2 combination. The late-1920s "Light Finish" versions sold for $42; the "Dark Finish" models went for $50. Additional butts, mids, and tips were available at extra cost—between $7.75 and $19. We'll take two tips, please—for what, $15.50?!

In 1930–31, Jim Payne dropped the lighter-colored series and some of the longer models and introduced the little No. 96, a diminutive two-piece 6-footer that weighed 1¼ ounces—theoretically. I had one that weighed 1⅝ ounces, so weights varied even with the best of the breed. I tried a 3-weight line on the dainty shaft, but it overpowered the rod. After discovering that it was "one o' those Smidge Rods #@!," the tiny Payne was never used; my eyes can't see a midge, let alone tie one to a tippet that I also can't see. When Jim introduced a new 6½-foot model that he had to sandwich into the lineup before the 7-footers, the little No. 96 became the No. 95.

The Payne two-piece 7- and 7½-footers were built in two actions, Dry Fly and Fast Dry Fly, handling line sizes 3- to 5-weight. Jim's 8-footers were Dry Fly models designated as Light and Fast, and carried the same line weights. Payne's 8½-foot two-piece models had three separate actions: Light Dry Fly, Slow, and Medium or Fast Dry Fly. An uncommon rod, the 8-foot, 10-inch model weighed around 5 ounces and

The Outdoorsman's Handbook, W. H. Miller, 1916

Throughout most of the company's history—into the 1930s or later—many Paynes carried the "E. F. Payne, Maker, Highland Mills, N.Y." marking.

filled the niche of a long rod needed for delicate casting. The longest trout models, the 9-foot Light Dry Fly and 9-foot Standard, allowed anglers to toss 4- to 6-weight lines. The shortest Payne, by the way, was the 4-foot, 4-inch Banty.

Other two-piece models included the superb Parabolic series, developed with John Alden Knight.

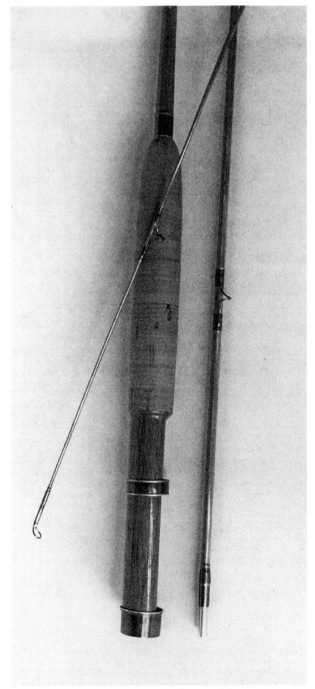

A true American classic, in the form of a 6-foot Jim Payne No. 96. One of the most coveted of trout rods, this ultralight two-piecer carries the older, 1930s "E. F. Payne" marking.

In 1936, Knight and cohort Charles Ritz went to Highland Mills with the parabolic idea, and Jim built the "first American rod in the true parabolic action." The design required pinpoint timing, so Payne smoothed the midshaft tapers to a semiparabolic curve. In 1951 Knight noted, "Mr. Payne modified the original pattern and produced a rod more acceptable for general use and for which, I understand, there is an ever-increasing demand." A vast understatement today. The rods were available in two lengths. The 7-foot, 1-inch Parabolic weighed 2⅞ ounces and handled a 3- or a 4-weight fly line. A larger 7-foot, 9-inch version took a 4- or a 5-weight line. Both rods featured a gunmetal black finish on the hardware. These powerful fly rods are for monied collectors who enjoy distance casting.

In today's market, three-piece Payne fly rods are—sometimes—within the average collector-user's grasp. For a short time in the 1930s Jim built a now scarce three-piece 7-footer that averaged a mere 3 ounces. Later Payne three-piece fly rods ranged from 7½ to 9½ feet and handled lines of from 3- to 7-weight in six different actions: Slow, Light Dry Fly, Dry Fly, Fast Dry Fly, Medium or Fast, and Heavy. As working collectibles, they're wonderful trout tools, exuding the rich aura of supreme craftsmanship. Today, many rods within this grouping are sold in well-used condition at reasonably low prices—for Paynes.

Jim also built some big-water three-piece classics. The 8½-foot Canadian Canoe Rod, designed for distance and large squaretails, ranged from 5 to 5¼ ounces and could drag a wet 7-weight line off the water, arch into one backcast, and present the fly again. The easy-working Canoe Rod was complemented by Jim's own versions of Leon Thomas's Streamer Rods. Payne Streamer Rods came in two sizes: an 8½-foot model weighing up to 5¾ ounces and a 9-footer reaching 6¼ ounces. With a 7- or 8-weight line, the big rods could handle streamers and more. In his 1951 catalog, Jim boasted that the model was "an ideal rod for Colorado spinners where large rainbows and steelheads are encountered." He didn't mention, of course, that Colorado spinners were frequently used with worms.

Early on, Jim addressed the needs of saltwater buffs. A unique Payne of 1930s vintage was the 10-foot

Tarpon Rod developed for A. E. Hendrickson. In the 1940s Payne introduced a line of Bonefish Fly Rods that incorporated a 1⅞-inch fixed fighting butt. The standard 9-footer probably handled an 8-weight line. And two 9½-foot bonefish sticks could throw 8- and 9-weight lines. All three models were built with corrosion-resistant guides and furnishings. In the 1950s the company still offered up to 11 Dry Fly Salmon models and eight different lengths in its Two-Handed Salmon Rod line. Frugal West Coast speycasters take note: The long big Paynes now sell for a fraction of the price of trout models, and many of these big babies were double built. A few four-piece salmon rods can also be found, although they don't seem to be a catalog item.

In 1963 Jim sold the E. F. Payne Rod Company to the old Gladding Corporation. He couldn't retire, however, and stayed on as the shop's master maker for another five years. The Gladding folks continued the line until the mid-1970s before closing the doors of the old Highland Mills operation.

From the 1930s until the end many Payne rods were sold through Abercrombie & Fitch, and the seller's stamp is so marked. The last A&F Paynes were similarly marked, but the wording was changed to "Made For" instead of "Sold By." When collectors heard about Jim Payne's death in 1968, there was a run on the remaining rods at A&F, and prices skyrocketed. There would never be another Jim Payne. The late Alfred W. Miller, the inimitable "Sparse Grey Hackle," said it best:

> All the old Leonard workmen who set up their own shops were notable for their rigid adherence to the highest standards of workmanship in spite of small incomes and, often, poverty. But I think Jim surpassed them all in his fanatical adherence to his own incredibly high standards of craftsmanship and his rigid determination to uphold the status of the Payne name as a tribute to his father.

HIRAM HAWES LEAVES LEONARD-MILLS

Within the history of the classic rod, no man has received so little credit for his accomplishments as has Hiram Webster Hawes. Starting his long career with his uncle Hiram Leonard in Bangor, Hawes was in large measure responsible for the building and design of many classic-era Leonard models at Central Valley. An accomplished tournament fly caster, Hawes applied his knowledge of distance casting to the design of several of the Leonard & Mills Com-

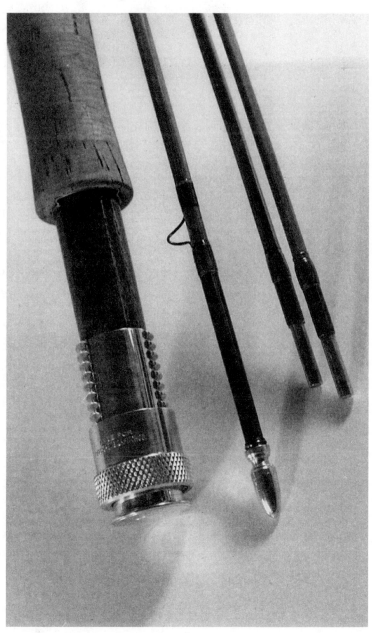

An 8-foot, three-piece Payne of later vintage "Made For Abercrombie & Fitch." Payne spared nothing, and appointed his fly rods with beautiful fittings, including the ferrule plugs. (The McGrath Collection)

pany's tournament sticks. When he finally left the Mills interests in 1909, Hiram Hawes; his wife, Cora; his son Merritt; and his mother-in-law, Lizzie Leonard, all moved to Canterbury, Connecticut.[7]

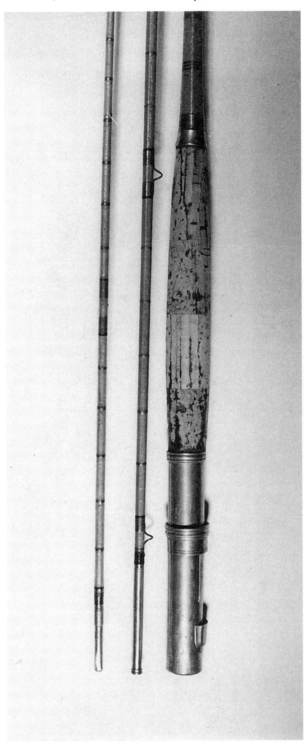

An early Hiram Hawes 10-footer, built around 1910 and sporting an old-fashioned full-metal seat. It's marked by Abercrombie & Fitch, New York.

One of the items that Hiram inherited was the original wooden Leonard beveling machine. In Connecticut, this last of the Mainers to leave Leonard-Mills started the Hawes Rod Company, and produced a small number—fewer than 1,000 units—of high-quality rods.

It's ironic that a man who helped build so many rods at Leonard-Mills would make so few under his own name, but so seems the case. So little is known about Hawes's output that collectors have been left hungry for knowledge. At a recent auction, a Hiram Hawes catalog sold for more money than most production rods. Many surviving examples feature a Wells grip, and the rods have flawless shafts. Hawes's wife and son were also winning tournament casters, and a few Hawes rods were built with tournament tapers, but the majority of his production was designed for delicate casting astream.

At Canterbury, Hawes spun the famous old beveler with waterpower, but his shop was far from being a factory. He did produce enough rods to attract the attention of Ezra Fitch, however, and many rods were marked "A&F, Sole Agents." The earliest Hawes creations, some made with Calcutta cane, were built without the use of finishing dies and were marked "The Leonard-Hawes Rod," a stamping that the Mills folks took legal steps to kill. Slightly later rods show the refinement of a knurling tool, as Hawes bought his own dies. These carry the standard "Hawes Rod Co., Canterbury, CT." or "Abercrombie & Fitch, New York" markings. The rods were built with serrated Leonard-style ferrules or with drawn Montague fittings. Some models had very wide $1\frac{7}{16}$-inch cork rings in the grip—a very unusual feature, which can aid collectors. With sometimes heavy, soldered-pocket reel seats and old-style grip checks, Hawes's rods looked much like Leonard's.

Hawes rods came in five different styles: Trout and Bass Dry Fly models, the "Featherweight" rods, the Tournament models, two versions of a two-handed Salmon Rod, and the ultra-rare Touradif—a six-piece pack, or "trunk," rod. Its entire length, when broken down, could easily fit into an Adirondack pack basket. While the regular 8- to 10-foot models were "quicker in action and more powerful for their weight than any other rod made," the little Featherweights were perhaps Hawes's glory. Made in lengths

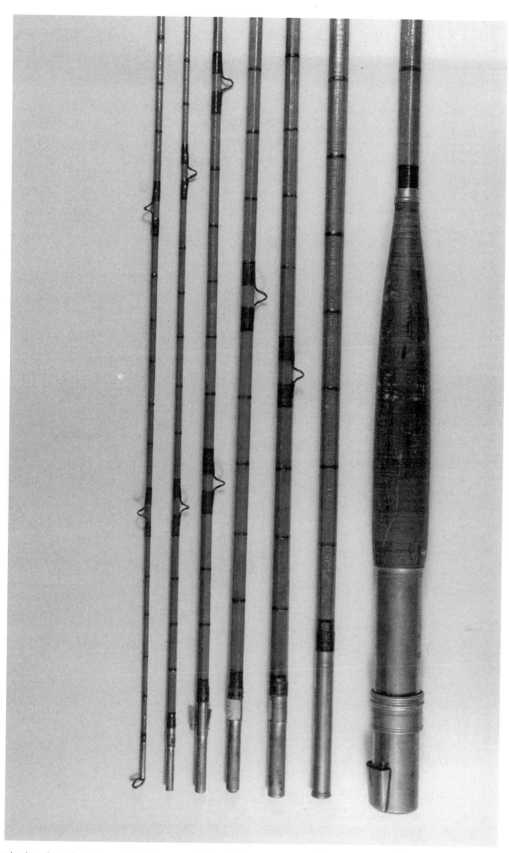

Built in a style that dates back to early-19th-century England, this Touradif pack rod by Hiram Hawes was built in six sections. The conservative full German silver reel seat, red intermediate wraps, and serial number X194 all indicate an origin in the very early teens.

of between 7 and 8 feet, these delicate buggers could cast a 2- or 3-weight line, weighed from 2 to 2¼ ounces, and were finished with cedar cap-and-ring seats. Tournament and Bass Rods were equipped with full

Bill Edwards, Ed Payne, Loman Hawes, and Hiram Hawes all learned their trade from the original master, to the independent decades when each of these craftsmen went out on his own and applied lessons

A Bit About Inflation

In these days of $350 production fly rods, it's depressing to learn that such things as a 1920s-era Payne sold new for a mere $50—especially considering that the same rod today might bring upward of $5,000. But was Granddad really getting that much of a bargain? Not if you adjust the figures for inflation. In 1994 dollars, Granddad's $50 Payne set him back the equivalent of $364—about what you'd pay for a quality production rod today.

Here's a handy yardstick for figuring out historical prices, which uses a figure easily manipulated by even the most mathematically inept—10 bucks. Prices are compared with 1994 dollars.

1980	1970	1960	1950	1940	1930	1920	1910	1900	1890	1880
$5.65	2.66	2.03	1.65	.96	1.14	1.37	.64	.57	.54	.55

German silver seats and carried a reworked Varney (pricey Montague) ferrule.

The standard Trout models were made with both styles of reel-seat furnishings: the 9-footer took a 6-weight line and weighed 4½ ounces. Rod features found now often vary with the piece's age. Later Hawes models, built from the late teens on, had flame-treated Tonkin cane and no intermediate wraps. After Hiram Hawes's death in 1929, Merritt continued his father's product lines for another 10 years or so. These rods were also bench crafted, and Merritt built only 315 examples; 220 pieces were stamped "Parker-Hawes, Meridan-Canterbury, CT.," and were sold through that fine old scattergun company.

I remember a time when Hawes rods went unrecognized and sat unsold on successive tackle lists. Those were the days. Hiram Hawes's legacy goes back to 1879, when he first began making rods with his famous uncle in the antique period. Merritt carried on the tradition into the late classic era. Today, any Hawes rod, by father or son, is a tangible link to our fly-fishing heritage. As the great maker's heirs, the Hawes family inherited several of H. L. Leonard's tackle artifacts. Two historically important items, a superb pair of early Philbrook patent reels, were donated to the American Museum of Fly Fishing by Elsie Hawes.

And so it goes. The chain was never really broken, from the journeymen years when Fred Thomas,

well learned, to the final link—the teaching of values, hard work, and great skill to the next generation of classic rodmakers.

THE NET RESULT.

Notes

[1, 4, 5, 6, 7]Keane, Martin J. *Classic Rods and Rodmakers.* New York: Winchester Press, 1976.

[2]Keane, Martin J. *Classic Rods and Tackle* catalog no. 67. 1995.

[3]Spurr, Dick, and Gloria Jordan. *Wes Jordan: Profile of a Rodmaker.* Grand Junction, Colo.: Centennial, 1992.

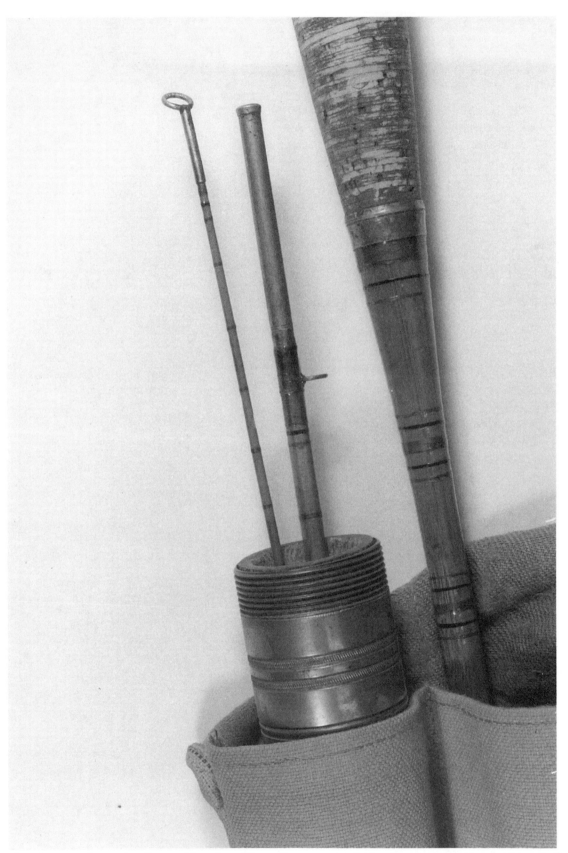

A rare product of one our finest rod craftsmen, this well-used George Varney example is stunning even today. The graceful swell of cane entering the 1/8-inch cork-ring grip isn't present on all Varney fly rods; many were built with a straighter shaft and Wells-style hand-piece.

9 Classic Eastern Tackle

The period between the 1870s and the 1930s was a time of steady expansion in the fly-tackle business, a time when the word "maker" was slowly replaced by the word "manufacturer." Indeed, large outfits such as the Terry Clock Company, the Montague City Rod Company, The Andrew B. Hendryx Company, and A. F. Meisselbach & Brother produced some of the most versatile fly tackle of the classic period.

In this same 60-year stretch, the U.S. Patent Office granted 260 or more patents to "improve fishing reels," and 370 patents for fishing rods. Of course, not all these "angling devices" were actually made. The patent office didn't differentiate between a screwball invention and a practical one, which resulted in an obtuse collection of tackle oddities. Pennsylvanian Elmer J. Sellers, for instance, built a bird-cage winch exactly 75 years after its time. Another chap made a reel of celluloid—hopefully accompanied by instructions to keep it away from open flames. Most of the newer stuff was practical, though, and—more to the point—often extremely affordable.

At the same time, fine craftsmen continued to cater to the white-collar professional class. Many rod-makers, often putting excellence before financial gain, made superb rods during these emerging dry-fly decades. One such maker, George Varney, produced a high-end line of rods while working for industry giant Montague. Long-standing companies such as Orvis and Leonard produced rods and reels that retained the high quality of the smith age.

HORACE GRAY & SON, PELHAM, MASSACHUSETTS

It was not by mere chance that the earliest production companies arose in the eastern United States. Take the case of Montague, an operation established by an old New England family in the smith age that continued right into the fiberglass years.

The small western Massachusetts town of Pelham had so few businesses that it never published a directory, so the Calvin Dwight Gray story is pieced together from tax and land transaction records associated with a waterpowered sawmill on Amherst Brook. By 1837 a mill belonging to Horace Gray had expanded to include a wood-turning operation. In 1858 Gray's 20-year-old son Calvin "became interested in making a better fishing pole," and set up his original Pelham rod shop in the turning mill to make fly and bait sticks from wood.[1]

From the start of commercial operations in 1860, Calvin's inexpensive turned rods were so well received by the average angler—a market that had thus far been largely untapped—that, in 1864, Horace moved upstream and bought the neighboring gristmill to enlarge his rodmaking operation. One of the original millstones eventually became a flagpole base, a purpose it still serves today. In the 1865 Massachusetts Census Calvin was listed as a "manufacturer," while his father was a "miller"—indicating that the family kept one nose to the grindstone. Tax records for the same year show that the new "Fish and gun rod factory" was valued at $800—some $20,000 in today's currency.

A circa-1895 view of America's first rod factory, in Pelham, Massachusetts. Manager and one-time owner E. P. Bartlett is standing at right, foreground. (Courtesy the Jones Library)

The Horace Gray & Son operation is the oldest known rod-turning factory in the United States, predating Thomas Chubb's business by half a decade. Calvin Gray proved to be a production-rod genius. From 1866 to 1871 new wings were added to the factory and the workforce grew; in 1872 two of the six Bartlett brothers joined the crew as wood stokers and runners. Eugene Bartlett displayed managerial skills and eventually became a foreman. His older brother Leander would become a rod designer and patent-monger.

Still in his prime, Calvin Gray died in 1873, and his widow, Louisa, sold his half of the business to Joseph G. Ward the following year. On January 1, 1874, his 21st birthday, Eugene Prentis Bartlett became superintendent of the company, and in July he married Jennie Ward, Joseph's daughter. At this time 40 people worked in the Horace Gray & Son rod factory year-round; the workforce grew to 60 employees seasonally to help fill the spring rush of orders. After a family feud, Leander Lester Bartlett left Pelham in 1882 and went to the town of Montague City, starting a larger factory to meet the ever-growing demand for affordable fishing tackle. To acquire the capital needed to build the new plant, Leander took on two partners, Charles Hazelton and Bernard W. "Barney" Farren. They formed the Montague City Rod Company prior to March 1886.

ESTABLISHED 1860.

CATALOGUE AND PRICE-LIST

—of—

FISHING ❖ RODS ❖ AND ❖ MOUNTINGS,

OF ALL KINDS,

WOODEN LANDING NET FRAMES OF EVERY DESCRIPTION,

MANUFACTURED BY

EUGENE P. BARTLETT,

AMHERST, MASS.

LANCEWOOD AND GREENHEART FURNISHED.

TURNING OF EVERY DESCRIPTION DONE TO ORDER.

The cover and a salmon rod from Bartlett's circa-1885 catalog.

THREE JOINT HEXAGONAL SPLIT BAMBOO SALMON ROD.

EUGENE P. BARTLETT, FISHING-ROD MANUFACTURER

Also in 1886, aging Horace Gray sold the last of his interests to E. P. Bartlett, who owned the major portion of the Pelham factory until he relinquished his controlling stock to Montague in 1889. Located on the Connecticut River, the new Montague City plant produced inexpensive split-bamboo fly rods. The best Montague models, some selling for $35, continued to be built in the old Pelham facility under the direction of Eugene. An 1885-era catalog and price list offered 250 styles of rod, priced by the dozen, and also included wooden landing-net frames.

In 1891, after the last of Thomas Chubb's many fires, Montague grew again. Smoke was still rising from the Post Mills, Vermont, structure when the Bartlett brothers arrived to purchase what was left. The "Captain" stayed on for about a year and helped set up new machinery.[2] Along with the original factories in Massachusetts, the rebuilt Post Mills factory produced a fantastic number of rods. The original Chubb label was retained until 1927 or later. Most of the Vermont rods were lancewood and ash jobbies in the fine old Chubb tradition. The rod-furnishings department also kept up with constant demand from amateur rodmakers.

Very little is known about early Montague six-strip rods, since they were marked by retail outlets only, either with shaft labels or reel-seat stampings. Bartlett rods built from 1875 to 1905 are rare. Some may have contained Leander's first ferrule patent, which employed an interlocking pin and slot for alignment. This idea, patented on March 9, 1886, is still used on some saltwater equipment, such as Aftco's Unibutt.

Montague rods built from 1906 to 1930 are easier to find. The higher-quality trout and salmon models

built at the Pelham plant usually carried Varney seats or Leander Bartlett's "new" German silver locking reel seat. Initially patented on September 25, 1906 (with later improvements on October 30, 1906), Leander's seat can be confused with Terry's version, patented in the same year. Quite often the Montague twist-locking seat is marked only with an arrow.

In 1934 the company admitted that "until less than ten years ago, Montague rods, famous all over the world, were not marked with Montague's name." Interpreting this reference to mean 1925, you'll notice a very interesting fact; 1925 was the year company manager E. P. Bartlett died. Exactly when marked models were introduced remains unknown, but the real catalyst of Montague's split-bamboo evolution arrived in the person of a former employee of Hiram Leonard.

GEORGE I. VARNEY, MAKER, MONTAGUE CITY, MASSACHUSETTS

Like Hawes, Thomas, Edwards, and Payne, George Varney was one of Leonard's top rod craftsmen, especially when it came to the fashioning of ferrules and soldered-pocket reel seats. Varney left Hiram in 1889–90, at the same time that the Kosmic-Walton crew departed. Varney ventured off on his own, however—with a new ferrule design.

George Varney patented the improved serrated ferrule on March 4, 1890, and began making them for the trade from his home in Central Valley. His "first model" ferrule was illustrated in Shields's *American Game Fishes* in 1892, accompanied by the note that one could buy the fixtures "under license." With an unidentified sibling, George Varney formed the partnership of Varney Brothers & Company. Its original rods appear in an E. K. Tryon advertisement appearing in an 1892 issue of *The Sporting Goods Gazette* featuring a complete angler's package: Redditch hooks, Martin fly lines, A. B. Hendryx reels, and

fig. 15.

Fig. 16.

*From J. H. Keene's chapter in
American Game Fishes, 1892*

Varney Brothers . . . ferrules? No, the Tryons meant Varney Brothers rods. These first Varney rods, built in Central Valley, are extremely rare.

At some point George and Brother moved to Poughkeepsie, New York. This was the address listed for his next patent (April 9, 1895), for a tapered, yet still serrated, ferrule. During the next year, Varney obtained partial rights to Burt Leeper's method of forming "seamless metal," patented on January 14. The brothers continued rod production in a small way, and some Poughkeepsie-marked models have been found. Any Varney Brothers rod is a rare find, however. George's major occupation seems to have been fashioning drawn ferrules for other rodmakers.

Shortly after 1900 Varney was asked to join Montague by Leander Bartlett, who was looking for a Leonard-caliber master maker. George accepted the position, moved to Massachusetts, and formed one of the most unusual business arrangements in the history of tacklemaking. Hired as Montague's chief rod-, reel-seat-, and ferrule-maker, George would also be allowed to produce his own handcrafted models, which didn't carry the Montague City Rod Company name but were signed "Geo. I.

Varney, Maker, Montague City, Mass." Even these newer Montague City–labeled Varney rods are scarce as hen's teeth today; very few have surfaced in any given decade. I've seen only three or four examples. Two were in Montague cloth-covered forms, each having a modified Wells cork grip that was rounded at the top, as later seen on the Red Wing. Another Varney rod (figure 9-7) is in the American Museum of Fly Fishing Collection in Manchester, Vermont. All these examples are too rare to cast.

George continued to oversee the manufacture of his seats and ferrules in Massachusetts under the Montague name. These were popular with a number of prestigious former Leonard craftsmen during their early business years, including Ed Payne, Eustis Edwards, and Fred Thomas. The Bangor maker used the Varney seat and ferrule, plus lesser Montague rolled-welt styles, until 1915 or later. Varney's reel seat was an important item, and many collectors are unaware that the graceful soldered-pocket fixture was his creation. It was best featured on his own products, but also appeared on early classic models built by several other makers.

As a total package, the Varney fly rod was a knockout. Aside from being an excellent metal engineer, Varney was a master in split bamboo. A casual glance at some early Varney rods, built in the itty-bitty-cork-ring era, shows the incredibly graceful and

accurate superswell just above the hand piece—a feature that even Hiram Leonard couldn't do better. In retrospect, George Varney was a most important

This Varney 9-foot fly rod, circa 1898–1900, sports the very familiar reel seat and loose-ring guides. The windings are similar to those found on a Fishkill salmon stick.

figure in the history of the classic rod, a man who has yet to receive proper reverence for his contribution to the trade.

Varney should also be credited with the design of at least two of Montague City's top-of-the-line fly rods. It's doubtful that he personally performed labor on these rods, but there's no doubt that it was Varney's eye for quality that allowed a few Montague models to ascend to a higher plane. These rods, overlooked by collectors for three decades, are just now receiving serious attention. Some collectors, in their long-standing worship of Leonard, dismiss the "super-Montagues" as orphans, yet the greatest Montagues were actually direct descendants of the Leonard ethic. As informed collectors recheck the roster of "great" rodmakers, we discover along with Hiram Leonard and Everett Garrison the name George I. Varney.

MONTAGUE'S MANITOU AND RED WING

Even today many collectors believe that Montague rods are crapola—until they see a nice Red Wing in a short length and excellent condition. These Varney-influenced rods have an elegance not found in lesser Montagues. And they're far rarer than a similar-sized Leonard. The two top Montagues, the Manitou and the Red Wing, are very collectible and will increase in value as collectors realize that there are no more left, at least not in excellent condition. Originally, "Red Wing" signified a light-line or trout rod; "Manitou" implied a powerful tournament-bass model. In the 1939 catalog, both rods were available in either trout or bass actions.

A few other models carried nickel-silver furnishings, and, although less collectible, they make decent and affordable casting instruments. These include the Trail, Flipline, Fishkill, and Splitswitch. The Fishkill represented the heavy end of the line, and the name can be found on quality saltwater models. In the 1940s the company produced four double-built, high-end fly rods; the Powr-Built series, which ranged from the No. 20 to the No. 50 (the numbers actually designated the selling price).

Next in quality and scarcity come the Splitswitch, Blue Streak, and Rapidan. While they're nice rods and also have German silver fittings, they seem to lack that Varney influence. These rods make up the bottom half of Montague's lineup of quality rods, and are generally good casting wands in lengths

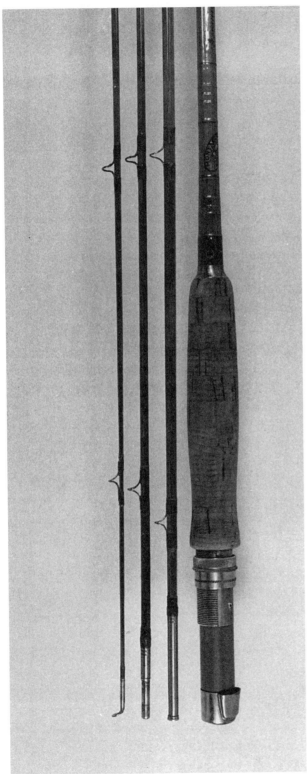

One of the top Montague models, this circa-1940 Red Wing has an acetate label and a red plastic reel-seat barrel. (The James R. Babb Collection)

of 8½ feet or less—somewhat heavy, perhaps, but with a soft action that's fine for early-season nymphing.

The grades below the Rapidan have little value, although some may fall into the "user" category. Such rods as the Flash, Rainy River, Sunbean, Clipper, *ad nauseum*, have plated-brass fittings. The cheaper the model, the thinner the ferrule walls. In fact, the lowest-grade Montagues, the infamous Clear Lake, Mt. Tom, and Highland, can bend their own ferrules after only a few casts. These rods are great for tricing up tomato plants.

Most marked Montagues were built between 1925 and 1952, all sporting a label on the lower shaft. Early round labels were post–E. P. Bartlett items. After Ocean City bought the company in 1934, an oval decal was used that included the model name; this was applied roughly from 1934 until 1941–42. After the war and up until 1949–50, a longer oval label in gold was used that also contained the model designation. The last decal just carried the words "Montague Rod & Reel Company."

The labels are the best way of dating classic Montagues. Although wrap colors were supposed to have been consistent, the company had upward of 500 employees, some of whom were color-blind or free thinkers. Wrap colors often verged upon the whimsical.

Trout-rod seat variations are legion, although most of the later ones were thread locking in some fashion. Early Manitous had a hard-rubber Landman spiral-locking seat. By 1939 or 1940, the better seats featured down-locking barrels—the Red Wing carried a seat with a red plastic spacer—but many cheaper grades had archaic furnishings that appeared to have emerged from a time warp. Old-style 1906 patent seats were used until the late 1930s, and were often seen on lesser models made of nickel-plated brass. Many midgrade classics had a Titelock seat, an extraordinarily long locking band that so lived up to its name that many reels had to be removed with a hammer.

Montague Manitou and Gaspé salmon rods often carried Chubb's reinforced female ferrule. One two-handed specimen, an unmarked 15-footer built for the Canadian market, had an extra midsection. Some later one-handed salmon rods carried black oxidized fittings. L. L. Bean's Light Salmon

Tournament Action Is What You Get in Any Fly Rod We Put "T. A." In

Famous Red Wing Game Cock Fly Rods (see cut) $27.50

Your Choice of 9 Grades, $12 to $41

YOU get it at its *best* in the Montague Manitou Tournament Fly Rod, with exclusive chemical tempering. Invaluable for its powerful casting with minimum effort, its better line control, its increased strength for fighting powerful, heavy fish. But be *careful* to get a *genuine* Montague split bamboo rod with *real* tournament action. Until less than ten years ago Montague rods, famous all over the world, were not marked with Montague's name. Many rods of inferior makes are still sold as genuine Montagues. Look for the name. It is your only assurance of the genuine.

Superior rods for every kind of fishing—in fresh water and salt water—$2 to $66. Ask your dealer.

Folders FREE. Mention Fresh Water or Salt Water

Montague Rod & Reel Co., Drawer A-K, Montague City, Mass.

World's Largest Manufacturer of Split Bamboo Rods

MONTAGUE SPLIT BAMBOO RODS

Field & Stream, *March 1934*

Rod, in a 9½-foot length, carried the Chubb ferrules in black. It was a very pretty Manitou-grade article, with all the eye appeal of contemporary Heddons. Good Montague cane, either flamed or rich brown, had tight glue lines, and the glue itself was strong stuff. Many abused Montague Citys that lack their original varnish have remained intact along their seams.

The best Montagues were built in lengths of down to 7 feet, and recently two short rods (no model name) of Manitou grade sold at auction for $400 each. Slightly longer rods in the same excellent condition bring about half that price. Ten years ago the Montague rod was dead meat, but savvy collectors are now discovering that *good* Montagues are tough to find. Excellent Montague buys include the early German silver–appointed, 1906-reel-seat rods, which almost mirror Varney's work; models in the higher grades built in the 1930s, such as the Manitou Tournament and Red Wing Game Cock; and, finally, the down-locking Red Wings and Powr-Builts of the 1940s. Remember, many of these models originally sold at sums substantial for their time: in 1934, up to $66, and in 1940, as much as $75—prices comparable to those of top-quality contemporary production rods such as those from Granger or Thomas.

MONTAGUE CITY'S BROOKLYN FACTORY

According to reel historian Jim Brown, Montague purchased the U.S. Net & Twine reel factory in 1899. The manufactory, located at 251 Classon Avenue, was retooled and started to produce a myriad of reels, from multipliers to cute little single-action fly models. The quality of the reels varied greatly, as did their rods, and some Montague models were designed to give Andrew "Five & Dime" Hendryx a run for his money.

The fly reels came in Trout and Salmon models, and the highest-quality items looked and felt a lot like the work of Julius vom Hofe. To this day many Montague City reels are confused with the products of their neighboring factory. The Trout was one of the company's classic quality reels, and is often attributed to Pflueger or Julius vom Hofe. It had hard-rubber side plates and a German silver rim, a sliding click button, and a balanced handle and foot—the expensive look.

The charming Trout reel was probably built in several sizes, always described by yardage capacity, and the rear bearing was adjustable. A telltale feature found on this model as well as others was the stamping "Steel Pivot Bearing," found on the back side plate. Many Trout models were marked only by the seller; these included Edward vom Hofe & Company, M. A. Shipley, and Stoddard's.

Montague reels were mass-produced for other jobbers, wholesalers, and retailers. A big retail-wholesale outfit, E. K. Tryon in Philadelphia, sold Montague reels under the Pennell label. The "Pennell" mark, named after English hookmaker-author H. C., was always associated with a top-of-the-line product. Virtually all Montague reels built prior to 1924 were stamped with the retailer's or wholesaler's address only. Sears carried a line of Pennell reels circa 1902; these appear to have been Montagues as well. Smaller retailers, such as Iver Johnson's in Boston, carried Montague reels right next to the

Classic Reels, Modern Lines

In the old days, reel line capacities were never really standardized. The antique models came in sometimes arbitrary numerical sizes: #1, #2, #3, and so on. Fritz Vom Hofe was one of the first makers to switch to the yardage system, marking reel capacities of 20, 40, 60, 80, and on up to 200 yards or more. In its day this was a helpful system, but today's plastic fly lines are much larger in diameter for a given weight than were their linseed-oil-coated ancestors. Here are some rough guidelines to help you choose the right reel.

A tiny 20-yard reel holds the forward taper and most of the belly of a 3-weight line; a 4-weight loads a 40-yard model; a 5-weight fits a 60-yard winch; a 6-weight fills an 80-yard model; and salmon reels marked 100 to 150 yards are matched with 7-through 9-weight lines.

Most classic production-grade reels cost less than modern new ones and they match vintage cane rods. A Montague Trout model hanging off a mint Red Wing "chust hlooks mahvlus."

Salmon Reel.
(About three-fourths size.)

An impressive and rare Montague salmon reel from the 1911 William Mills & Son tackle catalog.

look-alike Julius vom Hofe models. Like Montague rods, the fly reels are now collected according to their degree of quality and scarcity—two Montague attributes that seem to go together.

With a reel plant and three rod factories, Montague could wholesale like crazy; it even sold unfinished rod blanks, just as companies do today. In 1941 it claimed to be the "World's Largest Manufacturers of Split Bamboo Rods." A backcast to the early E. P. Bartlett days gives us an even more astonishing statement. According to Mark Aldrich, E. P.'s grandson, "By 1917, the company was making three-quarters of the fishing rods made in the world." As with the Montague rods, the reels are not scarce, although the better models are tough to find.

The Montague story must be told in two chapters. The original operation came to a screeching halt with the 1925 death of E. P. Bartlett. The company was then administered by the First National Bank of Amherst, which sold the stock to a brokerage firm in Springfield. The firm bled the company dry, and when the crash of 1929 hit Montague was in serious financial trouble. The seven-building Pelham facility was officially closed on June 21, 1931, followed shortly by the Vermont plant. By

Although it's a 40-yard-sized Montague raised-pillar reel, it has the wrong Montague box. The model here has metal side plates—not the rubber ones described on the Pennell label.

1932–33 the rodmaking facilities were consolidated at Montague City. In 1933 the reelmaking operation was moved to Philadelphia.[3]

The company name was changed to the Montague City Rod & Reel Company and stock was sold to the public. By 1934 the stock was controlled by the principals of the Ocean City Reel Company, namely Paul J. Johnson and his father-in-law, Louis M. "Lou" Moskowitz.[4] Ocean City used some Montague parts on its earliest reels, including bearing cups, retaining screws, and handles. One saltwater OC model, the Top-Sail, was built with a genuine MC handle, including the old-style wooden knob and all original MC screws. Ocean City eventually made "modern" fly reels, but as yet these haven't interested collectors.

Under Johnson's tenure, classic Montague rods were built in profusion. After the end of World War II, however, split-bamboo rod production decreased markedly—in part due to the embargo of Red China, which killed the supply of Tonkin cane, and in part to the rapidly growing popularity of fishing rods made from a wartime invention called fiberglass. The company appears to have ceased the manufacture of split-bamboo rods just after 1950, although it continued to produce fiberglass rods. When the company folded in 1955, owner Paul Johnson sold the "Montague" name to Brogan Industries, which built glass spinning rods.

SEWELL N. DUNTON & SON RODS

Sewell Dunton began working at the Montague City plant the year the Pelham facility closed; he was treasurer and manager. In 1954 he bought the remaining stock of Tonkin cane and the split-bamboo machinery and set up shop at 4 Fiske Avenue in Greenfield, Massachusetts, site of America's first cutlery factory. In the tradition of Horace Gray, the new rod company was called Sewell N. Dunton & Son. The Greenfield rods were a much lighter straw color than the chemically darkened Montague shafts. Labeled "Anglers' Choice" and also carrying the inscription "Genuine Tonkin," the most popular models were built in 7-, 7½-, and 8-foot lengths. The trimmings, reel seats, and ferrules were fashioned from good yet plain materials. Early Dunton rods, however, carried crappy bottleneck ferrules, a fixture

used at Montague for well over a decade. Other than these cheap, drawn ferrules, the rods were well made.

The Duntons also built a few short two-piece rods with finer appointments. One gem, the 6-foot

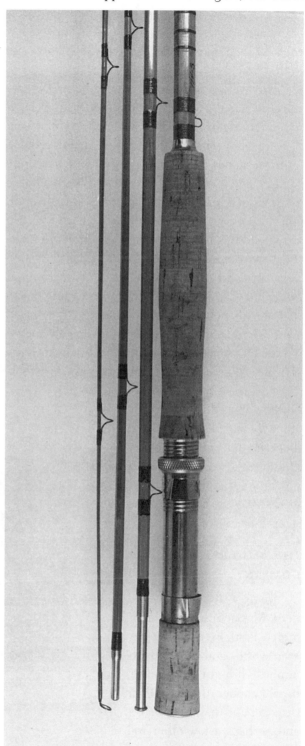

A Sewell N. Dunton & Son 8½-foot Anglers' Choice model, owned by noted author Frank Woolner of Shrewsbury, Massachusetts. The rod has a short permanent extension below the reel seat, popular for a few years in the early 1960s.

No. 161, had black furnishings and a superfine-style grip. At one time I used a nice 7-footer that worked a 5-weight line to perfection. This rod, like many late Anglers' Choice models, had a set of higher-quality straight ferrules, evidently introduced in the 1970s. Even rarer Anglers' Choice rods sported Super-Z ferrules, as Dunton tried to salvage his reputation late in the game.

The average Anglers' Choice came in a brown fiber tube and gray compartmented bag. Several models from the early 1960s, such as the 8-foot #114-S, were built with a short, 1½-inch nondetachable butt. Most rods carry forest green wraps, a carbide stripper, and a Wells grip. The Sewell N. Dunton & Son products are good, usable wands, usually available in popular 5- to 6-weight sizes. If an old-style Dunton ferrule has a tick or comes apart a tad too easily, it can be upgraded with a set of Bailey Wood's Swiss-style furnishings with little loss to original integrity and with a great increase in casting enjoyment.

In 1974 Tom Dorsey and Tom Maxwell bought out the Duntons and moved the machinery to Turners Falls, Massachusetts, to start the famed rodmaking partnership of Thomas & Thomas. Sewell Dunton retired to the sunny climes of San Diego. So ended the Gray-Bartlett-Montague-Dunton saga, leaving behind a vast array of related rods and reels that were built over the span of a century.

THE AMHERST ROD COMPANY

In 1947, three local Amherst, Massachusetts, businessmen—Paul French, Wesley Wentworth, and Arthur Davenport—bought a building, hired Leander Aldrich as their manager and master rodmaker and perhaps a few other rod craftsmen from the Montague City workforce, and founded the Amherst Rod Company.[5] Aldrich was one of E. P. Bart-

lett's grandsons and had grown up in the "Montague family." The new company's initial offerings, for the upcoming 1948 season, were three models of fly rods and two of spinning rods, all named after local towns. The little outfit built a modest number of midquality sticks until 1953. The model seen most often now is the Hadley 8-footer. Built in a three-piece design with two tips, fewer than 100 Hadleys were so marked.[6] However, the L. L. Bean Company of Freeport, Maine, sold a number of Hadleys under its house name, L. L. Bean Lightweight. This was a superb rod for the practical angler, with class and the ability to work a range of midsized lines.

Another model, the Berkshire, was appointed with German silver hardware, a Wells grip, and salt-'n'-pepper wraps tipped black. The Berkshire was built in the same length and style as the Hadley, but is seen now far less often. Like most Amherst rods, it carried a wooden-barreled cocabola-wood barreled reel seat reminiscent of the Edwards style. The third model, marked "A.F.R.Co." and "The Amherst," was a 4½-ounce 8-footer that featured a series of two delicate trim wraps located just above or below each winding at the ferrules—a very nice touch for a production rod. Additional models included the Pelham

Details of an 8-foot, 4½-ounce rod marked "The Amherst," showing a wooden-barrel down-locking seat and trim wraps whose design dates back to early-quality Montague models. (The American Museum of Fly Fishing Collection)

and the Scarborough.7 The rods—better than a Montague or a Dunton but certainly not the equal of a Leonard or a Thomas—were priced around $70–80.

As a company, Amherst suffered from overspecialization. The guy who made the blanks was a whiz, but the finisher couldn't keep up with production. In the face of sagging sales the company ultimately foundered, leaving behind a backlog of hundreds of glued-up blanks. The Amherst Rod Company assets were split three ways: One owner kept the cash, another received the machinery, and the third partner literally got the shaft(s)! Eventually, the equipment was bought by the Holyoke Machine Company.

Mark Aroner was the first to discover the Amherst bamboo repository, which was then brokered to Joe's Tackle Shop. In turn, the late Joe Sterling hired Mark to turn the blanks into finished fly rods, which were marked "Aroner Amherst" or just "Amherst." Mark noted that the 1983–84 Sterling association is "a constant source of embarrassment." You can now "cull" Amherst rods, original or Aroner models, by the color of the shafts. The best casting rods have light flamed cane; "weepy" models were glued from a lighter, almost Calcutta-blotched bamboo. In view of the extremely low output of this short-lived company—even with the addition of Aroner's production of "considerably less than 200 rods"—good Amherst fly models should increase in value in the near future.

THE CROSS ROD & TACKLE COMPANY

The Cross rod was born in Lynn, Massachusetts, just like its maker. Around 1919 young Wesley D. Jordan built his first fly rods from Calcutta culms supplied by his boss, William Forsyth, who owned a woodworking mill. The rods, made by trial and error by the old bench-planing process, were only a partial success. Wes Jordan soon discovered that the key to a good casting rod was matching its tapers to a particular size of fly line. His first exceptional model was built from the new Tonkin cane, a three-piece 8½-footer for an HDH line, now called a 6-weight double taper.8

Upon casting the new Jordan rod, W. R. Forsyth was so enthusiastic that he wanted to enter the big-time production game. To keep up with projected demand, Wes and his brother Bill built a 14-foot-long milling beveler. Forsyth wanted a name with local fame to be associated with the new rods, so he hired William Cross, a renowned Lynn-area rodmaker who specialized in wooden saltwater sticks. The business was now called the Cross Rod & Tackle Company but, as it turned out, William Cross knew squat about split bamboo, and in 1920 "wound up" as a rod wrapper and finisher. Less than two years later, Cross left the company and started kicking around in the shoemaking industry.

The Cross rod became a coveted instrument, thanks not to its namesake but to the superb skill and inventiveness of Wes Jordan. By 1924–25, the company offered 25 various fly models and produced 3,000 rods per year. The drawn ferrules on these rods, designed by Wes and Bill Jordan, were similar to the E. W. Edwards style, with the male cap cut and replaced by a soldered disk.9 The female was waterproofed by a swaged internal disk that wasn't soldered.

Rods were built in three grades—the double-built Cross and the single-built Forsyth and Essex. The early Cross and Forsyth rods can be identified by a German silver, serial-numbered medallion soldered to the full-metal reel seat on the side opposite the pocket. A second style of marking, found on the Essex, featured the model and company name stamped directly on the German silver seat. Saltwater models were offered in the Cross line, often triple built in the heavier sizes. Although any Cross rod in good condition is a usable classic, it's best kept as a collectible, as the rods were built in quantity for only six years. The company folded around 1925–26 when Bill Forsyth died. His widow sold the business, the Cross name, and the machinery to South Bend. Wes Jordan, the great Lynn rodmaker, went with the deal. What we collectors want to know is what happened to the 6,000 original Cross rods?

TERRY AND HENDRYX CONNECTICUT REELS

Two eastern companies, both located in Connecticut, manufactured more than just tackle. As a matter of fact, their tackle business, confined to reels and metalwork only, was probably just a sideline—albeit a sideline that allowed both Terry and Hendryx

to weather various financial cycles. In the end, though, each company was swallowed by larger ones.

Silas Burnham Terry, originator of timepieces and chronometers famous to this day, entered the tackle trade in November 14, 1871, with a three-part reel patent: There was an interior flange on the face plate to confine gearing, an adjustable pin on the rear plate to create a drag against the spool flange, and a reel foot made from one unitized piece of metal. The Terry reel was designed to be manufactured cheaply. The fly-fishing and single-action models were reminiscent of products built in Old Knickerbocker or even by Haywood or Ustonson in the old country. Some models even carried an old-fashioned rod clamp that was tightened Waltonian-style to the rod's reel seat.

The fly models had an internal click, a simplification of the drag idea mentioned in Terry's 1871 patent. This nonadjustable click pawl worked full time, like so many of the earlier New York fly reels. Terry models were made of brass and fitted with a wooden-knobbed single-crank handle. Nickel plating was an option. I have yet to see a rubber version of the Terry fly reel, but a March 1, 1884, advertisement in *The American Field* listed such an item. A readily identifiable feature of a Terry reel is the stamped reel foot, which was attached to each side plate by heavier shims that supported the screw threads. This unique foot, simpler in construction than those later designed by John Kopf or Andrew Hendryx, was usually fitted with a single screw on one side plate and two screws on the opposite plate. An odd little quirk.

Starting in 1871 the reels were built at the Terry Clock Company (established 1869) in Waterbury, Connecticut. Fly models probably accounted for more than 60 percent of the reel production. Reel

This angled view of the Silas Terry fly reel shows one of the foot shims, its standard wooden knob, and its single-crank style. This is a nickel-plated version.

manufacturing continued after Silas Terry's death in 1876; and either just prior to, or in, 1884, the plant was moved to Pittsfield, Massachusetts. The 1884 advertisement mentioned above also listed a Chicago business address. In 1888 the Terry Clock Company was restructured as the Russell & Jones Clock Company, which may have continued to make the reels until it went out of business in 1893.[10]

The Terry fly reels, smacking of early-19th-century design, were made for roughly two decades and probably in great profusion. Most examples were unmarked. Others had the Terry patent stamped on the underside of the foot, and a few had "Chapman & Son, Theresa, N.Y." fixed to the side plate. This well-known metal-bait company was one of the few identified retailers. The "toughest" Terry reels to find were those with the maker's name, and they're rarely seen. Overall, the Terry reel is not scarce, but plated examples in very good or better condition are difficult to find.

The other Connecticut company to produce fly reels as an offshoot of its major business was the Andrew B. Hendryx Company of New Haven. Simply stated, Hendryx products were for the birds. Chances are that most late-1800s parrotkeepers changed their newspaper in the bottom of a Hendryx bird cage. The famous Andrew Hendryx patent, granted on March 21, 1876, was for a simplified construction of aviary restraining devices.[11] This patent, often found on the company's reel feet, was accompanied by a second one, dated July 10, 1888, that protected the idea of a pressed reel spool.[12] The great majority of Hendryx fly reels were very inexpensive, stamped products—in fact, the cheapest reels in the world. The famous old mail-order house of "Monkey-Ward" sold a simple Hendryx reel in 1893–94 for 10 cents—the price of a good cigar.

Hendryx began making fishing tackle in 1878, and by the late 1880s the Andrew B. Hendryx manufactory was producing riveted, stamped reels at a volume that exceeded any other tackle company's; during one 12-year period, Hendryx spat out 2 million reels. Some models, those of interest to fly-tackle collectors, were rather upscale and included a rare German silver vom Hofe look-alike.

The quality fly-reel lineup also encompassed a series of brass and rubber revolving-plate models that had a stiff-upper-lip British look, replete with fancy

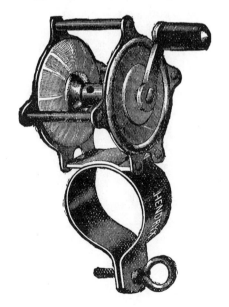

A small, inexpensive stamped fly reel with an old-style clamp that can be traced back to Venables's time. From the 1893 Andrew B. Hendryx Company catalog.

Perhaps the most "patented" reel in existence, this medium-grade Hendryx has seven patents stamped on its rear plate.

FOWLE, COZZONE, AND OTHER NEWARK MAKERS

Newark, New Jersey, spawned a number of reelmakers who were active from the end of the Civil War until World War II. We usually think of Meisselbach and Pliny Catucci, since they were the noted makers, but the Newark reel was built by a number of individual craftsmen, the earliest being A. S. Fowle and the last being one "D. Slater."

Andrew S. Fowle built a number of reels from the 1860s until around the mid-1890s. Surviving examples of his work are usually large, salmonesque models. Fowle was first listed in the *Newark City Directory* in 1853–54 as a "jeweler." By the late 1860s he was recorded under the heading of "machinist" and living at 68 Orchard Street, where he remained until 1889, the year that he may have "fished 'round the bend."[13]

Typical Fowle reels seem to have been constructed of German silver; their sizes were denoted by single-digit numbers, and they usually featured rear plates, feet, and markings similar to those typical of Kentucky makers. An 1880s-era raised-pillar reel, sold at a 1989 auction for $600, had the marking "A. S. Fowle" stamped in an arch shape; under this was an additional stamping, "No. 7." This salmon reel was 4 inches in diameter and had a "bullet" knob on a single crank and rollers located at the two double pillars. It was a knockout.

Another salmon model by this maker featured an additional marking, "Newark, N.J.," in a reverse arch. Combined with the "A. S. Fowle" stamping, it made a pattern shaped like a football. The reel's size, "No. 6," was inscribed inside the maker's name and address. The model had no external pillars. This German silver reel was designed as a revolving-plate model, again with plates measuring 4 inches. Additional Fowle reels and data should bring this talented Newark artisan to the fore as an important maker of quality products.

turned-brass pillars. The foot, however, remained the same riveted classic, stamped (usually illegibly) with the two patent dates and the word "HENDRYX" in capital letters; often it read as "ENDRYX" or "HENDRY." These Limey look-alikes were mostly salmon and grilse reels; the one example I examined was marked "3½," its actual size in inches. The reel was rather wide and could be taken down by holding the spool in place and turning the handle counterclockwise, a feature found on many similar revolving-plate models built on both sides of the Atlantic.

Good-to-great Hendryx fly reels are a scarce commodity today, while the lesser stamped models are worth little more than their original price. The reels were made until 1919, when Hendryx's tackle division was sold to the Winchester Repeating Arms Company, also of New Haven. Winchester continued to manufacture the cheap stamped versions, but the quality fly models were never made again. I'll trade you a thousand bad Hendryx reels for one good one! Overall, Hendryx made a lot more reels than Terry, although the two Connecticut companies can share the blame for starting the trouting career of many a collector's great-granddad. Terry and Hendryx were important fly-reel manufacturers, introducing the sport to the young and others of limited means.

John A. Cozzone probably started to make reels a little after 1915 in a garage on Kent Street in Newark. The business grew into the Cozzone Machine Screw Corporation (in business from around 1925 to 1940), which produced a quantity of hand-fitted reels as a sideline. Later, reel production expanded along with the popularity of the company's surf-casting models. Cozzone also produced a number of quality fly reels, most of which were marked by the maker, "COZZONE," in block letters inside an oval. No address was given. The Cozzone fly reels were designed to look like a Pflueger Hawkeye or a Julius vom Hofe product.

Typical Cozzones were constructed with hard-rubber side plates; the balance of the reel was of German silver. At least three sizes have been found, from "60 yd." to "100 yd." The back plate contained a sliding click and a removable knurled bearing cap, which could be adjusted over a Julius vom Hofe-style spider washer. Cozzone fly reels are uncommon—a shame, since they're beautiful little classics.

"D. Slater, Maker, Newark," is a stamping found on a fly reel dating from around 1900–20. Although obviously built by a small-shop Newark producer, little else is known. The 3-inch-diameter Slater fly reel was rugged and of high quality, fashioned from brass and aluminum, with double knobs attached to the revolving face plate. Perhaps we'll learn more about its origins in time.

A charming and well-designed John A. Cozzone fly reel constructed of German silver and hard rubber. It's marked "COZZONE," with no address. (Bob Corsetti photo)

NEWARK'S EARLY MEISSELBACH REELS

Montague and Hendryx made some rather pedestrian reels, but Newark's most notable tackle manufacturer-designer mass-produced reels that were just as cheap—and they had style. August F. Meisselbach (1865–1927) applied for a patent for his original Amateur reel in 1885, but began manufacturing it before the patent was actually granted, on February 23, 1886. The sons of German immigrants, Meisselbach, together with his older brother, William, began making the reels in a small machine shop at 13 Mulberry Street. (Through writings in the *Antique Angler* and his *A Treasury of Reels*, Jim Brown has documented the "Meisselbach trail," and I refer to him for production years and address changes.)

The first Amateurs were designed to be mounted on the fly rod horizontally, like those of Morgan and Billinghurst, and often carried a spring drag lever that kept tension on the spool until depressed by the angler's thumb. On November 23, 1886, Gus received a patent for an upright version of the Amateur. The reels were never marked as to the model, but are now found with one or both of these patent dates plus "A.F. Meisselbach & Bro., Newark, N.J."

One of the earliest Amateurs I've found is unmarked by its makers but has the first and rarest Meisselbach stamping: "PAT AP'LD FOR." It's a simple, 3-inch (bass sized) external-click reel, with no drag lever; it's counterbalanced and sports the earliest of rounded wooden handles. The model was also produced in a smaller 2¼-inch trout size, and in larger side-mounted trolling sizes.

With the introduction of the Meisselbach brothers' Expert in 1888 (jointly patented on February 5, 1889), the small company outgrew its old quarters and moved to 197 Halsey Street, not far from where Charles F. Murphy once lived. The Expert reels, which retained the spool within a single-piece circular frame, were made in several variations and were very popular. You'll probably see more of them than any other model of Meisselbach fly reel. The bass model came in narrow and

An early Meisselbach upright Amateur, marked only "PAT AP'LD FOR." It features a click mechanism on the front spool plate and post. This click was moved to the rear of the reel on the later Expert.

wide versions. The wider version was large enough for grilse, and drag could be added by pressing the frame against the spool. Until 1895 the reels were marked by the maker and carried the name "'EXPERT'"; later models are marked "expert" without quotation marks.

Popular as the Expert was, the next design was a near fizzle. The Allright was a raised-pillar reel with a solid back plate that covered the rear spool and click mechanism, but it was poorly balanced and didn't sell well. Covered by a January 14, 1896, patent obtained by the brothers, the Allright was a heavy affair. It remains one of the "toughest" Meisselbachs for collectors to acquire. It came in three sizes (#110, 120, and 130); the rarest is the little #110, which has yet to surface on the antiques market.

By contrast, at the same time that the first clumsy Allrights were introduced (1895) the Meisselbachs also introduced the charmingly original Featherlight. I fished this model in the narrow #270 version, fitted to a little 4-weight Thomas, for almost 20 years, and it's dear to my heart. The little Featherlight was one of the great mass-produced reels, sturdy enough to withstand the rigors of real angling. It was also one of the last models built by Gus and William. The senior brother retired shortly after the turn of the

century and was replaced by his son, William Meisselbach Jr.

At this time the Meisselbachs were located at 26 Prospect and carried a payroll of 10 employees. In 1911 the company had expanded to an extensive facility at 22–28 Prospect and carried a workforce of 75. Also at this time, the first model Featherlight was joined by an even lighter one-piece, stamped version. This lesser reel usually carried the 1896 patent, along with a newer December 27, 1904, patent obtained by Gus and William Jr. This lighter-than-a-Featherlight model was prone to streamside warpage, and good undented models are difficult to find.

LATER MEISSELBACH AND CATUCCI FLY REELS

A few years prior to the War to End All Wars, A. F. Meisselbach & Brother introduced another cheap version of the original Featherlight, the Zephyr, which had a one-piece rear plate and was made from lighter stampings. It was not marked by its maker, and was wholesaled for peanuts. The short-lived Good Luck fly reel, which had no front frame, was perhaps the last of the Featherlight-style models. Introduced at about the same time as the Zephyr, it was a fragile "palming" reel.

In 1917–18 the company was restructured as the A. F. Meisselbach Manufacturing Company and introduced its first departure from the classic ventilated-spool style: the Rainbow, a rip-off of Hardy's Uniqua, right down to the half-circle spool-retaining spring. The reels were offered in two sizes and finishes, the better being the satin finish. Later aluminum Rainbows had a large spool-retaining screw instead of the half-circle spring.

In 1920 Meisselbach discontinued its line of historic ventilated-spool models and within a year removed to Elyria, Ohio. Bought by General Phonograph and eventually swallowed by General Industries, the new Meisselbach Division produced the Rainbow until 1941. During this period, the Rainbow was sold by William Mills & Son as the Cresco.

The earlier and more important years of the original Meisselbach company saw the production of

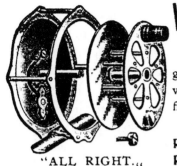

Maine Sportsman, *March 1901*

excellent and innovative examples of the reelmaker's art. The Amateur, Expert, and original Feather-weight models are not the only Meisselbach fly reels coveted by collectors, however. Our last great example, the wonderfully smooth Symploreel, resulted from the combination of Gus's expertise in manufacturing phonograph components and the invention of a new malleable substance.

A new enterprise known as the Meisselbach-Catucci Manufacturing Company was formed in 1910 as a direct result of the influence and association of two other men. The first was Leo H. Baekeland, a brilliant chemist who immigrated to the United States from Belgium in 1889. Baekeland developed the first successful photographic paper for Eastman and also, in 1909, a pourable synthetic rubber called Bakelite. Pliny Catucci, a mechanical wizard born in Rome, Italy, in 1869, helped create the jukebox and patented a number of important features incorporated into A. F. Meisselbach & Brother saltwater reels.

Catucci went into partnership with Gus in the same year that Baekeland started the Bakelite Corporation, and the two were among Leo's first customers. For all practical purposes, Pliny Catucci was the driving force behind the newer Newark company, which also manufactured machine gears and electronic components. By 1925 Catucci had devel-

oped a complete line of multiplying reels as well as five different fly models—all employing Bakelite as a major component. The screws, pillars, knob shafts, and feet were of German silver.

The Symploreel fly models, protected by Pliny's two patents granted on December 28, 1926, were made in three diameters, from 3 inches to 3¾ inches, and two widths, ¾ and ⅞ inch. Larger models had a genuine agate line guard encased in a German silver ring. The Symploreels were knockouts. Their sole fault appeared only if you accidentally dropped one on a rock! Catucci's fly reels were produced in Newark until 1931, when Pliny sold the company and design rights to the Bronson Reel Company in Michigan. In turn, the Bronson-Catucci products were built for another nine years. Perhaps among the smoothest turning of all classic production fly reels, the Newark Catucci models are now quite collectible.

HORROCKS-IBBOTSON'S "HEXI-SUPER-CANE" RODS

In 1940, Horrocks-Ibbotson claimed to have been making tackle for "over 125 years." Just a little white lie. It forgot to mention that its original products were kid gloves, made by George A. Clark. In 1863 Clark hired Englishman Hugh James Horrocks, who suggested expanding the product line by selling British fishing tackle. Edward D. Ibbotson showed up in 1894 as an errand boy. Clark, Horrocks & Company entered tackle manufacture in 1900, purchasing the stock of Trenton, New Jersey, rodmaker George Kemp. The next year it bought out the old J. B. McHarg facility. In 1905 it built a new factory at Whitesboro Street in Utica, New York, and four years later the firm became the Horrocks-Ibbotson Company.[14]

According to its 1914 catalog, H-I built rods so cheap that even Montague couldn't undercut its prices. You could buy an H-I velvet-covered case for a dollar, but the rod was only 90 cents! The cheesiest rods came with an "ebonized wood grasp," meaning birch painted black. More "expensive" models, $5 and $6 items, had the infamous "sheet cork grasp,"

The "Gogebic" Trout & Bass Reel.

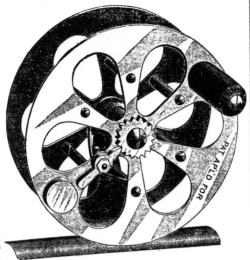

In presenting this line of Reels to anglers, we call attention to the moderate price and practical value over the majority of reels, as are pointed out in the following description:

They can be changed at will from trout click to free running bass reels. The large diameter of spool makes the multiplication equal to any quadruple reel made, and the open side discs permit of a free circulation of air, which dries line quickly on spool, and is of practical value in keeping any line in prime condition, thereby preserving it from rottenness, which often occurs on the old style closed disc reels where line is left on wet.

Before we leave Meisselbach, note the A. G. Spalding & Brothers "Gogebic" reel, introduced in 1889. This was the first Amateur offered for sale.

famous for peeling away from the wooden undergrip. Horrocks-Ibbotson made rods for people who were in a depression well before the Depression. Somewhere in my collection I have one of those 90-cent lancewood numbers, complete with light-gauge corroded brass ferrules. Lancewood was big on H-I's early agenda, and the company claimed to use only the real thing, genuine Cuban lancewood. It also mentioned that the Cuban Revolution made it difficult to import the stuff, and that the last person who'd tried to sneak some out of that country was depicted in Winslow Homer's watercolor, *The Gulf Stream*.

As for H-I's "better" rods, an angler had quite a choice. The best models, #2354 and #2353, were nicely rounded in African steel cane (poorly selected Calcutta), finished with a multitude of mostly green silk wraps, appointed with nickel mountings, and available in lengths of 8¼, 8½, 9, 9½, and 10 feet. This era's real collectible rod was probably the #2350 Special Extra Light Fly Rod, which came wrapped in three colors of silk; the 8¼-foot version weighed 3¾ ounces.

The Horrocks-Ibbotson Company produced a plethora of products from 1900 through the 1920s. Earlier rods were marked by a cellulose label on the shaft just above the grip—first with "Clark, Horrocks & Co." (1900–06), then the "UTK diamond" (1906–

19), followed by a bass and diamond and the trout logo (1920–29). Later labels dispensed with the fish and just had the diamond. Some antique-era models were also die-stamped on the reel seat.

The next phase in the Horrocks-Ibbotson evolution came in the early to mid-1930s, when the company became the first U.S. fly-rod manufacturer to impregnate its cane. An ad in the March 1934 issue of *Field & Stream* touted this accomplishment as a way to avoid "chronic curvature" (a set in the blank). The identity of the creosote resin used for impregnation was kept a secret: "The substance which fills the pores of the cane penetrates clear through." Sounds good so far, but reading on, we find that a "fine chrome vanadium steel core runs through the center from tip to butt." Yep. H-I had designed a Hardy-style shaft with a chunk of steel running through it. Even without the steel, its heavy creosoaked blanks were a sure ticket to a good case of tennis elbow.

Fly fishers looking for a durable rod welcomed the new Hexi-Super-Cane rods, however, and the various models were reasonably popular right through World War II. In 1939 and 1940 the Saranac, available in lengths from 8 to 9½ feet and weights from 4¾ to 5½ ounces, sold for $13. A slightly better model, the Golden Rod, came in 9- and 9½-foot versions; the smallest weighed 6 ounces and sold for $14. The Hi-Test fly rod was double built and even heavier than the Golden Rod. It came in 8½ and 9-foot lengths, the shorter version weighing 5¾ ounces, and was offered at just under $20. The top-of-the-line Cunningham sold for $30 and had quality German silver drawn ferrules. All of these models, like those that followed, were designed for popular 5- to 7-weight fly lines.

Late-1930s and early-1940s rods had a good Bakelite-barreled aluminum down-locking reel seat and a comfortable, full-cork grip. After World War II "modern" variegated plastic reel-seat spacers began to appear. Fly rods built from the late 1940s into the 1950s appear to have been the best of the Horrocks-Ibbotson models. The Vernley sold for $30, and the Cunningham increased to $36. Top-notch models

The Beaverkill, a Very Fine Lancewood Fly Rod, $4.25.

No. 6R8626 Our Special Beaverkill Lancewood Fly Rod, made up with solid reel seat below the hand, German silver mountings all engraved or milled to make a very handsome appearance. It has solid anti-friction guides, windings of black and red, beautifully clustered. Put up in three joints with an extra tip on a wood form covered with velvet. We cannot recommend these rods too highly. **They are perfection.** Ordinarily retail at $10.00. Weight, 8 ounces. Length, 8½ feet.
Our price...$4.25

If by mail, postage extra, 20 cents.

A "very fine" cheeze-de-twaux Horrocks-Ibbotson lancewood fly rod, as sold in 1902 by Sears. At 8½ feet, it only weighs 8 ounces. A half pound of daintiness.

included the Chancellor and Old HI's Hi-Test, at $40 and $50, respectively. Built in popular 8-, 8½- and 9-foot versions, these last H-I fly rods were also lighter, had no steel core, and are now usable classics. While most models were built in the standard three-piece style, the last of the Tonka Queens and the Tonka Princes came in two-section versions.

World War II-era rods had the first of the later H-I logos on their shafts. The logo changed two or three times, but the real identifier of latter-period rods is the presence of white-ink script, usually noting the model name. Today, Horrocks-Ibbotson split-bamboo rods from the white-ink era are still modestly priced and readily available tools for the budget-conscious angler.

THE CLASSIC DIVINE ROD

The old Fred D. Divine Rod Company survived into the classic era, and some very fishable rods were built well into the 20th century. Like Montague and Horrocks-Ibbotson rods, however, the newer Divine models varied in quality.

After Fred's grisly death in 1900 his widow, Ada, became the company's president and manager. Divine had been the firm's one and only traveling salesman, establishing

bonds with big tackle houses like Spalding Brothers and Abbey & Imbrie. Without Fred's amiable salesmanship, Ada decided to woo more dealers by offering inexpensive trade models, too cheap to pass up as profit-making items. The most frequently encountered lesser brand is the African Steel Vine, with rounded bamboo shafts similar to those on Horrocks-Ibbotson models; these were sold as the Schoverling,

FLY FISHING!
HORROCKS
IBBOTSON
Golden
Rod

Medium weight, low-priced, quality all-purpose fly rod. Hexi-Super-Cane bamboo. Durable—beautiful. Bakelite screw locking reel seat. 3 pieces—two tips. 9 ft., 6 oz.—9½ ft., 6½ oz.

PRICE **$14.00**

Outdoor Life, *April 1939*

Daly & Gales Saranac, and the Congress by the New York Sporting Goods Company.

Ada's brother and business partner, George F. McDuffee, gnashed his teeth while producing low-quality rods, but finally, in 1919, acquired controlling ownership in the business, now known as Fred D. Divine & Company. The ensuing classic McDuffee period has been chronicled by collector-author Mike Sinclair, and to tell the tale I've drawn upon his expertise.[15]

With 30 years of rodmaking experience under his belt, McDuffee brought The Divine Rod back to good health. He actually initiated the modern Divine line shortly after Fred's death, with the Divine Special, which followed prototypes built by Fred and went on the market in 1902. Designed with a dry-fly action, the three-piece model introduced the famous cherry-stained cedar reel seat, had signature wraps extending up the shaft in a 10/5/3 pattern, and came in lengths of from 9 to 10 feet. A 9-footer weighed 5½ ounces, indicating it carried a 6-weight line. The company also continued The Divine Rod, available in split Tonkin cane by 1910 yet still offered with wooden shafts for true traditionalists.

The first truly modern rod arrived at the end of the staid Ada Divine years. Designed by Frank Becraft, the 7½-foot Fairy Fly Rod was introduced in 1917. A delicate 2½-ounce model, it had a cork grip and seat complemented by a fancy brass sliding ring and cap. Mike Sinclair has praised this 4-weight rod, saying that it "casts with authority." As a sound, usable classic, the Fairy Fly Rod is the most collectible of all later Divine products.

Between 1920 and 1923, Becraft and fellow maker Frank Wolcott introduced more classics into the Divine line, including the Rainbow, the Trout, the Special Dry Fly (with oxidized fittings), a regular Special, the Tournament (9½ feet and 5¾ ounces, no intermediates), and the camouflaged Invisible. Available in lengths of from 8 to 10 feet, the Invisible Rod was finished in a nonglare dark green shaft and had dull green windings and oxidized hardware—all the better to fool trout with. It was specially made from the Standard model. The Standard and Trout were new markings for the original model, The Divine Rod. The 8-footer, whether marked as a Standard or an Invisible, weighed 4 ounces; the 8½-footer, 4¼ ounces; and the 9-foot model, 5 ounces. These were three-piece, medium-line-weight rods. The Special Dry Fly series handled heavier fly lines and represented the workhorse models; the Specials were designed for delicate work and the lightest of practical lines.

In 1923 Fred D. Divine & Company introduced its classic Golden West. This new model was only available in 9 and 9½ feet, and could be ordered made with the Silken process. Immediately, Pflueger got on McDuffee's case, claiming an infringement on the trademark of its popular Golden West fly reels, and the model name was changed to the Gloriwest. Like all Divine classics, the Gloriwest was built until the Depression killed the market.

A serial number, and the reel-seat stamping "THE DIVINE ROD," place this example in the 1900-09 era.

In 1927 master rodmaker and new general manager Frank Becraft unveiled four new models: the Celdimoc, Echo, Peacemaker, and Pathfinder. I believe that many of these last rods were built without intermediates; I fished an 8½-foot Celdimoc, so nonappointed, with a 6-weight line. For one season I used a tiny 5-foot, 9-inch Becraft two-piecer that tossed a DT5 to 70 feet. This short rod inspired me to design a synthetic version in E-glass of the same length, getting the line size down to a 4-weight.

Many split-bamboo Divines had rich chocolate-colored shafts complemented by yellow wraps. Others, like the models built totally of greenheart, had striking green wrappings. In 1910 these new Divines sold through Abercrombie & Fitch and weren't cheap (considering that an average day's pay was around $2). Six-strippers sold for $16, eight-siders fetched $20, and the varnish-soaked silk-wrapped

models commanded $30. In 1923 the retail prices were even higher: Gloriwest—$45; Fairy—$45; Special Dry Fly—$42.50; Rainbow—$37.50; and Invisible—$35.

Modern Divines were packed in a heavy muslin sack and a tip tube of Calcutta cane—later changed to aluminum. The latter tube contained another small partitioned sack for the two tip sections. Divine tip cases had a distinctive coarse-knurled flat cap.

You can roughly date classic Divines by standard serial numbers, which were initiated in 1900 and ran through 1909. Annual prefix letter codes "A" through "D" were added from 1910 through 1913. The "O" preceded the serial numbers until 1917. The next letter, "H," for some reason, seems to have carried into the early 1920s. Then the "R" prefix (aka "Rainbow") was used.

Field & Stream, *May 1920*

National Sportsman, *April 1923*

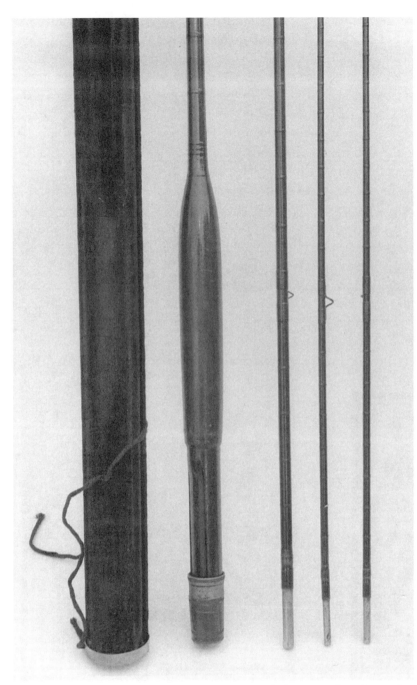

Built in the classic era, this Divine rod has greenheart shafts, a one-piece cherry grip and seat, and a later stamping with serial number 5365.

decal was used, perhaps until the end of production.

Today, antique Fred D. Divine fly rods are some of the most unique yet affordable collectibles made by an upper New York maker. Classic McDuffee production rods, in good fishable condition, are also enjoyable casting instruments. The makers of these early classics faded into the clouded legend of the Divine inspiration. Frank Wolcott retired in 1927. McDuffee remained as company president and treasurer until 1949. The following year the city directory no longer listed Fred D. Divine & Company, and a year later George F. McDuffee died at the age of 81. Francis F. Becraft, the great Divine rod designer, went 'round the bend on September 20, 1957.

ORVIS AND THE WES JORDAN ROD

No discussion of classic eastern tackle can ignore the evolving rods of the old C. F. Orvis Company. After the death of Charles F. Orvis the Orvis company declined steadily under the management of his sons, Albert and Robert, until it hit rock bottom during the Depression. The famous Orvis reel was discontinued around 1920, and the rods were then built from blanks and fittings purchased from Montague. Sewell N. Dunton, the treasurer of Montague, finally had to stop extending credit to the old Manchester company. In 1939 Dudley C. "Duckie" Corkran bought Orvis from Charlie's sons and hired Wesley Jordan to start up the idle rodmaking machinery.

After leaving South Bend, Wes had been operating a gas station in Lynn, Massachussetts. He jumped at the chance to build rods again, but by the time he had the Orvis machinery refurbished war had broken

Classic Divine fly rods can also be dated by their decals. Although the serial numbers remained on the seat, the "DIVINE ROD" stamping was discontinued around 1917 in favor of several strips of decals on the lower shaft. At first the decal read, "The Divine Rod, Utica, N.Y." From about 1926 to 1928 this original wording was retained but presented in an embellished Olde English style. Then a gold-foil Divine

out and Jordan wound up making not fish poles but ski poles, for the U.S. Army.

During the war, however, Jordan was able to begin experimenting with ways to improve the durability of what would become the new Orvis fly rods. Working with the Bakelite Corporation, he finally succeeded by heat-treating the six cane strips, gluing them up with a phenolic resin, and then having the Bakelite folks immerse the finished blanks in their liquefied product. Jordan applied for a patent for this process in 1946, but it wasn't granted until December 15, 1950. The Orvis rods built between these two dates were marked "Bakelite-Impregnated—Pat. Pend."

Jordan-Orvis rods built between 1940 and 1947 didn't have model names and were usually marked by the maker, along with "Pat. Pend." and the word "Manchester." The first actual model name, supposedly Duckie Corkran's idea, appears to have been the Battenkill, which arrived in the late "Pat. Pend." period between serial numbers 7,000 and 8,000. The Battenkill would be joined by the Manchester, Equinox, Shooting Star, Orvis Impregnated Salmon, Light Salmon, Superfine, and Deluxe.

The Shooting Star, probably Joe Brooks's favorite for casting in the brine, originally came with Aetna's Monel "foulproof" guides. By 1970 the Aetna guides were dropped in favor of a carbide stripper and large snakes. The Shooting Star tossed a 9-weight line (GAF), and came in three lengths: 8¾, 9, and 9½ feet. Shooting Stars were the first maintenance-free

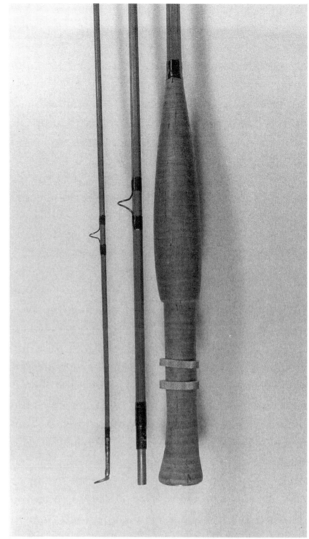

Wes Jordan's personal trout rod, marked "Orvis Impregnated, Battenkill Deluxe, Wes Jordan." It sports an Aetna "foulproof" stripper and second guide. (The American Museum of Fly Fishing Collection)

saltwater fly rods, and even today are fine tools for stripers, bonefish, and steelhead.

Lee Wulff was responsible for the development of the early short Orvis Deluxe models, and caught a number of respectable salmon on the smallest of the series. The two-piece Deluxe came in three sizes: a 6½-foot 2-ouncer, a 7-footer at 2⅝ ounces, and a 7½-foot model that weighed just under 3 ounces. The original Superfine was a one-piece rod that weighed 1¾ ounces. These early Jor-

dan models sold for exactly $100 and came in a plaid cloth sack inside a buffed aluminum case. While the rods were light, they weren't "midge" rods; they were designed to toss a standard double-tapered 5- or 6-weight line.

By 1955 the lineup included the 99, introduced on the 99th birthday of the C. F. Orvis Company. This was a no-frills two-piece line, ranging from 7 to 9½ feet in length. The 7- and 7½-footers handled a 6-weight line. The 8- and light 8½-foot models tossed a 7-weight, and the heavy-duty 8½- to 9½-foot 99s could throw 8-weight lines comfortably. Selling for $62.50 with one tip, the 99 was a powerhouse rod, actually built on some of Jordan's tapers that extended back to his Cross days. And for 24 bucks you could buy that extra tip.

With a durable finish that made them impervious to moisture and salt, Orvis impregnated fly rods were popular items. As competitors like Thomas and Heddon died off, the Orvis rod continued to please a new generation of fly fishermen. After 1960 the 6½-foot Rocky Mountain Fly Rod had a companion piece, the Rocky Mountain Spinning Rod. Both models were three-piece pack rods designed for western angling, put up in luxurious leather cases, and priced at $125. At about the same time the two-piece Midge joined the Orvis array, selling for $110 and designed for a delicate 6X or 7X tippet and a 5-weight line.

one of the first Orvis models introduced by Leigh Perkins, Corkran's 1965 successor to the business. At the same time the firm's name was changed to The Orvis Company, Inc. The forerunner of a series, the Wes Jordan Eight Footer had unusually dark cane and a thumb indentation on the grip; it worked down to the very butt while casting a 7-weight line. A 7½-footer for a 6-weight and an 8½-foot powerhouse for a 9-weight were later added to the Wes Jordan series, making a superb collectible lineup today. The Wes Jordan rods were packed in leather cases and had the original owner's name etched into the butt plate—great for collectors if it says "Jimmy Carter," but a little less great if it's "Joe Smith."

All late Orvis production wands are not only fishable but collectible, especially the Pending models and those named after the old caster himself. Wes Jordan, the master rodmaker, retired in 1971 and finally had unlimited time to spend on the waters with his second wife, Gloria. Unfortunately, he passed on in 1975. The Orvis split-bamboo rods are still being made, but—no offense intended—a good used model is a much better buy. New models in the Battenkill and low-priced Madison series are light sticks, handling lines down to a 3-weight. They're still great rods, however, and Orvis is the sole surviving American maker in the production split-bamboo business.

THE LEONARD ROD COMPANY

With the introduction of the Catskill series, popular through the turn of the century, the classic Leonard era began. The Catskills were the first "modern" rods marked "Leonard & Mills Co., Makers," a company guided by Ruben Leonard. In the post-1900 era, the factory was entirely owned by the Mills interests. The name was eventually changed to the H. L. Leonard Rod Company in 1927.

In 1915 Ruben Leonard introduced the Model 50DF, a fairly crisp 8-footer that featured a continuous cork grip and seat. The model evolved stiffer dry-fly action with the #4099 and the longer #4099½. The delicate—some say wimpy—

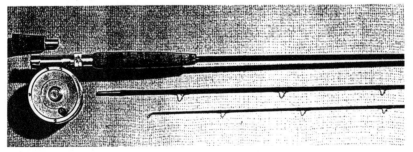

Orvis
Impregnated SHOOTING STAR Lt. Salmon & Salt Water Fly Rods

"Orvis Catalog," Spring 1970

The "new" Wes Jordan Eight Footer was introduced in 1966. For the first time in his life the great rodmaker had a rod named after him—almost 50 years after he'd begun working the culm. The rod was

Bogdan Salmon and Saltwater Reel

Sooner or later, every serious salmon fisherman owns a Bogdan. **Frame and spool are machined from a solid bar of aluminum,** perfectly fitted with no pillars to work loose.

Superb craftsmanship and engineering make this the ultimate fly reel for Salmon and Saltwater fishing. It is gold anodized to prevent corrosion under any conditions or climate. The ratio of retrieve is 2-to-1; the brake is smooth and easily adjustable. A double brake shoe drag has a wide range of seven stations from very light to very heavy, with click or silent action. Beautifully machined cross plate assures snug fit to reel seat. Large capacity and lightness make this reel perfect for any salmon or saltwater fly rod from the lightest to the heavier two-handed models. This fine reel is also available in lefthand models on special order.

F1662 — Dia. 3-1/4", Spool 1-3/8" wide, 11 oz. capacity WF9F plus 200 yards 20 lb. test dacron backing **$182.60**
F1663 — Dia. 3-3/4", Spool 1-3/16" wide, 13-1/4 oz., capacity WF10F plus 200 yards 20 lb. test dacron backing **$184.80**
F1664 — Dia. 3-3/4", Spool 1-3/8" wide, 14 oz., capacity WF10F plus 300 yards 20 lb. test dacron backing **$190.30**

As its top-grade offering, Orvis sold Stan Bogdan's reels in the 1970s. The prices started at $135 in 1970, and reached $193 in 1976. I wish I'd bought one.

rience, I know that it's far better to try for the fish 30 feet away than the one at 97 feet.

For a split-bamboo rod for general trouting—whether a Leonard or a rod from any other maker—choose a model that weighs less than 5 ounces. Bass and steelhead rods do require the extra power and weight found in Ruben Leonard's Tournament series and the classic workhorse rod, the Knight 99. John Alden Knight, "Mr. Solunar Tables," was a partial designer of this powerful steelhead model. The 7-ounce Knight 99 is still suited for northwestern rivers, although it kills at both ends.

We won't wade through a long list of the many variations and models of classic Leonard products; that subject has been well covered in print by Martin Keane and Len Codella. The vast spectrum of rods, from salmon tamers to dainty little midge models, is usually open to review at any given Lang or Broggi auction. The H. L. Leonard Rod Company made a fantastic number of quality rods. As testimony to the Leonard name the rods retain a high value, even though the market is flooded with them.

The Leonard rod remained a high-quality item throughout the company's history, with extreme attention paid to detail. Following Orvis's success with Bakelite impregnation, in the 1950s Leonard introduced a Duracane series built by Alan Sharpe in Scotland. The traditional models, which are on par with a comparable Thomas Special, are the most sought after today, however. In 1964 the Leonard plant burned to the point that just about everything—beveler, patterns, and stock—was lost or damaged irreparably. Hap Mills rebuilt, however, and hired Ted Simroe as master builder.

Although the postfire Leonard models held to the old high standards they were a new breed of rod,

Model 50 was also an offshoot of the original 50DF. Additionally, the model's action was modified into the Hunt Pattern, designed by salmon stalker Richard Hunt. This heavier 8-footer was complemented by black oxidized hardware.

Throughout its modern history, the Leonard Rod Company seemed preoccupied with sophisticated, specialized tapers. To some extent, the notion that all a fly rod had to do was deliver a fluffy hook with accuracy was forgotten. Reversing its original focus, the company moved from delicate to tournament models, producing rods that could shoot an entire fly line to hell and gone. Whether the average streamside angler really needed the reserve power found in the new Hunt and Tournament models is a moot point. Most fly fishermen are less than the rod they hold, and after an hour's worth of diligent flailing they usually start to dump casts. Speaking from personal expe-

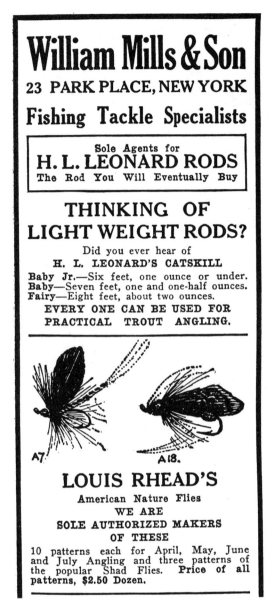
Field & Stream, *May 1920*

characterized by serial numbers and furnishings made by Research Engineering of Vermont. No offense intended, but the RE fittings were not made to the same delicacy as the older reel mountings. The knurling tool used on the seat bands was coarser, and the grip check had no knurling at all. Very modern, and very sterile. In 1978 the shop was sold to Johnson's Wax. Bill Alley took the helm as president, with Tom Maxwell of Thomas & Thomas fame as head rodbuilder. Production of a high order continued until 1985, when the Johnson folks, after reviewing the balance sheet, closed the

doors of a company that had built great rods for over a century.

Today, the prefire Leonards, built with the original machinery and sporting the traditional older furnishings, have the highest value. Postfire rods in identical condition and length bring lower prices, and therefore may be the best bargain for those who wish to fish a Leonard. Overall, similar rods built before and after the fire deliver the same fly with equal finesse. It's hard to imagine that a bad Leonard was ever built, and both Simroe and Maxwell made a superior product. If we count original Bangor-fashioned tools, the Leonard name appeared on quality products for 130 years. Outlasting Thomas and Payne, the company was one of the last of the old-time eastern tacklemakers.

THE ELMER SELLERS "CASTING BASKET REEL"

We finish our discussion of eastern tacklemakers on a somewhat lighter note, with a product that just didn't fall into any slot. Somehow the ass end of a chapter seems an appropriate place for a 20th-century dinosaur. After William Billinghurst's patents expired, Elmer J. Sellers of Kutztown, Pennsylvania, built very similar reels—only he made his reels in the mid-1930s. All the fly fishermen who really wanted a reel like this had died of old age some 30 years before.

Called the "Casting Basket Reel," this side-mounted curiosity was patented on February 13, 1934. The Sellers model came in several variations and, according to writer Albert J. Munger, it was made in the rear of Elmer's drugstore during his spare time.[16]

Field & Stream, *March 1934*

This postfire 7-footer is marked "H. L. Leonard Rod Co., Makers," with serial number 325, and remains a fine example of the old company's craftsmanship. (The McGrath Collection)

Various methods of attaching the handle were used, and Munger has found at least one folding version. The reel is mainly noted for its ventilated basket area and agate line guard. The latter looks like a reworked Esing's "line saver" or similar commercial rod top. Stamped from two pieces of metal as opposed to Billinghurst's soldered round stock, the Sellers basket was easier to produce.

I suppose that the round agate guard was an improvement over Billinghurst's model, but other than this one feature, the reel looked as though it had been cloned from the DNA of a Billinghurst found in a small piece of amber stored in a closet in an old house in Plymouth, Massachusetts. A few other quirks of the Basket Case Reel included a

Built about three-quarters of a century too late, this is the amazing Basket Case Reel. Probably not a big seller in its day, this model now commands a premium price on the collectors' market. (The McGrath Collection)

sliding click located on the face and an adjustable center knob that allowed the angler to vary the tension, or drag. This last feature was probably added for those coarse fishermen who trolled. Elmer's handiwork was a tribute to an original idea. Unfortunately, it wasn't his. Nonetheless, Elmer's reel is now a classic collectible, and seems about as rare as William Billinghurst's.

Notes

[1, 3, 5]Bigelow, Paul. *Wrights and Privileges*. Athol, Mass.: Haley's, 1993.

[2, 4]Kelly, Mary Kefover. Personal correspondence, 1977.

[6]Oliver, Richard W., Bob Lang. *Sixth Annual Summer Fishing Tackle Auction Catalog*. Kennebunkport, Maine, 1990.

[7]Aroner, Mark. Phone conversation, 1995.

[8, 9]Spurr, Dick and Gloria Jordan. *Wes Jordan: Profile of a Rodmaker*. Grand Junction, Colo.: Centennial, 1992.

[10]Vernon, Steve. *Fishing Collectibles Magazine*, Winter 1994.

[11]Herr, Harold G. "Doc." Personal conversation, 1989.

[12]Vernon, Steve. *Antique Fishing Reels*. Harrisburg, Penn.: Stackpole, 1985.

[13]Bodner, Mike. *Old Reel Collectors Association Newsletter*, Fall 1994.

[14]Sinclair, Michael. *Bamboo Rod Restoration Handbook*. Grand Junction, Colo.: Centennial, 1994.

[15]Sinclair, Michael. *Fishing Rods by Divine*. Grand Junction, Colo.: Centennial, 1993.

[16]Munger, Albert J. *Old Fishing Tackle and Tales*. Philadelphia: Munger, 1987.

Nathaniel Currier published a number of early fly-fishing prints, such as this 1862 example of an original painting by A. F. Tait. The angler in Brook Trout Fishing *is using a rod with a reduced butt fitted with a click reel.*

THE MALE BROOK TROUT [Solvelinus Fontinalis]

One of the most popular chromolithographs by Sherman Foote Denton, The Male Brook Trout *has superb color and charm.*

Antique brass reels have a special appeal. Left to right: a tiny raised-pillar model marked by Trowbridge, Boston; a size 3 ½ grilse reel by Hendryx; a small clamp reel by Allcock; and a raised-pillar model with Abbey & Imbrie's original 1878 logo.

This pair of New York fly reels could date back to 1864. Both are marked "J. C. Conroy & Co." At left is a German silver click reel; at right, a tiny "No. 6" multiplier made from plain brass.

Two perennial favorites from Julius vom Hofe. The square-tipped, raised-pillar fly reel (left) and rimmed model (right) are both early hard-rubber reels with single-crank handles.

The A. F. Meisselbach Manufacturing Company produced some charming fly reels. A sampler includes a later automatic made in Elyria (left); an early small Newark Amateur with a drag spring (middle); and an Expert, shown from the rear to reveal the open click mechanism.

Prior to the tackle patent period that began in the 1870s, the only reels large enough for salmon anglers were multipliers. Left to right: a big size 6 brass New York model by John Conroy; a Frederick Vom Hofe New York brass reel in size 5; and a huge, early hard-rubber-and-brass example believed to have been made in Birmingham, England.

Classic Atlantic salmon tackle. The reels are (left to right): an Otto Zwarg 3/0; an Edward vom Hofe 6/0, owned by William Mitchell; a smaller 4/0 Edward vom Hofe; and a #3 1/2 grilse model by A. Hendryx. The rods include: a 9 1/2-foot L. L. Bean Atlantic Salmon (foreground) by Montague; a 9-foot Leon Thomas Special (left); and a 15-foot two-hander by Leonard & Mills.

The chronology of the fly-rod reel seat from the 1860s to 1910.
Left to right: C. F. Murphy (1863); CB&M (1878); T. H. Chubb (1880); H. Pritchard (1881); C. F. Orvis (1882); William Mitchell (1883); H. L. Leonard (1885); T. H. Chubb (1889); J. G. Landman (1890); Thomas & Edwards (1898); and F. E. Thomas (1910).

Four 1898-1910 Maine-built fly rods. At left are a Dirigo with a standard signature wrap and an older Dirigo with a full German silver reel seat, both built by Thomas in Bangor. The two rods at right are a circa-1900 Isaak Walton and a rare circa-1896 Thomas & Edwards, both made in Brewer.

Although tough to find, early American fly rods do show up—occasionally. Here's an ultra-rare two-handed trout rod by J. C. Conroy (left), along with two lighter wooden models featuring dark-stained and Japanned butts, stained mids, and natural-colored lancewood tips.

Rods of the 1865-75 era. At left, a C. F. Murphy 12-footer in a green velvet form case. The center rod was built by Hiram Leonard in Bangor, circa 1875. On the right is an ash rod showing the defined "grip area" of the 1860s.

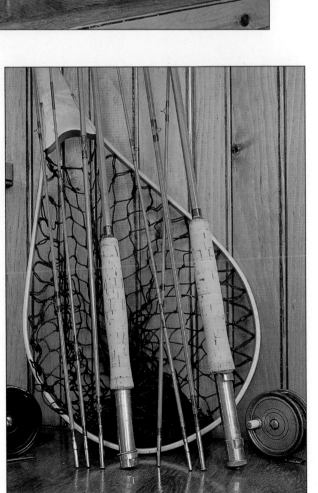

Two Granger fly rods owned and well used by the late author Frank Woolner. At left, an early Goodwin Granger 9-foot model with the original Granger reel seat. At right, an 8 ½-foot Wright & McGill Victory with the patent Phillipson-era seat. The matching reels are a Meisselbach Rainbow and a Sears J. C. Higgins.

A selection of vintage two-handed salmon rods includes (left to right): anonymous lancewood (1850s); ash Thomas Mack (1860); split-bamboo William Mitchell (1885); split-Calcutta Kosmic (1895); split-Tonkin F. E. Thomas (1915); and split-Calcutta Montague Manitou (1915).

E. W. Edwards was the first maker to "autograph" shafts as a way of marking a rod (left); Hardy followed suit. Thomas stuck to the old gunmaker's die stamp (center). The Montague Red Wing and South Bend #359 are marked with acetate labels. The rods are perched atop a nice 1920s willow creel, and the reels are a Pflueger Gem and a Union Hardware model.

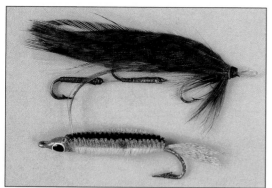

The new breed of flies. At top, a very early tandem streamer, attributed to Carrie Stevens, pattern unknown. Below, a 1940s streamer built in the "Optic" style.

Three antique #6 trout flies on their original 1885 patent "tension envelope." The patterns are (top to bottom): an Adirondack; a Silver Doctor; and a White Miller.

"The Silver Doctor...has been proved effectual under so many circumstances, and for such a variety of fish, that it is probably valued by American anglers more than any other salmon fly."
—Mary Orvis Marbury, 1892

Antique salmon flies are top collectibles. This is a Jock Scott, a pattern originated in the mid-1800s by Lord John Scott's "water bailiff," according to George Kelson. The fly was tied by the lassies of Forrest & Sons.

A classic-tackle sampler, from antique to almost new. One of the greatest creels yet found, this old beauty, fashioned from yucca and willow by an Arizona member of the Paiute tribe, has a little flap over the center hole to keep trout from jumping out. The rods include a matched pair of 6 1/2-foot Carlson Fours, a 7-foot Carlson Four Special, an 8-foot Gillum, and a fine Garrison (center). We also see an old patent Ross-White rod-reel combo, a fly reel by Edward Hewitt, and a Meek 44. (Photo by David Allen)

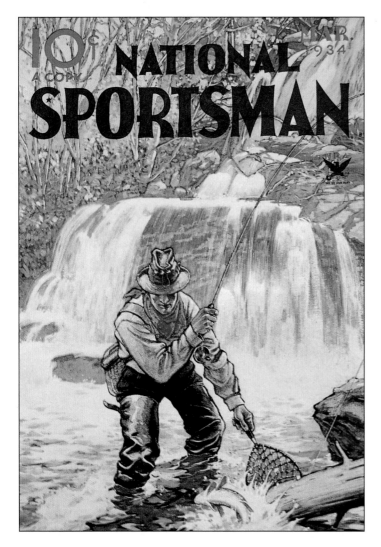

A. B. Parsons's March 1934 cover for the National Sportsman is a good example of the new breed of illustration art, using shades and colors beyond reality.

Displayed over fly trays from a large Abercrombie & Fitch chest are (top to bottom): a beautiful 8-footer by Dickerson; a recent H. L. Leonard Rod Company 6-foot Baby Catskill; and a 6 1/2-foot Maxwell Hunt. The reels include (clockwise from upper left): two rare Meek No. 44s; a Mills Crown; an Edward vom Hofe 1/0 Tobique; and an early Edward vom Hofe saltwater model. At center is a superb, and rare, Philbrook & Payne fly reel. (Photo by David Allen)

10 From the Heartland Westward

Not all the great fly tackle was produced in the East. From the Midwest to the West, manufacturers produced an array of rods and reels that satisfied both the needs of everyman and the tastes of the wealthy, from the cheap Shakespeare Featherweight reel to the superb Heddon "Rod of Rods." The lineup of tackle companies read like a *Who's Who* of Angling Americana—Pflueger, Talbot, Young, South Bend, Granger, Wright & McGill, Meek, Gayle—and stretched across America's Heartland, from Kentucky to Michigan to Colorado.

The easterners may have considered themselves the founders of the American tackle industry, but the heartlanders knew better, especially the Kentucky makers who'd been producing excellent fly reels since the early 1800s. Like thoroughbred horse racing, fly fishing came naturally to the Bluegrass State. Two top Kentucky reelmakers, B. F. Meek & Sons and Geo. W. Gayle & Son, produced some of the finest fly reels ever made. They were not the only well-known makers of casting reels who fashioned single-action models, however. William H. Talbot, a native of Missouri, ranks with them.

ANTIQUE NO. 1 & NO. 2 KENTUCKY REELS

The venerable little Kentucky winch must have been a popular fly reel; it was shown mounted on wooden and bamboo wands in several 19th-century paintings. But, with the exception of Jim Brown, Frank Stewart, Steve Vernon, and Michael Nogay—the founding president of the Old Reel Collectors Association—few collectors seem to have given this much thought. Today the tiny, almost toy-sized Kentucky models are considered nominal fly-fishing

tools, since most of the miniature reels were used on diminutive "trout rods." A trout rod, typified by the little "E. E. Eaton, Chicago"–marked Chubb pictured in chapter 4, was a very light, short, bait model with the reel mounted above the grip. Such rare, small bait rods matched the No. 1 and No. 2 reels. Even smaller Kentucky reels were built, including a size o and a oo. Many of these little winches were favored by midwestern fly fishers as late as the early 20th century.

Silversmith George Snyder, of Paris (Kentucky, not France), was the first maker of this style reel, beginning his limited career prior to 1820. The prototype for his original model was a multiplier made by Onesimus Ustonson. Today, Snyders are as rare as their British counterparts. Our next craftsman was Jonathan Fleming Meek, a jeweler in Frankfort (Kentucky, not Germany) who began making multipliers in the mid-1830s. He was later joined by his brother, Benjamin Franklin Meek. The gears for their early reels were made in Danville on a cutting machine imported from England.[1]

Other little Kentucky models were built by James L. Sage, a gunsmith formerly from Hartford, Connecticut. Sage was a devoted fly fisherman. James Henshall wrote, in the December 1900 *Outing* magazine, that Sage "showed me his fly rod and flies, all made and used by him as long ago as 1848; so that fly fishing for black bass was practised as early in Kentucky as in any other section of the country."

Our last maker of tiny trout reels was Benjamin Cave Milam, who began his career in 1834 at Frankfort, according to a statement recorded in *American Game Fishes* (1892).[2] Milam wrote that "o and oo are

Classic rods from Denver, and reels from Ohio: left, a 9-foot Wright & McGill Granger Victory and a Pflueger Gem reel; right, an early Good-win Granger 8½-foot prepatent model and a Meisselbach Rainbow reel, made in Elyria, Ohio.

fancy sizes, and too small for much heavy angling. Nos. 2, 3, and 4 are the sizes most used. We attach a click and drag so that our reels can be used for bait-fishing, fly-fishing or trolling." These solitary reel craftsmen would later form partnerships, including Meek & Milam and B. F. Meek & Sons. Meek & Sons outlasted the rest, eventually went public, and finally was purchased by Horton-Bristol.

The little Kentucky multiplying fly reel has an excellent pedigree, and remains at the zenith of the hand-built-reel maker's art. The single-action fly reels of 1900–16, discussed below, were built to the same high standards as this first generation of Kentucky reels.

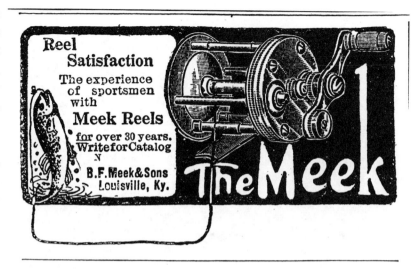

National Sportsman, April 1916

THE B. F. MEEK & SONS'S NO. 44

In 1898 J. H. Sutcliffe and Associates purchased the old Louisville, Kentucky, firm of B. F. Meek & Sons and restructured it as a production reel company called B. F. Meel & Sons, Inc. With Sylvanus Meek as master reelmaker, the new firm produced some of the best production reels ever built, including such popular models as the No. 25 and No. 3 casting reels. Its only single-action fly reel, the renowned raised-rim No. 44, was introduced around 1904 and continued in limited production until 1916, when the company was sold to the Horton Manufacturing Company of Bristol, Rhode Island.

Built entirely of German silver, the No. 44 came in two models. The earliest version had a flat back, a nonremovable rear-bearing cap, and trim bands on the rims. The second model had a raised back, an ivoroid handle knob, and foot screws that fastened through the front plate. Both versions had a counterbalanced handle that was retained by a large and often fancy oval-head screw.

Through the years various tackle auctions have promoted the No. 44 as a "handmade" item and extremely scarce, in one instance claiming it was built by Benjamin Cave Meek. In truth, these production reels sold for $15 until 1910. In 1911, due to "sufficient orders" and reduced "manufacturing

costs," the reel's price was dropped to $10.[3] This may well be when the model change occurred. Still, handmade or not, the Meek No. 44 was one of the most appealing fly reels ever built, especially to those who like a lot of German silver. It has sold for higher prices than either the Talbot Ben Hur or the Gayle Aluminum Trout Reel, and remains one of the most coveted reels of the classic era.

CLARENCE GAYLE'S KENTUCKY FLY REELS

George W. Gayle, a talented silversmith from Frankfort, Kentucky, built a few reels prior to his death in 1896, but it was his son Clarence who continued "Geo. W. Gayle & Son" into the present century. An excellent but unfortunately not prolific craftsman, Clarence Gayle produced a limited number of "Frankfort" and "New" Kentucky reels. He also made at least four single-action fly models, including some very inexpensive ones.

Geo. W. Gayle & Son built reels from 1897 until after World War II. As with so many manufacturers, fishing reels were only a sideline: The company also made a variety of items, including oil burners and precision atomic bomb parts. From the 1920s until the end Clarence built reels that incorporated "light" materials, such as aluminum, rubber, and Bakelite. After 1930 Gayle's highest-quality reels were marked "HAND MADE"; a few presentation-grade items were engraved "C. Gayle" in script.[4]

Looks Good; Don't It? It Runs Better

This cut shows the GAYLE "SIMPLICITY" No. 3 ALL BRASS Trout and Bass Reel, exact size, but it cannot convey to you the beauty of its action or the charm of its softly humming click that is made with just the right tension. This is a good, strong, well-made single-action reel, that can be used either as a F L Y CLICK or a FREE RUNNING CASTING reel, as the click is movable and can be thrown on or off instantly. Is made of BRASS, with dull SATIN or ROMAN finish. A reel that will please you. SENT POSTPAID TO ANY ADDRESS FOR $1.00. If you want to be sure of getting yours, order today, as we have only 5000 of these reels for this season's sale.

If you are not satisfied with this reel when you get it, return it to us and your dollar will be immediately refunded without question or quibble.

GEO. W. GAYLE & SON .. .. Frankfort, Kentucky
See our other "SIMPLICITY" Reels at your dealers, 25, 35 and 50 cents, and all good

National Sportsman, *April 1923*

The most common Gayle fly reel is the No. 3 Simplicity, which was made from 1925 until 1940 with steel or brass frames and sold for around 25 cents. At peak production, 100,000 Simplicity reels were produced annually; later, the model lost its market share to even cheaper Japanese copies. After the war, Geo. W. Gayle & Son produced a second inexpensive Simplicity—the aluminum No. 6; other than double knobs, it had essentially the same features as the No. 3. Selling for roughly $2.50, about 25,000 new Simplicitys were built. Most were scrapped and destroyed after Clarence Gayle's death in 1948.[5] The No. 6 reel is now somewhat collectible. At least one example, with its original round cardboard box, sold for $85.

Quality Gayle fly reels are rare items and eminently collectible. At least two versions have been found, dating from the 1920s and '30s: a German silver model with rubber knob and Bakelite side plates, and a mostly aluminum model with German silver accents. Both reels were built with a straight, counterbalanced handle. The aluminum version appeared in a 1937 Gayle catalog as the "Aluminum Trout Reel" and was available in two sizes. The Trout model, at 2¼ inches in diameter, sold for $30; the 4-inch Salmon winch was $75. These labor-intensive

reels were marked "G. W. Gayle & Son, Frankfort, KY," with "HAND MADE" stamped under the maker's mark.

Of the two aluminum models, only the Trout size has shown up at auction (1988). At least one Gayle rubber-Bakelite reel has been located as well. But it's evident, through a general lack of "sightings," that very few high-quality Gayle fly reels were made. I guess we know what that means, as far as value is concerned. All the superb fly tackle made by the Kentucky-Missouri school, "miniature multipliers" included, now brings much higher prices than contemporaneous New York–built collectibles.

THE TALBOT BEN HUR TROUT REEL

William H. Talbot was a master jeweler and Kentucky-style reelmaker from Nevada, Missouri. He entered the profession in the very late 1890s, and on January 22, 1901, patented his improved "shouldered" pillars, which strengthened a reel where the pillars met the side plates, an idea later used by William Shakespeare, who somehow circumvented Talbot's patent. Talbot was known for his superb, jewel-like multiplying winches, the most famous being the gold-accented "Fred N. Peet" presentation casting reel that sold for a record price of $7,000 a few years ago.

Sometime after 1901, William Talbot moved to Kansas City, Missouri, and expanded his reelmaking business. His only contribution to fly-fishing history was a singular model called the Ben Hur, a single-action reel constructed with aluminum side plates and the remainder of German silver. The Ben Hur had a graceful S handle, similar to those of Edward vom Hofe, and apparently was made in at least two sizes.

I've only seen the smallest Ben Hur—a striking Model No. 100, which had a diameter of 2⅛ inches. The reel was marked "Wm. H. Talbot Reel Co., Nevada, Mo." and had the 1901 shouldered-pillar

Forest and Stream, *July 28, 1900*

patent date. This particular specimen sold at Oliver's Fourth High Roller's Auction for $6,000 despite not actually being in the "Excellent" condition mentioned in the catalog. Scarcity equals high price, however, and no one knows how much time will pass before the next Ben Hur appears on the market.

PFISHING WITH PFLUEGER

Originally a bass-lure outfit, this Akron, Ohio, business was founded by Ernest F. Pflueger after he purchased a hook company that dated back to 1864. Reel manufacture began some time after 1900. By 1908 Pflueger's son Ernest A. had changed the name to the E. A. Pflueger Company. Around 1910 the firm became known as the Enterprise Manufacturing Company. Under the management of sons George, Joe, Charles, and Bill, Enterprise produced a plethora of Montague-style winches that were sold under three labels: "Atlas," "4-Brothers," and "Pflueger." Of the thousands of reels made, most were lesser-quality multipliers.

The early Enterprise fly reels, which often bore Montague City–style end caps, were mass-produced up until the early 1930s. The line included many stamped models and a very few higher-quality items. During the Depression, most of the cheaper models were dropped in favor of newer offerings. Perhaps the least expensive yet still collectible early Pflueger reel today is the Progress skeleton model—a direct rip-off of Meisselbach's original Featherlight. A raised-pillar click reel with a wooden knob, the Progress is usually seen in a 2½-inch size. The trick with this model is finding one with its nickel plating intact.

Moving up the early Pflueger fly-reel hierarchy, we reach the Four Brothers Delite, built in sizes from 40 to 80 yards. The Julius vom Hofe–influenced Delite was a pleasing little reel with hard-rubber side plates, an ivoroid balanced handle, and metal parts of nickel-plated brass. A popular "trade" reel, it sold through many companies such as Edward vom Hofe & Company of New York City and M. A. Shipley of Philadelphia. Like the Progress, the

One of Pflueger's finest earlier fly models, the Delite wasn't an expensive reel then—and still isn't. (Bob Corsetti photo)

Delite—with all its nickel remaining—is today most collectible.

Next was the charming Pflueger (with the Bulldog trademark) Hawkeye fly reel. The Hawkeye was similar to the Delite and came in the same sizes, but was of generally higher quality, with side plates and a handle knob of hard rubber, and all metal parts fashioned of German silver. The popular Hawkeye had all an aspiring trout angler really wanted—a click, a slotted bearing cap for oiling, and a couple minor patents (January 23, 1907, and April 3, 1923). It has sold for as much as $300 in excellent condition.

The last of the great early Pfluegers was the Bulldog-trademarked Golden West, introduced in 1915 when western angling was indeed golden. This most coveted of Pflueger fly reels had strong side plates made from German silver sandwiched between outer layers of hard rubber—an idea patented by Joe Pflueger on February 10, 1903. The Golden West came in a fancy padded box, like the company's best saltwater trolling reels, and had a nice, tacky, cloisonné German silver label. The model was usually serial-numbered and built in the same sizes as the Hawkeye.

Now we come to Pflueger's "new" models, introduced in the 1930s to replace the cheaper Atlas reels. The Gem was an excellent low-cost click reel with a whisper of German silver accenting. It remained in production for a decade or more. That perennial favorite, the Medalist, introduced as the 1590 series, was built with agate line guards encased in metal rings. A Medalist No. 1592, once owned by Harry Darbee, had two agate line guards. These rare original Medalists had sculptured pillars and spool arbors and a one-piece cast foot. In the good old days, a Medalist fit the reel seat of any top-quality fly rod.

Eventually the agate line guards were dropped and the foot was changed so that the reel no longer fit on a quality reel seat. By the postwar era the new 1490 series Medalist had lost much of its original appeal, though it remained a good, practical reel. Finally, in the 1960s, Shakespeare absorbed Pflueger and the model was cheapened further; eventually, it was made in the Orient. A few years ago it was reissued,

improved slightly with a palming ring. Original Medalists still in their variegated green Bulldog box are easily dated. The date is stamped in red ink on the bottom of the container.

Pflueger made other fly reels below and above the Medalist in quality. The old No. 576 Trump was a stamped aluminum jobbie for beginners. The No. 1554 Sal-Trout was, and still is, a fine low-priced reel. Ultimately more collectible, however, were the "big-fish" reels built at the end of the American-made Pflueger era. These included the No. 577 Single Action Supreme, a disk-drag salmon reel, and the larger Pflueger No. 578, built for saltwater fly rodding. The No. 578 also had the drag and line-stripping lever, which allowed the angler to bypass the drag while stripping out line.

SHAKESPEARE'S FLY-CASTING DEPARTMENT

William Shakespeare was a Kalamazoo, Michigan, watchmaker who started his tackle company in a one-room workshop in 1897. Thirteen years later, the William Shakespeare Company had grown to occupy a three-story factory that cranked out a phenomenal amount of tackle, particularly casting reels and bass baits. In production casting reels, Shake-

In 1916, the Golden West came in a velvet-lined box with a tool and grease. Today, this entire package would add great value to the reel.

Pflueger
FLY ROD Reels

**Pflueger
MEDALIST
$5.00 to $8.00
(3 sizes)**

**Pflueger
GEM $3.50**

**Pflueger SUPEREX (Automatic Reel, 2 sizes)
$6.00 and $7.00**

Field & Stream, *March 1934*

speare was a market leader, building the Standard Professional, the little Tournament, and the later Supreme, the first semimodern winches designed to toss bass lures great distances. Its fly-fishing production was negligible, although this was not exactly the picture it presented.

A story in the 1910 edition of *The Sporting Goods Gazette* featured three photographs of Shakespeare's "rod department," including depictions of "cutting wood for the construction of fishing rods" and the "finishing room." There was only one problem. In those days Shakespeare never made any rods, fly-fishing or otherwise. It purchased rods at jobber prices from real makers such as Montague, Heddon, and South Bend.

In 1924 Shakespeare offered four different models of the Ideal Rod, which were probably made by Montague. The three lesser models had plated-brass fitting and full-metal seats, of suspiciously Montague style. Shakespeare Ideal Rods in the lower price bracket included the No. 3310 Bass action and the No. 3318 Trout action. A slightly better model in the Trout action, the No. 3328, sold for $18. The best of

the offerings (probably comparable to the Montague Red Wing in quality), the Ideal No. 3338 Trout, had nickel-silver furnishings and sold for $24.

Sometime after 1926 Shakespeare added Heddon models to its rod lineup; these can be distinguished from the Montagues by the dark, flamed cane. During this late-1920s-to-1934 era, Shakespeare added model names to the numbers. Among the Heddon-crafted rods were the Deluxe, Intrinsic, Tillamook, Rangeley, Favorite, and Perfect. The lower grades, such as the Manistee and Trout River, were Montague built. Most rods had a "B" after the number, although the significance of this escapes me. One atypical 6½-foot two-piece model was built with a full-metal reel seat stamped "Shakespeare 1233-B-6½, Made In U.S.A., Pat. 1,410,906." According to these numbers, the patent was issued in 1924, but it wasn't recorded by Mary Kelly.[6]

Beginning in 1935 the South Bend Bait Company produced most of the later Shakespeare fly rods, in lengths from 6½ to 9 feet. By 1941 the Shakespeare fly-rod line was down to 11 models, the top 8 made by South Bend and the rest by Montague.[7] Most Shakespeare cane fly rods can today be roughly dated by their labels. Prewar rods had an oval gum label reading "Shakespeare, Honor Built, Honor Sold." After the war the words "Made in U.S.A." were added. Collectible high-grade prewar Shakespeare include the Superba Double-Built, the Superba, and the Deluxe, which originally sold from $50 down to $30. After World War II, Shakespeare offered only three split-bamboo models, all made by South Bend. Shakespeare had something new in store for American anglers.

In 1947 Shakespeare introduced the first fiberglass rods to the American market. The Shakespeare Wonderods, built using the Howald process (which produced a hollow glass blank) invented by Dr. Arthur M. Howald, had a gleaming white blank and a "snap-lock" reel seat, a carryover from late split-bamboo models that was used well into the 1950s. Models were available from 7 to 9 feet in length, all two-piecers. The top of the line was the Presidential series. Some rugged 9-footers, built for 8- and 9-weight lines, had a nondetachable 2-inch extension butt. That first year the Wonderod sold for $60. As competitors introduced glass rods of their own, the

One of several Meisselbach copies, the Shakespeare Featherweight was an inexpensive model sold in quantity. Not many are found in their original box, though.

price gradually dropped. Today these original fiberglass fly rods are becoming collectible. A pair of Wonderods owned by pioneering saltwater fly fisherman Frank Woolner appears in the color section of this volume.

Shakespeare did make a few fly reels, some marginal, some better models that are collectible and usable. The Kazoo was a Union Hardware–style skeleton reel made from rustable metal. The company's somewhat rickety version of Meisselbach's Featherlight was called the Shakespeare Featherweight. It was a copy of Meisselbach's last model, without the raised pillars, and held 60 yards of line.

The Shakespeare Russell trout reel was on par with Bristol's Meek series. For steelhead, the 3½-inch-diameter Shakespeare #1900 was a copy of Hardy's Perfect, including the ball-bearing shaft, adjustable drag, and so forth. This reel was produced until World War II. Like most vintage fly reels, it can be dated by the model number. Using the accompanying table, the No. 1900 Model HE, for instance, was built in 1936.

In the late-1960s-to-early-'70s period, Shakespeare made a steelhead-bonefish reel called the No. 1898 Purist Heavy Duty Fly Reel. It had a six-disk drag (similar to a spinning reel's), a nonreverse lever, and a "built-in click" on the drag. I suppose this was a lot handier than installing the click yourself. Tackle hype can really wear down one's brain.

SOUTH BEND AND CROSS FLY RODS

An old company located in the city of the same name, the South Bend Bait Company entered the fly-fishing game in the early 1920s, making a few reel models and selling fly rods built by Eustis W. Edwards at Winchester. In 1925 Edwards left Winchester, and the quality of rods deteriorated significantly. South Bend then turned to Mrs. William Forsyth, the widow of the man who'd owned the Cross Rod & Tackle Company in Lynn, Massachusetts. In 1926 South Bend entered into the rodmaking business, acquiring not only Cross Rod & Tackle but also, in one

SHAKESPEARE 1898 PURIST H.D. FLY REEL. 6-D drag. On-off non-reverse lever and built-in click on the drag (signals a running fish), tackle the toughest tarpon, coho salmon, chinook, or almost any other big gamester. Will hold any level or tapered fly line plus 200 yards of 20-pound braided line backing.
1898 Shakespeare Fly Reel — Net $24.88.

"Finnysports Catalog," 1972

Cracking the Shakespeare Model Code

1	2	3	4	5	6	7	8	9	0
K	J	H	G	F	E	D	C	B	A
V	U	T	S	R	Q	P	N	M	L

of the great coups in the business, the services of its master rodmaker, Wesley Jordan.[8]

Wes Jordan moved to Indiana, set up the old Cross machinery, and built one of the largest rod-making operations of its time. In 1927 he patented his modern screw-locking reel seat, assigning the rights to South Bend. In time the company demanded that Jordan increase his production of lesser rods, some of which were wholesaled to Sears, Roebuck for 83 cents. Within a few years, the plant was building rods day and night. Two styles of fly models were built. The lesser, designated South Bend Rods, gave Montague a run for the small dollar. The Cross Rod line was intended to compete with Heddon, Granger, and Edwards.

Early rods were made with German silver reel seats and ferrules, and featured the oval South Bend medallion. Cross Rods came in lengths from 7 to 15 feet, and were available in both double-built and single-built versions. The double-built rods were not nearly as heavy as cane laminates from other makers, such as Hardy. Rod shafts came in an array of colors, from light and burnished to a rich dark brown, and

even a "green finish." For windings there were 14 standard "main colors" and 14 tipping colors. Custom windings were available in any combination of the 28-color palette. The most popular colors were brown and jasper.

The little Single Built Cross Sylph, a two-piece 7-footer weighing only 2½ ounces, was a carryover from the old Lynn days. Perhaps the greatest example of Jordan's talent was the 9-foot One-Piece Fly Rod. If found today, this would be a great collector's item. Most models, however, were made in the standard three-piece configuration. The 1927 "South Bend Cross Catalog" offered eight different actions, all single or double built. In addition to the two models mentioned above, the line included the Dry Fly (8 to 10½ feet), Dry or Wet Fly (7½ to 8½ feet), Salmon Wet Fly (13 to 15 feet), Salmon Dry Fly (9½ to 10½ feet), Tournament (9 to 9½ feet), and Bass Bug (8 to 10 feet). It's amazing that, with so many models offered, so few surface on the collector's market. All Cross Rods were put up in a bag and a labeled aluminum case.

By the 1930s the Cross line was rather modern looking, with models like the 9-foot No. 166 Dry or Wet Fly having Bakelite and German silver Jordan reel seats. Just prior to World War II, many models

A Model 264 South Bend Cross rod. The double-built models are uncommon today, especially in shorter lengths. This one is an 8½-footer. (The American Museum of Fly Fishing Collection)

DOUBLE BUILT FLY RODS

No. 30 Trout—No. 31 Bass or Steelhead—No. 330 Dry Fly

Cross Craftsmanship The finest rods produced by South Bend. Constructed of selected butt-cut, mountain grown Tonkin cane, double-built by the craftsmen who make the famous Cross Rods. Three-piece fly rods (with extra tip). "Lite-Lock" end-locking reel seat of ivory plastic with Colonial Black hoods. Shaped cork grip of best quality. Bamboo finished with alternating black and natural flats. Stainless Steel chrome plated stripping guide, others tool-hardened steel and Perfection tip top. Contrasting black and white silk winds. Ferrules are nickel silver, chrome plated, hand welted, serrated and waterproofed. Packed in sateen bag and No. 280 aluminum case. Made in three actions and various lengths:

Rod	Action	Grip	Length	Weight	Line	Price
30	Wet Fly or Trout	6 in.	8½ ft.	5 oz.	E or HEH	$38.50
			9 ft.	5¼ oz.	D or HDH	
			9½ ft.	6 oz.	D or HDH	
31	Bass or Steelhead	7½ in.	8½ ft.	5½ oz.	D or HCH	$38.50
			9 ft.	6½ oz.	C or HCH	
			9½ ft.	7 oz.	C or GBG	
330	Dry Fly	6 in.	8½ ft.	5 oz.	D or HDH	$38.50
			9 ft.	5½ oz.	D or HDH	
			9½ ft.	6 oz.	D or HCH	

No. 119 Trout—No. 120 Bass or Steelhead—No. 319 Dry Fly

Carefully selected materials and highest quality workmanship characterize these double-built three-piece (with extra tip) split bamboo fly rods. Thread-locking reel seat made of yellow plastic with Colonial Black hoods. Shaped cork grip. Bamboo finished in natural color. Stainless Steel stripping guide, others tool-hardened steel and Perfection tip tops. Yellow and black silk winds. Ferrules of nickel silver, chrome plated, waterproofed, welted and serrated. Complete with No. 280 aluminum case. Made in three actions as follows:

Rod	Action	Grip	Length	Weight	Line	Price
119	Wet Fly or Trout	6 in.	8½ ft.	5 oz.	E or HEH	$33.00
			9 ft.	5¼ oz.	D or HDH	
			9½ ft.	6 oz.	D or HDH	
120	Bass or Steelhead	7½ in.	8½ ft.	5½ oz.	D or HCH	$33.00
			9 ft.	6½ oz.	C or HCH	
			9½ ft.	7 oz.	C or GBG	
319	Dry Fly	6 in.	8½ ft.	5¼ oz.	D or HDH	$33.00
			9 ft.	5½ oz.	D or HCH	

The 1942 line of South Bend Cross Craftsmanship rods. The 8½-foot No. 30 is the delicate rod, tossing a 4-weight line. Weighing 2 ounces more, the 9½-foot No. 31 is the powerhouse model, carrying an 8-weight.

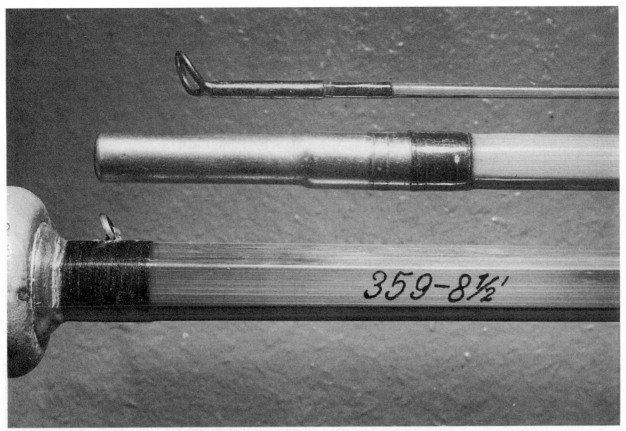

Although a lower-end South Bend model, the Model 359 has a crisp action, making it ideal for modern dry-fly fishing.

were dropped and model designations were changed. Also at this time, Wes Jordan left South Bend and returned to Lynn. In 1942 the company offered top-of-the-line Cross Double Built Rods in lengths from 8½ to 9½ feet in three separate actions: the Model 30 Wet Fly Or Trout, the Model 31 Bass Or Steelhead, and the Model 330 Dry Fly. These various models had standard salt-'n'-pepper wraps and "Lite-Lock" up-locking reel seats with an ivory plastic spacer and "Colonial Black" hoods. Some blanks were built with dark and light cane in alternating strips.

Three lesser postwar models with the same actions were the Model 119, the Model 120, and the Model 319. These rods sold for $5 less and had upchuck-yellow plastic reel-seat barrels. Both rod grades had serrated nickel-silver ferrules and came in a "sateen" bag and a "No. 280" aluminum case. The Cross Rod was discontinued around 1946–47. Today they're very uncommon and priced accordingly.

The modestly priced South Bend rods were built on into the 1950s. Like lesser Montagues, lower-end

South Bend had nickel-plated brass ferrules that would bend if an angler hooked a stump—or a large fish. The better models had German silver ferrules and dark flamed cane, but were cursed by a common problem of the new Plastic Era: tacky reel seats. All South Bend models and grades came with two tips in a bag and a fiber case.

ALMOST CRACKING THE SOUTH BEND CODE

The South Bend model code used in the 1940s and '50s isn't easy to follow, but a good place to start is the two- or three-digit number written on the butt of the fly rod's shaft. The number is followed by a dash, then the rod's length. Models 29 and 290 came in 7½-foot lengths only. The Model 23 grade came in lengths of 8, 8½, 9, and 9½ feet, while the other grades were available in 8½ to 9½ feet. Models with a dry-fly action used three-digit numbers.

The shaft markings of a postwar Model 359 are illustrated to help clarify all these wicked numbers. When shopping for a classic South Bend, determine whether you want a rod that tosses a 6-weight line, like the 359, or one that carries something a bit heav-

Model	No. of Pieces	Action	Construction	Price
30	Three	Trout	Double-built	$35.00
31	Three	Bass	Double-built	
330	Three	Dry Fly	Double-built	
12	Three	Trout	Double-built	$22.50
13	Three	Bass	Double-built	
312	Three	Dry Fly	Double-built	
23	Three	Trout	Single-built	$17.50
24	Three	Bass	Single-built	
323	Three	Dry Fly	Single-built	
46	Three	Trout	Single-built	$15.00
47	Three	Bass	Single-built	
346	Three	Dry Fly	Single-built	
59	Three	Bass	Single-built	$10.00
359	Three	Dry Fly	Single-built	
57C	Three	Bass	Single-built	$ 6.25
357	Three	Dry Fly	Single-built	
290	Two	Dry Fly	Single-built	$10.00
29	Two	Stiff Dry Fly	Single-built	

ier. The Model 59, slamming a 7- or 8-weight, would be good for windy days and bass or landlocked salmon. To aid in delicate presentations on the Little Lehigh, the Trout action grades, originally priced at $15 or higher, will toss a 5-weight. That's about as delicate as South Bend got. Now selling for modest prices on dealer listings, appropriate South Bend fly rods are excellent casting tools, giving hours of pleasure on the stream.

Most models found today are of the lower-cost, single-built style, originally selling for $7.50–12 in, for instance, 1941. That year, South Bend introduced its new patented "Comficient grip," an early form of what we now would call an ergonomic handle. It was available on three models. The following year, six models featured the new grip. Most 1942 models sported the Lite-Lock reel seat. During that year, prices ranged from $10 to $16, while double-built versions commanded $22. These early-1940s rods were available in three-piece models from 8½ to 9½ feet, in

Hunting & Fishing, *February 1942*

Bass, Trout, and Dry Fly actions. The South Bend Model 51, however, was a two-piecer that had lighter, honey-colored cane and gold wraps tipped with black. This appears to have been a better rod than most examples built during this era. This two-piece rod came in standard lengths plus a little 7½-footer.

THE MARVELOUS HEDDON RODS

This old Dowagiac, Michigan, company was founded in 1902 by James Heddon as a bass-lure manufactory. By 1910, Heddon was making a number of bait-casting rods and advertising "Dowagiac Baits and Rods." After James Heddon's death in 1911, the company was restructured, becoming known as James Heddon's Sons. The Heddon fly rod was born in 1924, and quickly became synonymous with quality. Today, the many Heddon grades rank alongside those fashioned by Goodwin Granger and the Edwards family as the best production classics ever built.

From the start Heddon wholesaled to the largest tackle houses in the biz, including Spalding (Chicago), Tryon (Philadelphia), L. L. Bean (Freeport, Maine), and Lyon & Coulson (Buffalo). These retailer-marked rods were very popular, often selling for more than comparable rods marketed under Heddon's label. At first Heddon produced only one model, the No. 35, selling, appropriately enough, for $35. In 1926 the No. 2 was introduced, selling at a corresponding price and called the "Bill Stanley's Favorite."[9]

Original Heddons had a German silver sliding-band seat. Soon, a few rods were fitted with a new screw-locking reel seat, patented on April 12, 1927, by Charles Heddon and Jack Welch. On most rods the seat barrel was cedar. In the 1930s the barrel became Bakelite; this was later replaced by plastic when that material came into vogue. In 1927 Heddon also introduced the No. 14, perhaps the best buy of all the Dowagiac fly rods.

By the 1930s the company was making specialized models for steelhead and bass. Heddon suggested that its Regular Bass rod (8½ to 9½ feet long) be used in conjunction with—what else?—Heddon Bass Bugs. The Power-Plus steelhead model, introduced in 1938, was built in lengths of 9 and 9½ feet

and weighed around 6¼ ounces. Heddon's largest early rod, the Extra Duty, was intended for salmon and built in the same lengths as the Power-Plus. It was not the same rod, however; it had larger ferrules and more backbone for hefty lines. On special request, Heddon made the salmon rod in a 10-foot length, which weighed over 7 ounces.

Early Trout models were built as 8-footers (5⅛ ounces), 8½-footers (5¼ ounces), and a larger 9-foot streamer model that weighed 5½ ounces. The darling of the line, and the most collectible of these earlier Heddons, was the diminutive Featherweight model, made only on special order. This "fairy wand" was "ideal for those who wish to get the most sport out of their fishing." We could use one of these rods today! The two-piece Featherweight was built in two sizes: the 7-foot No. 31, and the 7½-foot No. 51. Both weighed about 3½ ounces, and they sold for $35 and $50, respectively.

Beginning in 1939 the model designations, always inked on the shaft, included the butt ferrule size. You can determine the action of a three-piece Heddon rod at a glance using the sizes listed on page 196 from the 1938 Heddon catalog. In the 1940s Heddon started to produce many of the models that collectors now drool over. Introduced in 1948, the Model 1000 "Rod of Rods" was built by Heddon's master builder, Sam Anson. About 400 Model 1000s were made, each accented with gold wraps and plated-gold furnishings. Even the anodized aluminum rod tube was gold.[10] The gold theme later showed up in the Models 125 Expert and 115 Premier, which were sold through Sears, Roebuck. The next-highest-priced Heddons were the Models 60 Deluxe-President and the 35 Peerless, each built in lengths of 8 to 9 feet and weighing from 3½ to 5½ ounces. These Deluxe models were superb factory rods.

Perhaps the most popular Heddon rod, then and now, was the Model 17 Black Beauty. Built in lengths of from 7½ feet and up, the Model 17 had black ferrules, a black Pyralin reel seat, and black wraps tipped with orange. The Black Beauty is today a great casting rod in average used condition, and very collectible in excellent condition. A grade above the Black Beauty was the evolved Model 20 Bill Stanley's Favorite, which was built in the same lengths as well as in a scarce 7-foot, two-piece Featherweight ver-

Heddon Bamboo Fly Rods -

You will be proud to own one

Mr. Irvin S. Cobb, nationally known writer, says: "That Rod! The slickest, smartest rod I ever saw—a typical Heddon product."

A Heddon Fly Rod is a true fisherman's pet, his pride, his sweetheart. It is something to treasure. Seldom will he lend or sell it, and even after years of faithful service he still keeps it for the memories of finny battles won and lost. Just to grasp one and see its grace and strength, and feel its balance, live quick action and suppleness, makes you realize that Heddon Rods are the last word in Fly Rod construction, and truly "Rods with the Fighting Hearts."

Bass, etc., on Fly Rods is Thrilling Sport

Many anglers now realize the sport that can be had in fishing for bass with a fly rod—also for crappies, etc. Remarkable catches of large bass are made with Heddon Bass Bugs, etc., when fish will touch no other lure. Heddon "Bass Weight" and "Power-Plus" Fly Rods are world-famous for Power and Action. "Dry Fly" Rods are stiffer in butt and mid-section and have a quick-acting or "fast" tip. "Wet Fly" Rods are whippier throughout. As most anglers now prefer the "dry fly" action, this will be sent, unless "wet fly" is specified. "Dry Fly" Rods are suitable for both methods of fishing.

A New Special "Power-Plus" 6¼-oz. Rod

Steelhead fishermen and others have asked for a rod a little heavier than our "Bass" weight, around 6 ounces, and are much pleased with this new weight and its nice balance.

"Featherweight" Model — 7 and 7½ ft.

Made on special order, this "fairy wand" of a rod is ideal for those who wish to get the most sport out of their fishing. An ideal lady's rod. Weight about 3½ oz., 7 and 7½ ft. Two-piece construction and extra tip.

No. 31 — $35.00	No. 51 — $50.00

Five Types of Fly Rods

Five standard types are now made, the ferrule sizes of which are given below; also on special order a "Featherweight" Rod.

1. "Standard Trout" Fly Rod, No. 2 Ferrule.
2. "Regular Bass" Fly Rod or Dry Fly Trout Rod, No. 2½ Ferrule.
3. "Power-Plus" for extra service, Steelheads, etc., on 2¾ Ferrule.
4. "Extra Duty" Fly Rod for Salmon, etc., No. 3 Ferrule.
5. "Extra-light Trout" Rod, No. 100 Series, No. 1½ Ferrule.
6. "Featherweight," Nos. 31 and 51, No. 0½ Ferrule.

Length and Weight of Standard Fly Rods with Screw-Locking Reel-Seats.

Note: Weights will vary ⅛ oz. from that shown. Rods with plain wood reel-seats ½ oz. less.

Kind	8-Foot	8½-Foot	9-Foot	9½-Foot
"Trout", No. 2 Ferrule.........	5⅛ oz.	5¼ oz.	5½ oz.	Not Made
"Bass-Trout", No. 2½ Ferrule..	Not Made	5½ oz.	5¾ oz.	5⅞ oz.
"Power-Plus", No. 2¾ Ferrule.	Not Made	Not Made	6⅛ oz.	6¼ oz.
"Extra Duty", No. 3 Ferrule...	Not Made	Not Made	6⅜ oz.	6¾ oz.

10-ft. Rods in "Extra Duty" Weight made on Special Order

From the 1938 "Heddon Catalog"

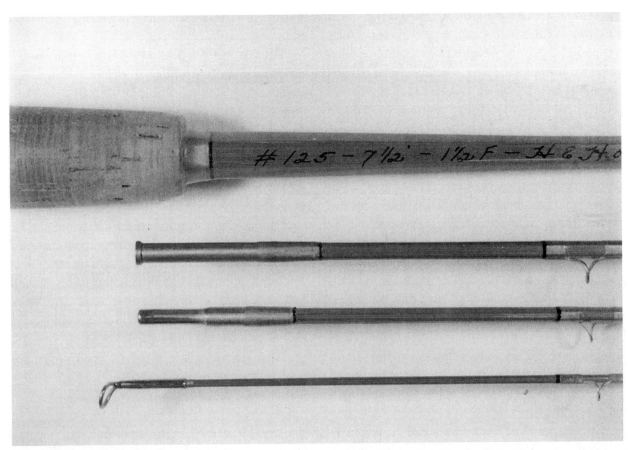

The Heddon No. 125 fly rod in a 7½-foot, 2½ ounce size. The hardware and wraps were colored bronze/gold. (The American Museum of Fly Fishing Collection)

sion. Stanley is best known as the New Age inventor of the weedless hook. I assume he fly fished for bass and used the little 7-footer on bluegills—another weed-loving species.

Throughout Heddon's rodmaking history its master craftsmen—a group that included Harold Drueding and James Hedges—built some fairly obscure models, including a saltwater version, the Model 19 Riptide, which had chrome-plated nickel-silver ferules and other marine-resistant features, and the rare Model 60 Deluxe Salmon Rod. But Heddon also built rods for the regular guy, such as its bottom-of-the-line Model 10 Blue Waters, which remains a good fishable stick. The Models 13 Lucky Angler and 14 Thorobred were a cut above the entry-level Heddon, giving the average angler a proper rod at value prices.

For production rods, Heddon offered a magical mixture of delicacy, power, and sheer eye appeal. From the top of the line to the cheaper grades, Heddon rods were built with a distinctive rich brown Tonkin bamboo, usually in three pieces with an extra tip. Heddon ceased split-bamboo rod construction in 1958. Recently, collectors have recognized the inherent value of these striking instruments and prices have skyrocketed. A Heddon collection might encompass the multitude of retailer-marked models bearing famous old names, including Folsom Arms, Lyon & Coulson, the Webber fly folks—even Paul Young.

THE PAUL H. YOUNG COMPANY, DETROIT, MICHIGAN

The fly rods of Paul Young probably have the greatest collector value of any true production rods. Their superb craftsmanship is equaled only by that of the very best American makers. But Young was an innovator and an expert caster as well as a great bamboo artist. He began making rods around 1924 or '25, when he was about 35 years old. Before that he had been a taxidermist, a fly tyer, and, most important, a complete angler. After a number of bouts with a variety of game, from trout to redfish, Young realized that a fly rod needed both delicacy and power. Using compound tapers, he designed fly rods that wedded these two attributes. By 1927, when he issued his first catalog from his shop on Detroit's Grand Avenue, Young listed "four models in sizes from 7½ to 9 feet in length," designed with this concept in mind and made from light-colored Tonkin cane.[11] These compound-tapered rods were the first real innovation in cane since the early days of Fred Thomas and associates. American fly fishermen fell in love with the Paul H. Young rod—and the price was right.

The demand for Young's product was overwhelming, and in the space of two years he became a major player in the production business. By 1929 the little Detroit shop couldn't keep up with orders, and Paul turned to Wes Jordan at South Bend and to Heddon for contract blanks built to the precise Young tapers; the blanks were finished at the Grand Avenue shop. As the business grew, the entire line of fly rods changed; Paul was searching for the elusive perfect rod.

In the 1930s Paul Young was one of the few makers to take the Depression in stride, introducing the optimistically named Prosperity Rod, a model that sold for around $10—perhaps one of the greatest buys that Granddad ever had. At this time Young also offered a number of more expensive PHY models, including the Ace ($22.50) and the Dry Fly Special ($42.50). The 1935 Paul H. Young Company catalog also carried rods with snob appeal, including models by Payne, Thomas, and Edwards, all selling for $50. Of course catalog items included flies, leaders, lines, and reels by Hardy, Pflueger, and Bristol-Meek. In

the late 1930s Paul's son Jack entered the rodmaking trade.

After World War II the Paul Young rods entered a new era, and the Midge, Perfectionist, and Parabolic series emerged. The Martha Marie rods also appeared at this time; these were named for Paul's wife, an avid and well-rounded fly caster who loved to fly fish in both salt water and fresh. In 1953 Bob Summers started working with Paul and Jack, and he remained with the Youngs until 1972, learning every aspect of rod construction and finally becoming a superb maker in his own right. During the postwar period, Jack developed new milling machines, and after his father's death in 1960 carried on business in the finest production facility in the trade. In 1969 Jack relocated to cosmopolitan Traverse City. Working with his son Todd, Jack currently makes a few rods a year.

Throughout its tenure, the Paul H. Young Company made its own reel seats and ferrules. In addition to finished rods it sold raw blanks and hardware. One rep, Gene Bullard, of Dallas, Texas, began selling shafts and furnishings to amateur makers. As far as I know, Gene was the first person to cater to the modern amateur rodbuilder trade. He offered Young blanks for less than $30. As a pioneer in the Young network, Gene was a caring and dedicated supplier, and long after he retired he'd phone me "just to see how things were going in Maine."

Postwar Young rod sections were glued with a phenolic resin that was impervious to water. The finished shafts were dipped in a preservative coating of Bakelite, making them ideal for rugged fresh- and saltwater angling without the heaviness that full impregnation would impart. Ferrules were serrated; those used on light models were made from aluminum, and on larger fly rods, from nickel-silver. Paul didn't like three-piece rods, because of the possibility of ferrule breakage, so his popular models were all two-piecers to lessen the necessary evil. He did offer four heavier models, 9 to 9½ feet, in the old three-piece configuration. They had no model designations.

The finished Paul Young products, representing a production of less than 6,000 rods, are today premier collectibles

Outdoor Life, April 1940

I couldn't afford to buy a Paul Young rod for an illustration, so I purchased this leader package instead. It's a very nice leader, made from Spanish gut, and it saved me $597.

and include a number of superb light-line models. The Midge, at 6 feet, 3 inches, weighs 1¼ ounces and handles a 5-weight line. The Driggs River is a 7-foot, 2-inch, 2⅞-ounce gem with plenty of tip power, taking a 6-weight line for short casts and a 5-weight for distance work. The Perfectionist is a 5-weight rod, a 7½-footer weighing 2⅔ ounces that's ideal for small dry flies and nymphing. Another 7½-footer, the Nymph Jr., rings in at 2⅞ ounces and takes a 5-weight line, either double tapered or weight forward. At the same length and, at 3⅓ ounces, just a bit heavier, the Martha Marie is an excellent rod for close quarters and pinpoint casts, taking a 5-weight for distances and a 6-weight close in.

Paul Young's larger models include the Standard Dry, Parabolic 15, 16, and 17 (these numbers indicate ferrule sizes in ⁄64-inch increments; the Parabolic 15 carries a ¹⁵⁄64-inch ferrule), Boat Rod, Texan, Texas General, Nymph Special, Bobby Doerr, Florida Special, and Powerhouse.

Perhaps the most exotic of all Young fly rods are those in the Parabolic series, which evolved from his original experiments with compound tapers. The rods work all the way down to the grip, but their slow pace masks considerable power. The Parabolic rods are often found with two different tips, one designed for a true parabolic action and the other with a lighter action for dry-fly work. The Para 15—at 8 feet and 3¾ ounces the smallest of the series—handles a double-tapered 6-weight line when used with its 4⅝⁄64 tip for general angling; with its ⅝⁄64 tip, a weight-forward line for bugs and streamers. The Para 16 is an 8½-foot, 4½-ounce rod for 7-weight lines. The Para 17, although the same length, weighs an additional ½ ounce and can slam an 8-weight bug taper to the backing. An amazing rod.

Today, Young rods are not cheap. Examples in excellent condition command four-figure prices; those in lesser condition still bring over $500. All later Young rods have flamed cane shafts and serial numbers and are well marked. These easily identifiable casting treasures were built until the late 1980s. The Paul Young influence still can be seen in the immaculate work of Bob Summers, who carries on the Traverse City tradition for a new generation of discriminating anglers.

GOODWIN GRANGER AND WRIGHT & MCGILL

The fly rods of Goodwin Granger have long appealed to collectors and anglers alike, but in recent years their maker has become something of an icon. The Goodwin Granger Rod Company was started in 1918 as a small rod shop in Denver, Colorado. It's been said that Granger developed his original and superb tapers by trial and error—and in virtual isolation.

The original rods did have a rather Leonard-esque look, however, with their medium-brown cane and German silver reel seats with soldered pockets and simple sliding bands. Granger rods always seemed to work "easy," and, while the longer sticks were probably designed for large, windy western rivers, easterners soon discovered that these rods made excellent upcountry streamer models. Frank Woolner, author of *Trout Hunting* and my mentor, was a devoted Granger addict who fished an original, prepatent 8½-footer and a later 8-footer dating from the Wright & McGill era. Unlike many outdoor writers, Woolner was noted for using utilitarian tackle. He had no use for "that fancy stuff."

Original pre-1925 Grangers, with the old-fashioned reel seat, carried at the top of the seat between the knurled rings, a grade and maker name such as the "Granger Special," the lowest grade. Other early rods, increasing in grade, included the Deluxe, Premier, and Tournament. From 1925 through '27 or early '28, the reel-seat stamping read "Made By Goodwin Granger & Co., Denver."[12]

Around 1924–25 Granger rods began appearing with the model name applied to the shaft by a gum-rubber stamp—unfortunately for collectors, as this faded over time. The inked stampings included, from lowest to highest grade, the Colorado Special, the Denver Special, the Goodwin Rod, and the Granger Rod. The reel-seat stamping used from 1928 through 1938 stated, "Made By Goodwin Granger Co., Denver." Usually the earliest rods had classic "old red" wraps, tan trims, and the full German silver sliding-band seat; some featured snake-guide strippers. I used one of these early Grangers, a dark straw-colored 9-footer that, when refinished, took a 6-weight line, and found it to be a great early-season

landlocked salmon rod. Unfortunately, not many of these early rods have survived.

Goodwin's untimely death in 1931 handed the Granger banner to Bill Phillipson, Granger's top employee. Phillipson developed the various later models, the elliptical hammer handle used on the two-piece rods, and the famous 1938-patented uplocking nickel-silver reel seat. The seat showed up around 1934, first stamped with "Pat. Pend." and, later, "Pat. Apr. 12, 1938," along with the patent number.

Later Grangers were built in eight grades, listed here from the best to the most modest: the Registered, Premier, Deluxe, Favorite, Aristocrat, Special, Victory, and Champion. Through the history of the Goodwin Granger Rod Company, at least nine models were built, ranging from 7 to 10 feet in length. Finding a rod shorter than 8½ feet or longer than 9 feet isn't easy. Models can be identified by the original hanging tag (often missing), or by a rod's length and weight. A two-piece 7-footer weighing 3 ounces, for instance, would be a No. 7030 (i.e., 7´0˝, 3.0 ounces). Models longer than 7 feet were all standard three-piece rods.

A circa-1940 Granger flyer recommended old-style (HEH, HCH) double-tapered line sizes for its various models. The following models are matched to our modern fly-line nomenclature, developed in the 1960s by the late Myron Gregory. The No. 7633 (7´6˝, 3¾ ounces) tosses a 5-weight line. No. 8040 and No. 8642 also throw a 5-weight. Nos. 9043 and 9050 take a 6-weight. Nos. 9053 and 9660 carry a 7-weight, and the No. 1062 takes a steelheader's 8-weight line.

You can determine an appropriate line weight by actually weighing your Granger on an accurate scale, such as the one your postmaster uses. Each model was supposedly available within all the Granger grades, from Registered on down to the Champion. Champions shorter than 8½ feet are scarce buggers, however.

In 1946, Wright & McGill, also of Denver, made arrangements to lease the Goodwin Granger factory; five years later they bought it. For a time, Phillipson remained as

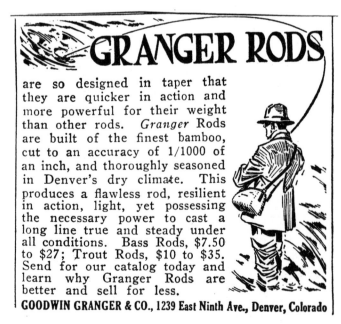

One of Granger's earliest ads, this appeared in the May 1920 issue of Field & Stream. Granger *began extolling the virtues of a "dry climate" early on.*

This 8½-foot, early prepatent model, marked "Champion" on the shaft, was Goodwin Granger's personal fly rod. (The American Museum of Fly Fishing Collection)

Granger lineup was priced as follows: Victory, $25; Special, $30; Aristocrat, $40; Favorite, $50; Deluxe, $60; Premier, $75; and Registered, $100.

W&M got a lot of miles out of Goodwin Granger's patent reel seat; it was fitted to glass rods until the early 1970s. In 1951, Wright & McGill began to simplify the wraps on Granger rods, eliminating some tipping and signature wraps. A year later the company introduced the impregnated Water Seal rods. These had no design or trademark relation to the Grangers other than the patent reel seat. Water Seals were built in two grades; the better model was the FA. Rods so designated had gold-tipped burgundy nylon wraps, and were put up in a burgundy bag and aluminum tube. The slightly lesser model, the Water Seal FB, had the same-colored wraps but was packed in a gold sack and anodized tube. The rods were labeled "Wright & McGill, Water Seal Hand-Made Super Rod," on the lower shaft. The Water Seals were built as late as 1960, according to rod historian Mike Sinclair.

After its starring role in *A River Runs Through It*, the original Goodwin Granger rod has taken a big upward swing in popularity within the last few years—which is only fitting. It was and is a pleasurable casting rod. Wright & McGill Grangers fall close behind, not only as collectibles but as good working rods. The Water Seals are an anomaly. Although the most recently made, they seldom show up on dealer lists or at auction. That's something to think about. They may well be a good long-term investment.

BILL PHILLIPSON RODS

William Phillipson came to the United States from Sweden in 1923 and a few years later went to work for Goodwin Granger. Phillipson became one of the finest rodmakers in his adopted country, first as the man who successfully redesigned the Granger rod and much of the company's production machinery; second as the head of Wright & McGill's production; and, finally, as the owner of his own

the master rodmaker, and the quality of the Granger rods remained high. These products were ink-stamped with "Wright & McGill" and the grade level; for instance, "Granger Victory." One of W&M's first actions was to change the reel-seat stamping to "Made by Wright & McGill Rod Co., Denver, Colo., U.S.A." A year later, Wright & McGill replaced the Champion grade with the Stream & Lake, which featured red nylon wraps.

After World War II, Bill Phillipson left Wright & McGill to start his own company, but the Granger models remained remarkably unchanged until production ended sometime in the 1950s. The later

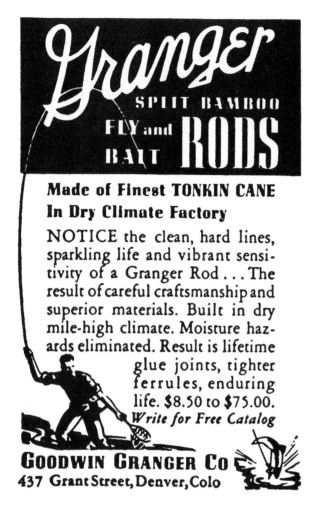
Outdoor Life, *April 1940*

excellent rod company. A pioneer in modern rod-making, Bill was quick to adopt new technologies, and was probably the first to construct blanks using the new resorcinol glue, and to wrap rods with nylon thread. His highest-grade rods, however, carried the traditional silk wraps.

The Phillipson Rod Company began operation in Denver in 1946, producing around 7,000 rods annually—about the same schedule Bill had followed at Granger and Wright & McGill. Although later Phillipson rods were impregnated, those built during the first decade received their browntone color from a chemical wash, probably ammonia, similar to the Granger and Montague processes. Phillipson's newly designed milling machine produced a rod section every 60 seconds, with a minimum of human labor, and was so precise that the strips

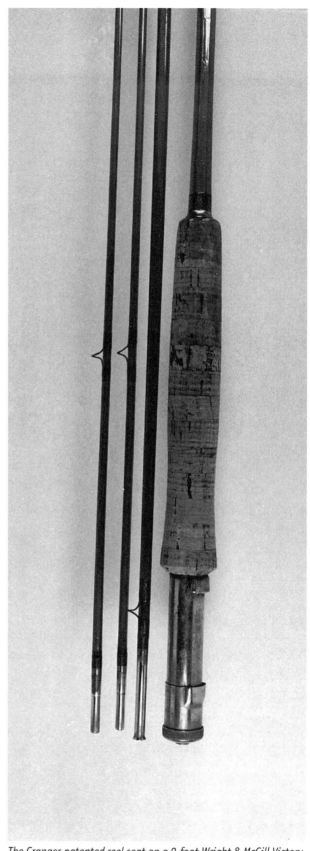

The Granger-patented reel seat on a 9-foot Wright & McGill Victory. Owned by Frank Woolner, author of Trout Hunting, *it saw service in Massachusetts and Maine.*

needed only slight trimming before being glued up. The finished blanks were fitted with tapered nickel-silver ferrules and quality down-locking reel seats made of plastic and aluminum

The inherent castability, frugal price, and honest value of a Phillipson rod just shone, and Bill was proud to produce a production rod that handled like the labor-intensive models built by the old masters. He also felt that his local high-and-dry climate was ideal for bamboo construction, keeping excess moisture from the culms and stirps until they were finally glued up. As Bill once said, "In Denver, the year-round average relative humidity registers only 39 percent, compared with 78 percent registered on or near the seacoast." To stabilize the Tonkin, Phillipson developed a process that provided the equivalent of 20 years' worth of seasoning, and all manufacturing steps were carried out in a controlled environment.

Regardless of model, the same grade of bamboo was used in all the rods, from the cheapest to the most expensive. In addition to a number of "value" rods, often private-labeled by dealers across the country, Phillipson sold a variety of original cured-cane models under his own name. The original 1946 line included, in a general order of modest to pricey: the Pacemaker, $25; Pow'r Pakt, $37.50; Paramount, $50; and Premium, $75. The Paragon, an economy model selling for $20, was introduced in 1949. Three years later the Smuggler ($25), a pack rod, was added to the Phillipson line.

The Phillipson Rod Company's biggest seller was the Pacemaker, which, though modestly priced, was a prized casting instrument. It was built in lengths of from 7½ to 9 feet, and originally had lime wraps tipped yellow; these were chanted to a brighter green with black tipping around 1952. Pacemakers and other modestly priced models had a black-threaded plastic reel seat. Like the entire Phillipson line, Pacemakers are easily identified today by the maker's name, model, length, and appropriate line weight—all written on the shaft just above the winding check. The Phillipson winding check, by the way, was similar to—although a tad smaller than—the Granger check. This check style, in ever-smaller sizes, can be traced back to Edwards and the old partnership of Thomas & Edwards.

In 1953 Bill started to make his famous impregnated Peerless rods. The modern resorcinol glue, readily identifiable by its purple color, kept the Bakelite resin from entering the blank's interior pores; the impregnated rods weighed only slightly more than the varnished versions. The Peerless, perhaps today's "fishingest" Phillipson, bore the hammer-handle grip and came in the same weights and lengths as the Pacemaker. The 7½-footer weighed 4 ounces; the 8-footer, 4½ ounces; the 8½-footer, 5¼ ounces; and the 9-foot rod, 6 ounces. In 1972 a modification of the

Genuine split jar cork grip cut with grain to keep moisture from entering pores. Shaped to give 100% natural contact with fingers.

Hard drawn solid nickel silver ferrules with inserted moisture seal. Each rod individually hand fitted.

Nylon guide wrap-plugs—stronger than silk.

Scientifically tapered on precision machines accurate within 1/1000 inch.

Phillipson Action — gives light delivery and controlled line curve.

Phillipson FISHING RODS ARE MANUFACTURED IN COLORADO'S DRY CLIMATE

Colorado's dry climate is ideal for working with moisture-sensitive bamboo. That's because the moisture content of bamboo has a very definite bearing on the resiliency, spring, and strength of the fibrous cellular structure of this choice INDO CHINA imported Tonkin cane. In Denver, the year 'round average relative humidity registers only 39 percent, compared with 78 percent registered on or near the sea coast. The moisture content of bamboo stored in Denver soon changes to harmonize with the natural climate, and reaches its highest point of efficiency. This moisture content never changes while the imported Tonkin cane goes through a special curing process which is equivalent to 20 years of natural seasoning. The moisture content stays the same while the cane is being split, straightened, glued into sections, and finished. That is why the ferrules of your Phillipson rod will not work loose. And that is one reason why your Phillipson rod will hold its shape, and retain its original action and backbone even after years of hard service.

Screw-type locking reel seat with metal butt cap fused between. plastic—holds cap forever. Pat. Pend.

What's new in Denver. How about making rods in Colorado's dry climate? Naww! Then how about Bill Phillipson sneaking cane out through Red China's back door?—through Cambodia or Vietnam, then called Indo China.

Peerless was introduced, the Dry Fly Special. The Special took a line one size heavier than did the corresponding length of Peerless. Peerless models built after 1963 had a torpedo grip and a large anodized winding check.

Bill Phillipson's finest rods were the Premium and Paramount models, which were built lighter than the bulk of his work and had the hammer handle. The rods were fitted with aluminum threaded reel seats that had a plastic spacer. Premium models were wrapped with black silk tipped gold and black. Paramounts had salt-'n'-pepper silk wraps tipped gold and black. For fans of the two-piece light rod there was the Preferred, which came in either a 7-foot or a 7½-foot length and tossed a 4-weight line. Starting in the 1960s Phillipson began making his excellent fiberglass rods, which are now gaining collectible status. In 1972 Bill sold the Phillipson Rod Company to 3M Corporation. I believe that the last impregnated bamboo Peerless models were built until 1974, when the veteran rodmaker left the factory for a much-needed and permanent angling vacation.

OBSCURE HEARTLAND RODMAKERS

The "big boys"—Heddon, Granger, and Phillipson—were not the only Middle Americans to produce the long wand. A number of small-timers should also be noted, just in case you run across one of their signed products. In several cases, these lesser-known makers actually started their trade a decade or more prior to the larger companies.

THE G. E. WHALING & SON COMPANY, CLEVELAND, OHIO

A maker of handcrafted fly rods, G. E. "Dad" Whaling built his models on the matched-cane principle, as advertised in a 1916 edition of the *Sportsman's Handbook*, "Each rod is made entirely from the same cane [culm]. You can readily understand what this means when no two canes have the same strength and flexibility. For a rod made up of strips from different cane is certain to take that dreaded set and hold it." This early reference may actually predate a construction style made famous by the Edwards family.

From his small shop at 800 Champlain Avenue in Cleveland, G. E. Whaling produced rods that were hand split, heat-straightened, and "finished to rigid specifications" in a classic six-strip style. The maker seemed intensely proud of his product and abhorred the common factory-made rods of the period. As he said, "Do you want one of 'them things' ground out of a sausage machine by the thousand, or a master rod made by master workmen?" Hmm, gee, I think I want a Whaling Good Rod.

H. B. CLARK, BELLAIRE, MICHIGAN

It's interesting to note, in H. B. Clark's ad in the April 1916 issue of the *National Sportsman*, that he made only one style of fly rod. Talk about specialization. Evidently, H. B. Clark's "dry fly rods" were built from hand-planed six-strip bamboo and finished with quality German silver waterproof ferrules, and agate strippers and tops. From the basic Clark info, we know he tried for a crisp dry-fly action—not easy to achieve with a heavy agate tiptop.

OLD HICKORY ROD & TACKLE COMPANY, BANGOR, MICHIGAN

If you should ever run across an antiquated fly rod marked "Old Hickory," you have this Bangor company and its owner, Mr. Kinney, to blame. Not only did it make rods outdated before their time, but it lived in the wrong Bangor. There's nothing better than a good hickory tuna rod, but let's face it: In fly-fishing circles, 1923 was not a big year for wood. And for the advertised price of $50, an angler could buy a nice Thomas, Leonard, or Payne. Kinney's ad read, "Rods for all kinds of fishing." It takes all kinds.

"WHALING GOOD" RODS are made in all wanted styles but only one quality, the best that can be made. When a better rod is built—Dad Whaling will make it.

The Outdoorsman's Handbook, W. H. Miller, 1916

National Sportsman, *April 1916*

And so ends a chapter in fly-fishing history. In time, the great Middle American tackle companies such as South Bend, Heddon, and Phillipson turned to fiberglass, and small firms like Whaling, Clark, and Kinney fell by the wayside. Split bamboo was in danger of falling by the wayside, too, just as had ash and lancewood of the previous century—and Kinney's 1920s hickory. Only a handful of dedicated American craftsmen would be left to carry on the cane tradition. Quality reels, like handcrafted rods, however, would be available as long as fly fishermen were willing to pay for them.

National Sportsman, *April 1923*

Notes

[1, 3, 4, 5]Vernon, Steven K., and Frank M. Stewart III. *Fishing Reel Makers of Kentucky.* Plano, Texas: Thos. B. Reel Company, 1992.

[2]Shields, George O. (ed.). *American Game Fishes.* Chicago: Rand McNally, 1892.

[6]Kelly, Mary Kefover. *U.S. Fishing Rod Patents and Other Tackle.* Plano, Texas: Thos. B. Reel Company, 1990.

[7, 9, 12]Sinclair, Michael. *Bamboo Rod Restoration Handbook.* Grand Junction, Colo.: Centennial, 1994.

[8]Spurr, Dick, and Gloria Jordan. *Wes Jordan: Profile of a Rodmaker.* Grand Junction, Colo.: Centennial, 1992.

[10]Keane, Martin J. *Classic Rods and Rodmakers.* New York: Winchester Press, 1976.

[11]Spurr, Dick. Introduction to reprint of *More Fishing—Less Fussing, 1956 Paul H. Young Catalog.* Grand Junction, Colo.: Centennial, undated.

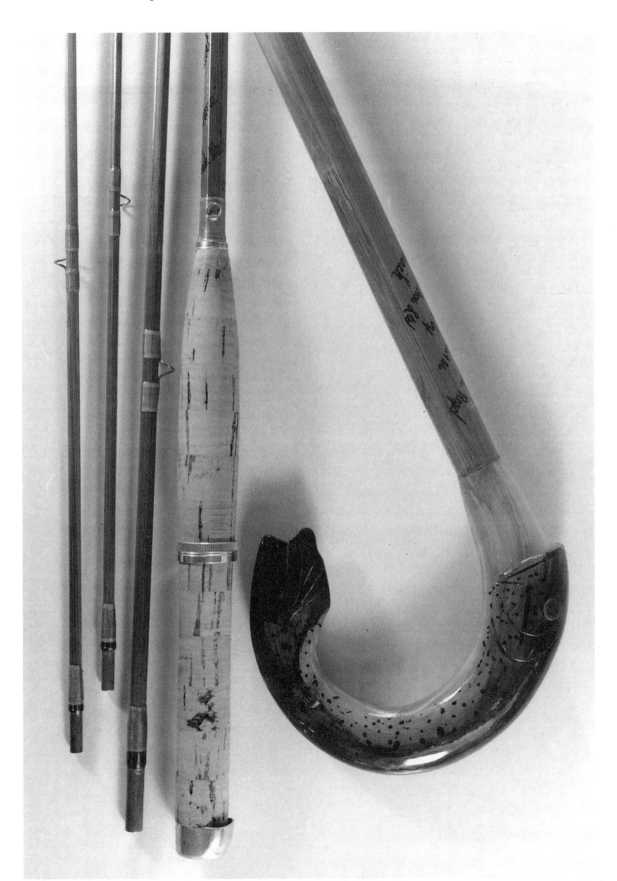

A beautiful little 7½-foot Hardy Marvel fly rod. This 2¼-ounce wand is complemented by a rare Hardy Brothers split-Palakona bamboo walking cane. (The McGrath Collection)

11 English Fly Tackle

We left the British tackle scene in chapter 2 during the American Civil War. At that time reels were plain single-action brass jobbies and the rods had just moved into a four-strip style. Most British tackle exported to the United States came from London makers such as Ustonson, Bowness (with sometime partner Chavalier), and the Farlows. After the war, two large companies from Redditch and Alnwick began exporting to "the colonies": Samuel Allcock & Company and the Hardy Brothers. Their early six-strip rods, newer brass stop-check and revolving-plate reels, flies, and myriads of accessories remain accessible and—sometimes—affordable collectibles.

English tackle of the late 1800s, the result of an entertaining marriage between old thinking and new manufacturing ideas, had a character quite unlike the products of the New World. This quaint charm was exemplified by Farlow and early Hardy rods, with their patented twist-locking ferrules—the Empire's way of overcoming an inherent distrust of friction—and their unusual "American Pattern" stop-latch reels. These odd products of the Empire are still modestly priced on the tackle market.

Anglo tackle found its way into American hands through the major importers of Thomas H. Bate & Company (later as William Mills & Son), Andrew Clerk & Company (later as Abbey & Imbrie), and Bradford & Anthony (later as the Dame-Stoddard-Kendall progression). Most of the products were "trade" items, unmarked by their makers. Reels in particular were often marked by the importer, leaving the actual maker unknown. After the turn of the

century, most English items were marked by their maker, and collecting becomes easier.

For two centuries a steady stream of rods, reels, and natty little gadgets flowed from the Isles to satisfy the whims of Anglophile anglers in the new world. Today, almost any tackle auction displays a substantial quantity of British tackle.

This chapter makes no attempt to chronicle the long list of English makers and their products, but concentrates instead on export tackle made mostly between the 1880s and 1960s, giving production years based on Graham Turner's *Fishing Tackle: A Collector's Guide*.[1] High on the list are Hardy reels, which have captivated anglers around the world, but other makers such as Allcock and Farlow played important roles as well.

A fair amount of imported English salmon tackle found its way into Maine and points south by crossing the Canadian border. At the same time English items imported by American tackle houses often traveled north to the Miramichi and other salmon rivers. This century-old interaction is a boon to the growing number of collectors specializing in Empire tackle. Not all tackle was destined for Atlantic salmon. Many English trout rods and reels were shipped to the States and eventually showed up on rivers as far west as Oregon and Washington.

EARLY ENGLISH IMPORTS

A surprising amount of lowland tackle arrived in American hands between the late 1700s and 1850, when our own rod-and-reel industry was still in its formative years. Makers such as Haywood and Ustonson exported the earliest examples found in America.

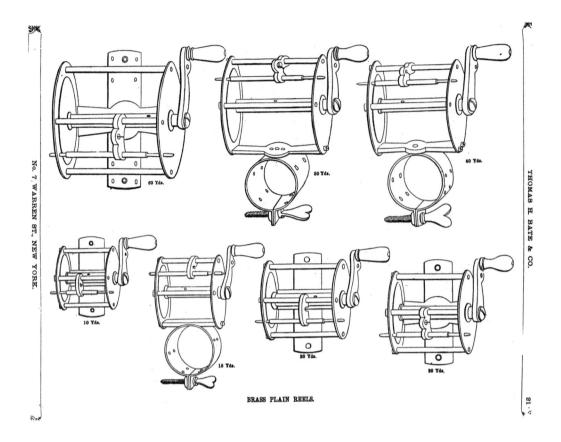

Allcock stop-latch fly reels, as illustrated in an 1867 Thomas H. Bate tackle catalog.

Ustonson reels are the most coveted examples of the smith age. *Fishing Collectibles Magazine* publisher Brian McGrath, who has fondled some of the rarest tackle built on either side of the pond, considers an Ustonson reel one of the two ultimate collectibles, the other being a Phillippe rod.

The list of early British "toughies" would also include the odd little Archimedian Reel and a number of wonderfully made folding-handle reels built in London. The wooden Nottingham reel, which appeared at midcentury, is a rare find in excellent condition. From the late 1700s until at least 1867, a small quantity of British multiplying fly reels landed on American shores. They were carried by T. H. Bate and other New York houses, and their market probably extended into northern New England via Bradford's. The old Boston tackle establishment imported a good quantity of tackle from Great Britain, including some rare rods built by William Henry Alfred & Son of 54 Moorgate Street, London. These were "inside-out" split-bamboo models, made circa 1856–58.

THE FABULOUS ARCHIMEDIAN REEL

The rare Archimedian Reel came from the inventive mind of Frederick Skinner of Sheffield. Skinner "registered" (patented) his design on April 25, 1848, making it the earliest known patent fly reel in existence—built some 11 years prior to Billinghurst's first American example. The Archimedian Reel may also have been the first winch with a variable tension device. This early form of drag featured a tempered wire that could be pressed against the rim of the spool by turning an adjustable knurled knob.

The drag system was not the only historic feature of the reel's design. Skinner also built the winch as a side-mounter and gave the spool (or "drum") a series of round ventilating holes—putting to rest the notion that these two design innovations were first addressed by Morgan James (side mounting) and Charles F. Orvis (ventilation holes). In one fell swoop, our Mr. Skinner produced a reel so adventurous that it was never duplicated. The Archimedian Reel falls second only to the Ustonson reel in rarity. Few collectors would pass it up while awaiting the rarest of the rare.

An innovative Archimedian Reel exhibiting Skinner's early drag and knob. Even though the spool-shaft nut isn't original, this antique remains provocative.

WOODEN NOTTINGHAM REELS

The Nottingham reel may not be as old as Friar Tuck, but it traveled along the same roads. There's disagreement about the first maker of these "centre-pin" wooden reels—whether Joseph Turner of Pomfret Street (according to Henry Coxon, 1895)[2] or Samuel Lowkes at Upper Parliament Street (as claimed by A. Courtney Williams, 1945).[3] Other makers included John Bickerdyke, who developed the Nottingham's wire line guard, and William Brailsford. Nottingham reels were first produced just prior to 1850 and were built well into the 1900s.

This charming wooden winch, with its brass reel foot and back strap and traditional twin handles, was not exclusively the tool of fly fishers. In fact, it was most popular with bait anglers. For example, Brailsford wholesaled much of his original production to Walter Wells, who owned the Nottingham Worm Farm. Another chap, George Holland, took a Brailsford reel to London, where it was a big hit, and whence it eventually traveled the Atlantic to America.

Through the decades the Nottingham was improved. In 1885 a Sussex Street dealer named West designed the first silent check. David Slater, who became the dean of centre-pin reelmakers, built advanced models with hard-rubber back plates and a patented (1883) line guard. Most of the Nottingham reels were unmarked, but a few were stamped by the maker, or seller, or both. Recently, I examined a 2⅞-inch-diameter model that at first seemed unmarked. Under magnification, however, the underside of the foot showed individual letter stampings: "Richardson * 77 * Finsbury Pavement." Henry Richardson, a London tackle dealer at 77 Finsbury Pavement from 1885 to 1890.

The centre-pin reel looks good in any antique reel grouping and is a favorite collectible. It had its faults, though. The wooden back often cracked at the screw areas, the varnish wore off, and the nonexistent bearings needed constant greasing to keep going. Like any other collectible, a Nottingham in excellent condition is a killer, while those in lesser shape can drop all the way down to the curiosity level. The centre-pin Nottingham was perhaps the original "palming" reel. Larger sizes could take a fly line and backing and could handle hyperthyroid game like sea trout.

The wooden Nottingham reel, introduced around 1850, had a long production life, taking it out of the rare category and keeping current prices reasonable. Note the cracked rear plate, invisible on a "wall hanger."

FOLDING-HANDLE REELS

Another prestigious early English style is the folding-handle fly reel. A number of artisans made these, but the identity of the original designer is unknown. A variation fashioned by Ainge & Aldred had the earliest 126 Oxford Street address, indicating that the folding crank was developed before 1856. Some makers built this style of winch as a multipier, with a gear-box cover on the right side plate. All had a locking disk on the crank handle just above the folding ivory knob. William Gowland, Giles Little, and James Jones, all active from the 1850s until 1890, made single-action fly reels with a similar locking, folding handle.

The off-and-on partnership of Chavalier and Bowness produced a folding-handle reel with a slightly different folding mechanism: The handle could swing back to lock into a notch on the side-plate rim. All folding-handle models appear to have originated from London makers, and have today a better-than-average collectibility. At least one other English winch falls into the "glorious antique" category: the stop-latch reel.

ALLCOCK'S RODS AND REELS

Many years ago, when I first started collecting tackle, I stopped at Dan Winchenbach's place in Waldoboro, Maine. Dan was one of the original "antique squirrels," and the outbuildings of his sagging farm were stuffed with worthless gewgaws sprinkled with just enough good items to be worth investigating. That's where I discovered, dangling from a twisted coat hanger, six English brass reels that

A folding-crank reel built by "Chavalier Bowness & Son"—a scripted marking used from 1866 to 1882. This exquisite all-brass model has an ivory knob and appears to be unused. (The American Museum of Fly Fishing Collection)

Still advertising in Forest and Stream *in the late 1880s, Giles Little was a pioneering split-bamboo craftsman and one of the known makers of the folding-handle fly reel.*

Dan had found in the basement of a long-gone Augusta hardware store. The reels had never been used and still carried their original lacquer coating.

Four were marked by Abbey & Imbrie, and, through the aid of Mary Kelly,[4] I established that they had a circa-1878–83 stamping. The unique sliding bar that popped into a hole on the single-crank handle told me I'd found some rare "stop-latch" reels—and in excellent condition. A great find.

The maker's origin and production period of this style of stop-latch reel remained a mystery, although reel historian Jim Brown found identical models in an 1867 Thomas H. Bate & Company catalog. Described as "brass plain reels," they were sold in sizes from 10 to 50 yards. The English source for these reels finally appeared in an 1887 Samuel Allcock & Company catalog, where they were described as "American Pattern, Plain with Stop."

Marked with the original Abbey & Imbrie logo, this nickel-plated Allcock stop-latch reel has an atypical hard-rubber knob, made in quantity in New York and shipped to Allcock for assembly. The nickel plating was an 1880s option, costing an extra 25 cents.

I consider the Allcock stop-latch reel one of the most "English" of tackle items imported during the post–Civil War period. Allcock produced it from approximately 1865 until 1887 or perhaps a little later. Soon, revolving-plate models and even a large raised-pillar salmon reel joined the stop-latch in the Allcock lineup. Perhaps the most popular of the Redditch

products, the revolving-plate reels are now often found entirely unmarked, and many were copied by American makers such as Julius vom Hofe and Andrew Hendryx.

Polycarp Allcock founded the company in 1803 to manufacture hooks. Through the decades, Allcock went into reelmaking, and by the late 1870s was making split-bamboo rods. In 1881 these were billed as "Hexagonal Built or Split Cane." Many Allcock rods, in particular the company's two-handed salmon behemoths, were exported. In the 1880s Samuel Allcock's operation was one of the largest in the tackle biz, with 400 workers stamping out reels and grinding out rods at a feverish pace. This has ensured the survival of a lot of Allcock tackle, although rods are tough to find in shorter lengths and in good or better condition. For this reason it's the winches, in brass and "ebonite" (hard rubber), that have garnered the most collector attention.

CHARLES FARLOW & COMPANY

Founded by Charles Farlow in 1840, this historic London tackle business, one of many located on the Strand, exported a substantial amount of tackle to North America. Its rods carried the common locking ferrule, similar to a design patented by Hardy. In fact, the rods can be mistaken for Hardys at a distance,

While this revolving-plate reel is marked "S. Allcock & Co., Ltd., Makers," many identical versions were not. A stamping adds a great deal of value to any piece of fly tackle.

having similar blued furnishings and uncountable intermediate wraps along the shaft—just in case the glue let go during a match with a fresh-run salmon.

Built from the 1880s until well after the turn of the century, the bayonet-ferruled Farlow rod was testimony to the company's absolute distrust of cane, glue, and friction-fit ferrules. With all their multiple wraps, massive ferrules, and lengths extending to 17 feet or more, these rods were some of the heaviest ever built. Along with heavy rods, Farlow also exported reels, accessories, and flies, particularly salmon flies. Collectors often run into Farlow reels, from the early single-action brass winches, to all-brass or brass-ebonite revolving-plate models, and finally to modern aluminum styles. Those products, marked by the maker, can be roughly dated: "C. F. 221, The Strand" (pre-1852); "Chas. Farlow Maker, 191 The Strand, London" (1852–84); "Chas. Farlow & Co., 191 The Strand, London" (1885–94); "C. Farlow & Co., Ltd.," with the same address (1895–1906); "C. Farlow & Co., Ltd., 10 Charles St., St. James Sq. SW and 191 The Strand, London" (1907–15); and the "10 Charles St." address (1916–25).

Charles Farlow & Company, Ltd. built a new, larger factory in Croydon in 1911, and later products (1926–57) had a Panton Street address. In 1957, the last controlling member of the family sold the business to the Westley Richards Gun Company, and reels built from 1958 to 1964 were marked "Burton Street." The ownership was then transferred to the Drayton Group and J. S. Sharpe of Aberdeen. Farlow is still selling tackle at Pall Mall, near Trafalgar Square.

The most sought-after Farlow reels are the all-brass models with the original Strand address, built with ivory single-crank handles and, sometimes, a raised rim. Later reels, many designed during Charles F. Farlow's tenure during the teens through the 1930s, included the Aluminum Salmon and Trout and Kelson models, which now are not only collectible but also very usable. Perhaps the most sought-after modern Farlow rod is the Elf, a little two-piece rod for a 5-weight line, which weighed 3¾ ounces and was 6 feet, 10 inches long. Another tiny brook rod, the Ultimate, was a one-piece, 5-foot, 10-inch rod that weighed a mere 1½ ounces. Designed by Lee Wulff and Alan Sharpe, the Ultimate could handle a 6-weight line and was quite popular in America. The Lee Wulff Ultimate also had a sister rod, the two-piece, 6-foot, 2-ounce Wulff Midge.

FOSTER'S STEEL RIBBED ROD

Founded in 1833 by David Foster, this old Ashbourne, Derbyshire, company, Foster Bros., was greatly expanded by the second generation. The Foster brothers, David and William H., came to the fore in the 1880s, first with their Steel Centered Rod and,

A circa-1880 Charles Farlow fly rod. Its loose-fitting blued furnishings were wrapped with wire loops. The loops were lashed together with thread to keep the rod in one piece. These were the worst ferrules ever produced, and many other English makers shared the blame.

later, their Steel Ribbed Rod. By 1895 this wicked derrick was "IN GENERAL USE THROUGHOUT THE CIVILIZED WORLD." One Foster photograph showed two mechanical engineers applying successively heavier ingots to one end of a Steel Ribbed Rod section secured in a massive vise. The Fosters even elicited testimony from the clergy: "The rod you have ribbed is very much improved" —the Reverend G. R. Peak, Bloxwich, near Walsall.

The epitome of steel ribbing was the rubber-handled Little David, built in lengths of 9½ to 11 feet. This patented (1892) design remained popular as long as the average British fly fisher remained convinced that heavy was heavenly. As far as I know, the Foster Brothers built rods only. Other items in their catalogs included "upright wing flies," Malloch's reels, and silk lines. A few Foster-marked items surely will surface in the United States, and collectors should keep an eye out for the rod that challenged the world.

NEW STEEL CENTERED FLYROD

The Fosters' Steel Centered Rod, from Shields (1892), was the immediate forerunner of the infamous Steel Ribbed version. Note the built-in reel.

J. GILLETT & SON

John Gillett began a rodmaking and retail career in 1865, first at 115 Fetter Lane, London. In 1876 he moved down the street to 40 Fetter Lane, where he and his son finished their tenure in the trade. Most marked tackle—a star-backed Nottingham reel, for example—had the latter address. After 1906, the firm became J. Gillett & Son, and, judging from the accompanying 1911 ad, it was into split bamboo. I have no idea of the origin of Gillett's great claim to seniority: "Established over 300 years." It sounds impressive, though.

HARDY BROTHERS

Originally a gun and cutlery maker, the Alnwick firm of Hardy Brothers was founded in 1872 by William Hardy, who was joined shortly thereafter by his brother, John James. Within a decade the company began producing quality fishing tackle. Most notable of these early products were the Gold Medal fly rod, introduced in 1885, and the Perfect fly reel, which showed up five years later. Additional models proliferated into a lineup that made Hardy Brothers the most popular tackle firm in the British Isles.

I owned a fine old Hardy Brothers mortised, swelled-butt rod for a number of years. A powerful 11-foot sea trout model that had probably been a grilse rod in Maine, it had "Lockfast" ferrules and a "Wee-

J. GILLETT & SON,

40, FETTER LANE,
—LONDON, E.C.—

The Oldest House in the United Kingdom.
Established over 300 years.

GILLETT'S CELEBRATED

SPLIT CANE Lock Joints, Cork Grip, &c.,

TROUT RODS 9 ft. to 11 ft., **42/-**

SALMON and TROUT RODS built to order.

"For general excellence I do not think there is a better rod on the market."—H. T. SHERINGHAM, Angling Editor of the *Field*.

Telephone: No. 5356 Holborn.

The Salmon & Trout, September 1911

jer" reel fitting, both of which were patented in 1881, as well as loose-ring guides and a serial number of 4014. Built around 1881–83, the rod was a beautiful example of English craftsmanship and still in very good condition, yet it was listed in successive editions of my catalog before it sold.

Even much newer Hardy fly rods, such as the Palakona and Deluxe Palakona series, are generally underappreciated in the United States. Astute collectors recognize a few rarer models, such as the little Fairchild and the Casting Club de France. All collectors, however, know well the incredibly popular Hardy reels. I have seen exotic specimens bring startling sums at national auctions. An American collector offered £5,000 sterling for an original pre-1891 Perfect. Other coveted items include the various gaffs, nets, tools, and myriads of gewgaws that the company has produced over the past 120 years.

Hardy rods changed appreciably in 1911, when the company introduced the ubiquitous reverse-locking reel seat, which set new standards for practicality and good looks. That same year Hardy introduced the Deluxe and Casting Club de France models, the latter probably designed by or at the insistence of John James Hardy. These delicate Casting Club rods are some of the finest British rods ever produced. J. J. Hardy used one—a little 7-foot, 2⅝-ounce jobbie—to set a French distance casting record at Bois de Boulogne. The company called these two-piece rods "The Lightest Practical Fishing Rod In The World." The elegant Fairchild, developed for S. W. Fairchild of New York around 1913–14, was a light yet powerful rod with jeweler's-quality fittings. Available in lengths of from 8 to 9½ feet, the series's largest rod weighed only 4¾ ounces!

On a light-rod roll, around 1922 the company introduced the mighty Marvel series, which featured a skeleton reel seat with a fixed and sliding hood. The 7½-footer in this series weighed a mere 2¾ ounces. Herbert Hoover owned a later Marvel and loved it. Starting in 1925, Hardy Brothers began building fly rods with new milling machines and introduced several models that established it as the largest rod manufacturer in England. The Halford Special was popular in the United States, as were the revamped Deluxe and the Fairy. The latter rod became famous after Ernest Hemingway fished an 8⅓-foot, 5-ounce model in Wyoming. Hemingway's Fairy is now in the American Museum of Fly Fishing.

As the decades passed Hardy Brothers continued to upgrade its selection of rods, and modern examples such as the Phantom and Palakona Superlight have found a following in America. Today, collectors

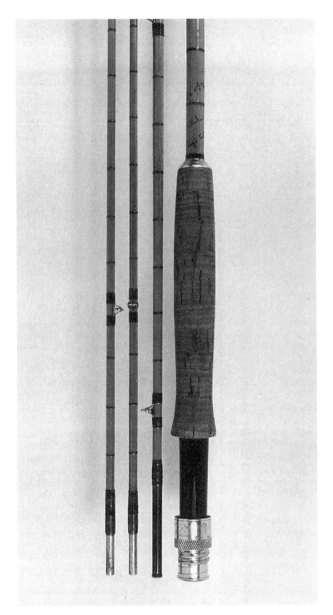

The Hardy Deluxe Palakona, shown here in a 9-foot version, was one of the company's first rods to sport the "modern" classic look. The reel-seat style was borrowed by American makers, including Thomas and Payne. (The Hillary V. Heaton Collection)

are beginning to take a new look at some of Hardy's exotics, and these shorter rods are catching on. The reels, of course, caught on long ago.

Some of the earliest Hardy reels should pique the interest of the antique-tackle collector, including a number of Birmingham revolving-plate models, a raised-pillar winch, several Nottinghams, and even an old-fashioned single-crank, fixed-click reel. From 1880 until 1921 the various Birminghams were offered in 12 sizes (2¼ to 5 inches), including a model with a hard-rubber back plate; perhaps the most coveted is

In the Empire's finest lingo we have here the Return of the Steel Centre Rod. Hardy didn't bother to list the weights of these Palakona models. From The Salmon & Trout, *April 1914*

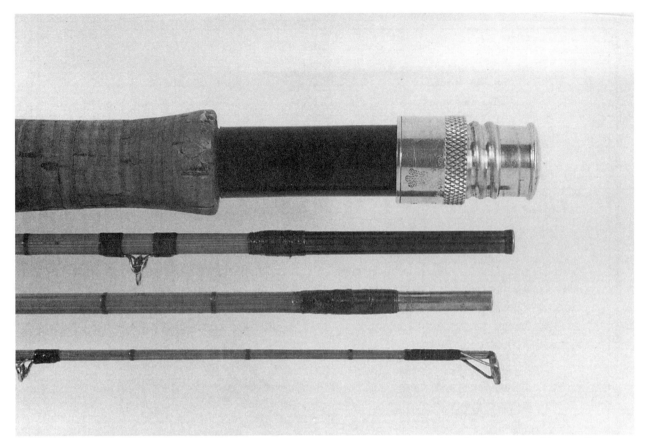

Hardy's classic up-locking reel seat was introduced on the 1911 Palakona series. Note the multitude of close-spaced intermediate wraps along the shaft. Labor was cheap in prewar Britain.

the Special New Pattern, a German silver reel made only in 1885–86. The American-style raised-pillar, or "External Pillared Crank," winch had a protective rim and single crank. It was not a cataloged item, but was made for a short period during the 1880s. The fixed-click reel, called the Bronzed Crank, was available from 1880 until 1909. Most of these early Hardy models carried the oval logo that read "HARDY BROs, MAKERS, ALNWICK."

HARDY'S PERFECT REEL

Starting in 1888 Forster Hardy, a third brother who ended up as head of the reel department, applied for two British patents for a process that used ball bearings to produce a smooth, long-lasting reel. The prototype Perfect was actually built prior to the acceptance of the second patent, on January 13, 1891. This brass reel, built between 1890 and 1892, had a 120-degree arched bridge over the adjustable click mechanism and perforations on the rear of the spool. The next model, an ivory-handled Transitional Per-

The original all-aluminum Hardy Perfect Narrow Drum fly reel, built from about 1902 until 1920–21, when the Mark I and II series replaced it.

fect, had no line guard and came in several variations, including a version with a nonperforated spool. Late Transitionals have the left-hand-threaded spool also found on the uncommon 1896 Perfect. This model retained the early click mechanism and was in production from 1894 until 1897. A few examples have been found that had the newer alloy frame but retained the original brass spool.

In 1898 the mostly brass 1896 Perfect was replaced by the first of several Wide Drum models. These retained the brass spool face yet were built largely from aluminum to cut down on weight. In production from 1898 until 1908, these reels begin to approximate the "modern" Hardy look, and have smaller ventilation holes in the rear. True aluminum models started to appear in 1899, still with the old click mechanism. These reels were superseded by the 1905 Perfect Wide Drum, built until 1921. Most Perfects built during the developmental years—prior to 1905—were marked with the "Rod in Hand" logo.

The narrow-spool, all-aluminum-alloy Perfect was introduced into standard production in 1897–98, only to be replaced almost immediately by the Brass Faced Narrow Drum, made from 1899 to 1901. The next year a third model narrow reel was introduced and built until 1921. To quote Billy Crystal, the reel was "mostly standardized, but not totally standardized." In fact, the additions and omissions of new clicks, old clicks, and even ball bearings make collecting the various Perfects an almost consuming affair. In 1921–22 the several models of Mark II wide- and narrow-spool Perfects were introduced; these continued in production until 1966. Throughout its history, the Perfect was made in various sizes, from 2¼ to an amazing 5¼ inches in diameter.

Hardy also produced a rare early Silent Check version from 1908 until 1910, and the somewhat rare raised-pillar Perfect, called the Bougle, from 1903 to 1922. Both reels had the early ivoroid handle. The older Bougle was replaced by a Mark II model in 1923; production continued until 1939. Other collectible Hardys include the original Uniqua (six models, 1903–59) and the St. George, which came in Trout, Junior, Silent Check, Salmon, and rarer two-handled Salmon versions. The St. George (1911–83) was probably the first reel to feature an agate line

guide as a standard fixture. A lesser reel, the St. John, is still in production.

Of all the Hardy reels, my personal favorite is the Flyweight Silent Check, which was imported by Abercrombie & Fitch in the 1930s and matched rods tossing 3- and 4-weight lines. The Flyweight and its larger cousins are still in production, I believe, but the Silent Check is history. Hardy's Cascapedia (1932–39) is mentioned on page 124 in conjunction with American multiplying salmon reels. The staggering selection of antique and classic Hardys could keep a collector busy for a lifetime.

H. MILWARD & SONS, LTD.

The firm of Henry Milward's, Washford Mills, Redditch, began as a small needle- and hookmaking foundry sometime in the 1700s. By the mid-1800s the company had become one of Great Britain's finest tackle establishments, winning 30 medals at various expositions and international trade shows between 1853 and 1922. The Redditch rodmaker also marketed a number of reels under its own label. A typical example was the 1920s-vintage Milward Flycraft, a 3-inch-diameter aluminum reel built by another maker, as yet unknown.

Milward rods from the 19th century are now somewhat scarce but often quite interesting. The Holland, a six-piece wooden trunk model, for instance, was built with two butt sections, eight midsections, and three tips. Split-bamboo Milwards are easier to find, although—as is usually the case with rods of this era—finding them in good or better condition can be difficult. Rods from the turn of the century tended to be large; some trout rods were 12 feet long. Many of the H. Milward & Sons rods were built in rather odd sizes, such as 8 feet, 11 inches or 10 feet, 7 inches—so close to the half-foot marks, yet so far.

Many post-1900 rods were put up in an excellent brass-capped aluminum tube, including such models as the Flyover (two piece) and Flycraft (three-piece). The most popular Milward rod was the three-piece Flymaster, which was built in lengths of from 6 to 9 feet. Very few Milward rods seem to have been imported into the United States, but a slew of later models made the crossing to Maritime Canada and are now beginning to filter into the Lower 48 through the back door. Milward products were good,

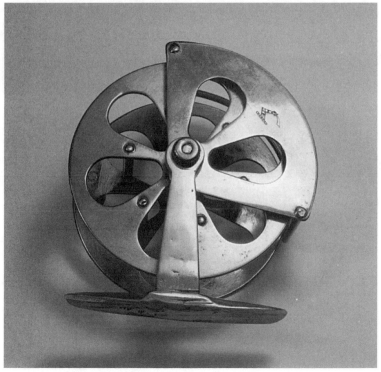

A unique Milward fly reel with a lot of panache from Gus Meisselbach. Milward went the extra mile, adding a rugged line guard to its model. Judging from the rear-view photo, the Milward reel's guard could be adjusted to either left- or right-handed use.

midquality items. Would that more American makers had built aluminum rod cases to equal those from Washford Mills, Redditch.

Other English tackle firms such as Ogden Smith, producer of the excellent Exchequer fly reels and long salmon rods, exported tackle to the States in a minor way. In more recent times the firm of J. W. Young produced a number of modern Orvis-marked aluminum reels that found a ready market with U.S. anglers. Other Young reels, such as the Beaudex and Pridex, were popular items in a number of U.S. catalogs, including that of the Paul H. Young Company. And some 30 years ago, Hardy Brothers built the original Scientific Anglers™ System reels, enameled gray with a palming rim.

In the future, antique rods and newer Hardy and Young reels will become more collectible as New World anglers look back to their British roots. The lowland reel was an excellent tool from day one, as were the English rods as modernized in the first quarter of the present century. Flies were another item, but for economic reasons they were usually tied in Celtic lands to the north and west.

We'll leave the old shore on a light note. In 1913 we find a London taxidermist named E. W. Little. Possibly related to our old Giles Little of Fetter Lane, E. W. stuffed his fish with papier-mâché, not a bad idea. One can smack a noisy angler over the head with a 3-foot molded-paper salmon and not really hurt it. Or the angler either.

The Salmon & Trout, *July 1913*

Notes

[1, 2]Turner, Graham. *Fishing Tackle: A Collector's Guide.* London: Ward Lock, 1989.

[3]Brown, Jim. *A Treasury of Reels.* Manchester, Vt.: The American Museum of Fly Fishing, 1990.

[4]Kelly, Mary Kefover. Personal correspondence, 1977.

12 Celtic and Canadian Makers

High above the English lowlands and westward to the Emerald Isle, the heirs of the ancient Celts carved their own niche in fly-fishing history and the art of tacklemaking. Much of what we know of classic Atlantic salmon angling, of spate water and the speycast, was born in the Celtic and Pict-Norse regions before the Renaissance.

The salmon, "lord of all freshwater fish," lived throughout Europe long before the Celts arrived. By 1830, however—about the time that large quantities of fly tackle began to become available—the pollution from burgeoning human populations had made major salmon runs in thickly settled areas but a memory. Only in the wildlands of the thistle, shamrock, and coal could significant runs of salmon still be found.

Celtic tacklemakers must have been around in the 1500s or even before, but there's a great dearth of information about them until the early years of the 19th century, when pioneer producer-exporters like Martin Kelly and John Forrest are recorded. The angling literature of Great Britain is, perhaps not surprisingly, Anglocentric. Even Graham Turner's recent book on British collectible tackle (*Fishing Tackle: A Collector's Guide*) gave only one line to Kelly (who wasn't even mentioned in the index) and 20 words to Forrest—out of 388 pages.

But the Celtic tradition carried on across the Atlantic in Canada, where the salmon also reigned supreme. In Saint John, New Brunswick, a number of fine craftsmen with links to Ireland, Scotland—and Hiram Leonard—emerged from the cabinet-making trade to become the last rodmakers on God's Green Earth to uphold the honor of Saint Christo-

pher's ancient spliced rod. (In early Celtic church art, Saint Christopher was often depicted at the river. At first he carried the baby Jesus, but later medieval depictions show him with a fishing rod.)

Celtic Fly Tackle

Collectors need only look at the fabulous multicolored illustrations in the *Book of Kells*[1] to see the same fanatical attention to detail that also shines in Celtic fly tackle—whether it be a rod, reel, or the mind-bogglingly intricate artistry of a classic salmon fly.

Unfortunately, the Celtic tackle artisans have remained obscure and their tackle has, for the most part, eluded collectors. It's only in the pages of *The Art of Angling, As Practised in Scotland*,[2] a slim volume published in Edinburgh in 1836 by poet Thomas Tod Stoddart, the first Scot to write an angling book for his countrymen, that we learn of the contributions of the Celtic makers.

DUBLIN'S MARTIN KELLY

For Irish tackle, Stoddart referred anglers to Martin Kelly and "Murray," both of Dublin. Of these two River Liffey makers Murray remains a mystery, but Kelly grew into "Kelly of Dublin" and later "Kelly & Son, Dublin." The Kellys are among the oldest known tackle producers. A few very small (1¾ to 2⅜ inches in diameter) brass "spike foot" Kelly reels have been found that date back to the early 1800s. The Kellys also built a number of impressive brass salmon

THE

ART OF ANGLING,

AS PRACTISED IN

SCOTLAND.

By THOMAS TOD STODDART, Esq.
AUTHOR OF "THE DEATH WAKE," AND OTHER POEMS.

SECOND EDITION.

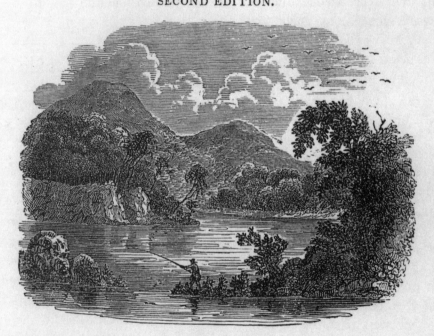

EDINBURGH:
W. & R. CHAMBERS, 19, WATERLOO PLACE;
AND ORR AND SMITH, LONDON.
1836.

Frontispiece from the original edition of Stoddart, 1836

Salmon anglers on the Conway in Wales, from an 1841 engraving in The Fly-Fisher's Textbook, by Theophilus South. Although just as powerful, Celtic rods weren't as stiff as English versions.

reels of up to 5 inches in diameter, which have brought solid prices in recent years.

Martin Kelly was also known for his excellent hickory and lancewood salmon sticks. In the first American edition of *The Compleat Angler* (1847), Bostonian George Washington Bethune wrote that "Kelly is said to be the best rodmaker in Europe." In 1869 Genio C. Scott wrote that as salmon rods, "the hickory ones made by Martin Kelly, of Dublin, are preferable to any that I have seen except for the split bamboo."[3] It's obvious that Martin Kelly rods were important in the 19th-century American tackle trade, and it's likely that at least some of these early Dublin products have survived. The full Kelly story has yet to be told, but one thing is sure: From a collector's viewpoint, the early rods and reels made by the Kellys are on par with a Murphy rod or an Archimedian winch.

THE CASTLE CONNELL ROD

The Kellys are the only recorded major Irish tacklemakers, but others plied that trade on the Emerald Isle. The Irish hookmaking industry dates back to the iron age. Unsung artisans produced an old hook pattern in Limerick, on the River Shannon, that remains popular today. Another famous Limerick-made hook was developed by O'Shaughnessy and is still a standard for saltwater flies. The original O'Shaughnessy began making hooks around 1790, but it was his son, Robert, who brought the steel to a fine art sometime before 1828. As Europe's finest hookmaker and an expert fly tyer, Robert O'Shaughnessy was active until midcentury or later. His ads in the *Limerick Chronicle* touted his flies, his hooks, and the now-famous collector's item, the Archimedian Minnow.

Rods, too, were made at an early date. A very early Irish fly rod, built and used throughout the Isle, finally got a name after rodmakers began to concen-

TO ANGLERS:

JOHN ENRIGHT & SONS,

Fishing Rod & Tackle Makers

Castle Connell, near Limerick, Ireland.

Reels and Lines to match balance of rods. No agents. All information respecting fisheries on the Shannon in this locality, as also catalogues of our greenheart rods and general assortment of tackle, to be had on application.

Forest and Stream, *August 11, 1881*

trate in Limerick—and in the nearby village of Castle Connell.

Early Shannon rods were probably made for local gentry, but by 1850 the rods had fans in Canada and the United States. The Castle Connell rod characteristically had a full bend to the butt and a slow casting action. Its two sections were joined by the ancient scarfed splice. Unlike some screw-joint Irish fly rods (c. 1840–55), the spliced style was cheap to build. John Enright was one of the first known makers in Castle Connell. His rod shop evolved into John Enright & Sons, which advertised its wares in *Forest and Stream* from August 1881 until January 1882.

In the 1850s, as noted by Edward Fitzgibbon,4 the Castle Connell style of fly rod was not high on the preferred-tackle list of British anglers, who preferred a stiff-action rod. Yet Castle Connell rods, in both trout and salmon models, were popular items with Fitzgibbon's "colonial descendants," and New World collectors can expect to find the style from time to time. The Castle Connell rods were interesting examples of the days when long fly rods were left in one piece to hang on dank castle walls, never far from clear, cold waters. In the end, the Castle Connell rod lived on in Canada until 1900. Later in this chapter the model's Maritime Provinces champion, Joe Dalzell, will explain its unique action.

IRISH FLY TYERS

Over time the obscure early Celtic fly metamorphosed into a thing of beauty in Ireland and began its travel across the seas. Landing first in Scotland, it competed against the somber Highland flies. Stoddart called it the "gaudy Irish fly," and defended the style because it worked when duller patterns wouldn't. In Stoddart's eyes, Tweed anglers who ridiculed the bright fly were "prejudiced clodhoppers."

Stoddart wrote: "There are no special rules for the composition of this fanciful lure. A general one seems to enforce the introduction of the golden pheasant's feather under the wings." The feathers of the golden pheasant—just one of many birds and ani-

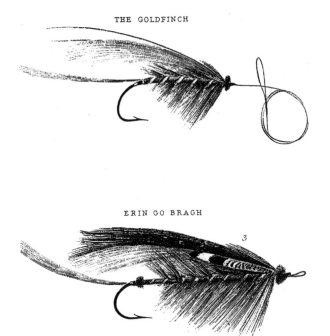

THE GOLDFINCH

ERIN GO BRAGH

Two gaudy Irish flies, from the frontispiece of Edward Fitzgibbon's 1853 edition. The Goldfinch (top), "a standard fly for grilse-fishing in the Shannon," has wings made from eight or nine golden pheasant toppings. The Erin Go Bragh, with features that evolved into the classic Scottish salmon fly and the later American streamer, is built with feathers from nine species of birds.

mals brought back by the Empire's vast oriental trade—appear to have been the first exotic material to land on tyers' benches.

From the time of the Dame's Twelve, flies had copied real insects, anything from a mayfly to a dragonfly. The Celtic tyers introduced a new idea—bright colors to inflame a fish's battle thirst. From Scotland the gaudy style traveled to England. In 1853 Fitzgibbon showed the "Goldfinch" and the "Erin Go Bragh," complete with golden-pheasant crests, etched from examples by William Blacker, the dean of London fly tyers (as well as a rodmaker). The gaudy fly may have been healthy, but the original tyers weren't. The Potato Famine (1845–49) devastated Ireland. Many Irish fly tyers either starved to death or emigrated to the United States and Canada.

The 1880s saw a slow rebirth of the Irish fly-tying industry, and flies were exported to America. For the next three decades the trade expanded, hitting an all-time high between 1915 and 1920. The Irish flies were both well tied and—important in America—very inexpensive, not surprising given the financial and political isolation of the Emerald Isle. Irish tyers, mostly women and perhaps children as well, worked long hours for little money.

The new Irish flies were mostly trout patterns, tied for and sold by American retailer-importers such as Herb Frost, Thomas Conroy, James Marsters, and others. In most cases the flies were offered to sportsmen as house specials, unlabeled bargain flies competing against nearly identical American products; the flies' Irish origins were rarely divulged. A few Irish producers had the brass actually to package and advertise their wares, however, and these flies are now a great find for collectors.

THE WHITE BROTHERS, OMAGH

The White Brothers exported flies to the United States before Ireland became a republic. Typical of the genre, its flies were tied in North American styles by colleens, mostly in the town of Omagh. "American patterns copied," said the Brothers' 1915 ad in *Field & Stream*. Tied on 4-inch gut snells, the flies were offered at an amazing 25 cents per dozen—an outrageously low price even in those days.

This Omagh firm also sold a general line of tackle, mostly of domestic origin, including an

Field & Stream, *April 1915*

enameled 30-yard "Waterproof" silk fly line for a dollar. Of course it was a level line, no taper, but what more could an aspiring fly fisher expect for a dollar? When I was young I used these so-called "waterproof" lines, coated with successive layers of linseed oil over the braid. They sank after the third cast and, if left in a car trunk on a sunny day, would melt into something like flypaper. Ah, the good old days.

L. KEEGAN, DUBLIN

The firm of L. Keegan also specialized in exporting trout flies from the Emerald Isle. In 1921 it was located at Three Inns Quay, Dublin, and placed an advertisement in the May issue of *Forest and Stream*: "Guaranteed Irish make and tied here on the premises." There was again a reference to American fly patterns, although Keegan would tie flies from "any special pattern supplied."

L. Keegan charged 60 cents a dozen for its flies, indicating a better product than White's. Keegan's flies were "tied on best tested hooks; eyed or taper shank and looped gut (4½ inches). Any size from No. 8 to 12." Although I haven't run across any fly packs marked by Keegan or the White Brothers, they must still exist, locked away in leather-covered tackle boxes from the era. You should keep a sharp eye cocked for paper envelopes containing these Irish treasures. And remember that original Celtic patterns exist as well. Gaudy Irish flies, looking somewhat like

Blacker's Goldfinch, are prime antiques and worthy of framing.

SCOTTISH FLY TACKLE

From Ireland it's a day's row to the Scottish coast, a crossing the gaudy Irish fly made sometime in the early 19th century. Prior to this, Scottish salmon patterns were homegrown and drab. An example used as late as 1847 was built simply from orange mohair, pig's wool, gold ribbing, red hackles, and a wing from the red-tailed kite. The Scots borrowed exotic ideas from the Irish attractor pattern, and began squirreling away bird skins in closets until the classic Victorian salmon fly emerged like a butterfly.

From midcentury until the publication of George Kelson's massive volume, *The Salmon Fly*,[5] in 1895, the patterns became more complex and intricate, and along the way they picked up personality. One of the early flies, the Jock Scott, was named after a River Tweed boatman. Scotland also gave birth to a number of traditional trout patterns. Foremost among them, and a favorite of Stoddart, was the Professor's, named after its creator, John Wilson (1785–1854) of Edinburgh University. Today we call it the Professor.

Fancy Scottish salmon flies are supremely collectible—and expensive. Scottish fly rods and reels, however, are moderately priced on today's classic-tackle market. William Rawson, ironically an Englishman from York who moved to Edinburgh in 1796, was one of the first known rodmakers in Scotland. His Leith Street business became William Rawson & Sons in 1801. Sons John and William each started their own rod shops in 1814 and 1816, respectively, and continued their father's trade until around 1850.

Thomas Tod Stoddart's favorite rodmaker was "the late Mr. Phin," also of Edinburgh. William Phin began making rods at his North Bridge address around 1810, dying, to Stoddart's consternation, around 1831. Mr. Phin was close to an institution and the demand for his excellent rods and flies encouraged his widow to continue the shop. She was succeeded by C. A. Phin, who carried on the business from midcentury until the mid-1890s.

Two uncommon collectible rods of the 1880s: At left is a two-piece, 10-foot spliced rod of Canadian origin with a small Meisselbach Amateur reel; at right, a John Forrest light, three-piece, 10-foot trout model with a raised-pillar Allcock reel.

A number of Scottish makers produced brass fly reels, some even embellished with rosewood about the side plates, such as those of Paton & Walch of Perth. Other reelmakers included William Brown of Aberdeen, John Gow & Son of Dundee, and Arthur Allen of Glasgow. Most of these Scottish reels were of the revolving-plate style; the great majority were salmon models.

FORREST OF KELSO

As Scotland's most recognized maker, John Forrest had many imitators. A New York tackle dealer even lifted his name and that of his town (see page 256). Knowledgeable anglers could see the Forrest through the trees, however, and bought the genuine article. A number of signed examples of Forrest of Kelso's impressive and massive salmon rods have survived, showing up at a half-dozen major auctions. The first John Forrest product I ever examined was a wonderfully light, 10-foot, mahogany-colored lancewood fly rod, which I used for a season or so before tucking away. The rich brown rod was unsigned and had loose-ring guides and brass seamed ferrules; I was sure it had drifted westward from the Isles, but I didn't know the maker at the time. Its unique features included a swaged butt cap plus brass sliding- and fixed-reel bands, which were not round, but swaged on one side to fit a reel's foot—not unique, but very British.

I kept the rod in the back corner of a closet for more than a decade, until I found its big brother at the 1988 Oliver's Summer Auction. This rod, a 16-foot two-handed salmon rod built by and marked "Forrest, Maker, Kelso," matched the first rod's mahogany brown color and brass handle fittings to a T. It was a fairly typical Forrest, with two mids and three tip sections. The grip area was cord wrapped, although many Kelso-built examples are found with plain handles. I had identified the maker of the little 10-footer.

John Forrest began making rods on the banks of the River Tweed around 1836–37; any rods from this early era would today be a great find. His original address was probably Wood Street, Kelso. At some point the business branched out into the reel and accessory market as well; it eventually became Forrest & Sons, a name it used as late as 1967, when it was at 35 Royal Square. By 1912 or earlier, Forrest established a store on Oxford Street in cosmopolitan London, giving the company access to the prime English market.

Forrest also exported a quantity of trout and salmon flies to William Mills & Son, Dame, Stoddard & Kendall, and many other U.S. dealers. The American fly was frequently built from dyed turkey feathers—far short of the real macaw, bustard, and other exotic feathers used by the Scots. Remarking about Forrest's salmon flies, A. G. Wilkinson, our itinerant fisher of Canadian waters, mentioned his tack for the River York in 1876: "I provided myself with a Norris greenheart and a Leonard bamboo in the way of rods, and with an assortment of proper flies made by Forest [sic] & Sons, of Kelso, Scotland. Not but that excellent flies are made in our own country, but the composition of an artificial fly is an art. . . ."

Forest and Stream, *May 24, 1888*

FORREST & SONS
KELSO, N.B.,

FISHING RODS

AND TACKLE

J. FORREST
24, Thomas St., Oxford St., LONDON, W.,
Catalogue on Application.

The Salmon & Trout, *June 1912*

early rods had swaged reel fittings, indicating that they were built in the drop-hammer era. The salmon reel seats often had a partial brass covering and the guides were loose ring. Nice touches included two extra tips, an extra mid, and handsome ferrule plugs, some made from mahogany. The lower shafts were often "ebonized" (Japanned black), while the tips were plain lancewood. Other rods were completely stained a rich mahogany brown and varnished. Later versions had rather small English snake guides and agate strippers and tiptops. Many of these salmon rods carried spliced joints.

It's difficult to date late Forrest & Sons products because their design changed very little over a span of many decades. British anglers still fish with 15- to 17-foot Kelso-built models. Early Forrest salmon rods had wooden butt buttons; later versions sported rubber caps and cork grips. Irrespective of era, the shaft material was usually greenheart heartwood (often called bethabara) and/or lancewood. Often built with a splice, the rods pinpointed Forrest as Scotland's champion of the classic 400-year-old Bernes style.

The salmon reels are today far scarcer than the rods, and appear to be solidly of 20th-century origin, made of aluminum alloy. At present, Forrest's wooden rods are real bargains on the collectible market. The time for big Scottish salmon rods has yet to arrive, but when it does, these excellent examples of Celtic craftsmanship will rise to the top of the list.

Throughout the history of the firm Forrest & Sons majored in the Atlantic salmon fishery, minoring in trout rods. Most Forrest rods I've seen can be dated from the 1880s to the 1920s or beyond. Even

MALLOCH'S SUN & PLANET

Born in the town of Almondbank, Scotland, in 1853, Peter Duncan Malloch spent his formative years observing nature as well as fishing the River Almond. By his early teens this self-taught

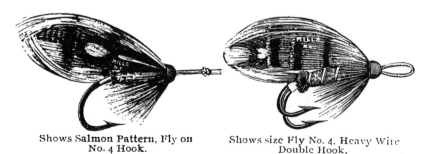

Shows Salmon Pattern, Fly on No. 4 Hook.

Shows size Fly No. 4, Heavy Wire Double Hook.

Scottish salmon flies, from the 1912 William Mills & Son catalog.

angler had mastered the nomenclature and natural history of most Scottish birds and fish and had become an amateur taxidermist. At the age of 14 Malloch was building his own rods and tying productive flies. When he was 22 P. D. Malloch, with his brother James, started a small taxidermy studio and rod shop on High Street in nearby Perth.[6] Malloch, considered the best salmon fisherman of his time, became famous for his monumental volume, *The Life History and Habits of the Salmon, Sea-Trout and Other Fresh Water Fish*, published in 1910. Malloch was the first to invent and market a working spinning reel—the Malloch Sidecaster. However, Malloch's gift to the fly-fishing fraternity came in a most unusual reel that became known as the Sun & Planet.

This unique fly reel had planetary gears: a large inverted gear that ran the circumference of the reel and a smaller gear that, when the handle was turned, circled inside it. When a fish took line from the reel, the cow-horn handle rotated as well. The Sun & Planet, intended for the salmon angler, was well designed but rather hard on the cranking fingers, since the gear ratio was actually a mechanical disadvantage. I know of two nice blued-brass Sun & Planets that came from Maritime Canada, and several others have been offered through American auctions. For a time Abercrombie & Fitch sold the reel as its Perfection Salmon Reel; the 1909 A&F listing offered it from sizes 2¼ to 5 inches, priced between $12 and $25.

Malloch also produced a number of other reels, including revolving-plate models of up to a 6-inch size. Other styles varied from a small brass Birmingham-style trout winch at 2¾ inches to a brass-and-ebonite salmon version at 4½ inches. His rods are tougher to find, although he built a number of them. Malloch's legacy also includes stuffed brown trout and salmon, aptly displayed in shadow boxes and proof of his accurate knowledge of the subtle coloration of both species.

Perhaps the greatest of Malloch's accomplishments came secondhand, as seen in the pluck of an 18-year-old ghillie's daughter who caught the muckle fish of Glendelvine. Miss Ballantine was fishing the River Tay on October 7, 1922, when she successfully hooked, played, and caught a 64-pound cockfish on a Malloch rod and reel.

Peter D. Malloch's rugged Sun & Planet reel, copied but never equaled, is a handsome collectible.

The family business continued until 1964, when Miss Maggie Malloch retired at the age of 82. Under new ownership, Malloch of Perth continued until 1981.[7] Although the Mallochs are gone, their historic tackle remains to remind us of great days on mighty waters. By the way, Miss Ballantine's "cockfish" still stands as the British-record salmon.

JOHN S. SHARPE AND THE SCOTTIE ROD

John Sharpe was yet another rod craftsman who started his precision career as a gunmaker, apprenticing with Davidson in Aberdeen. He then moved to England and worked in a fishing-tackle manufactory until World War I. Following the war, he returned to Scotland. In 1920, with his sons John and Alan, he established J. S. Sharpe of Aberdeen. From a small business Sharpe's grew into a major British tackle company, first with its rods but soon with its reels, telescoping gaffs, folding landing nets, and other salmonid-dispatching accessories.

By the 1930s J. S. Sharpe had become the largest tackle concern in Scotland and begun exporting to India, New Zealand, Canada, and the United States. The two most recognizable styles of the famous Sharpe rods were the Aberdeen and the Scottie. John Sharpe also manufactured a unique line of aluminum Gordon and Scottie reels, which ranged from smaller trout models to 4-inch salmon reels. But

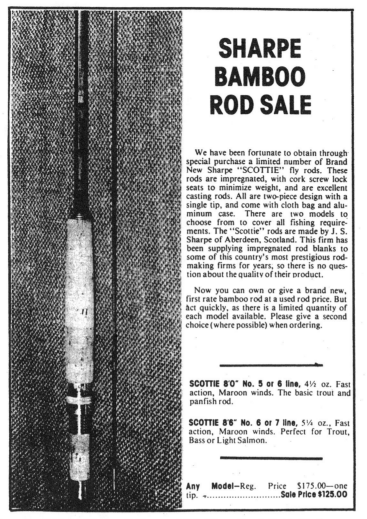

From the 1980–81 "Thomas & Thomas Fall & Winter Classic Tackle List"

it was the award-winning rods (a first-place medal at the Madrid International Exposition in 1932) that were the backbone of the company.

Collectors have a wide range of models to choose from, starting with early pre-model-name sticks with old-fashioned spliced joints and continuing to modern six-strippers of uncanny casting ability. Sharpe stopped production during World War II, but in 1946 it incorporated, retooled, and introduced its variously named series of rods. At this time it became known as J. S. Sharpe Fishing Tackle, Ltd., and began impregnating its blanks with the Orvis Bakelite process. These blanks, built to different tapers, also served as the foundation of the Leonard Duracane rods.

The Aberdeen series ranged from 14-foot, 10-inch two-handed salmon models down to a 9½-footer (6¾ ounces) designed for a 7-weight line and a 9-footer (6 ounces) for a 6-weight. The Aberdeens had a down-locking metal reel seat and were built in three sections with an extra tip. Perhaps the most recognizable rods were those in the Scottie series, which were imported into the United States until the early 1980s. The Scotties ranged from 14-plus feet to a dainty little 7-footer; most were built as two-piecers with an optional extra tip. A fine all-around rod was the 8½-foot (5½ ounces) Scottie for a 6-weight line.

Other Sharpe two-piece rods were particularly charming. These included the venerable Fario Club, an 8-foot, 5-inch rod for a 5-weight line with a parabolic taper designed by Charles Ritz; the Eighty Eight, which was 8 feet, 8 inches long, weighed 5 ounces, and tossed a 6-weight line; and the smaller Eighty Three, which weighed 4½ ounces and took a 5-weight. All these rods featured locking reel seats and a cork barrel.

The Featherweight series were fine rods, fitted with cork reel seats and a German silver cap and sliding band. They came in four models, ranging from a little 7-footer (3¾ ounces) and a 7½-footer (4 ounces) for a 5-weight line to the 8- and 8½-footers for a 6-weight.

Models from the Fario Club through the Featherweights were designed along Charles Ritz's staggered-ferrule principle, which placed the nickel-silver ferrule below the rod's center and closer to the power of the butt. Although a staggered-ferrule rod looks like hell when it's apart, the design gives it more lifting power than a conventional two-piecer, and it also helps remove casting vibration, making the rods both accurate and powerful.

Throughout its history the Aberdeen company collaborated with well-known anglers to produce specialty rods. In the 1970s Sharpe teamed up with author-riverkeeper Frank Sawyer to produce the Nymph Special. Another design, intended for itinerant brown trouters, was the Wilson-Sharpe, built in the odd length of 8 feet, 3 inches. Partially the brainchild of writer Dermot Wilson, the Wilson-Sharpe tossed a 5-weight line.

During the 1970s. Sharpe made an intense effort to increase the popularity of its products in the United States. The campaign never really worked, however, even though the lower-grade models, such as the Scottie, were priced as low as $125. American rodmaker Thomas & Thomas purchased a small shipment of Scotties in 1979, put them on sale at $50 below retail, and still had the same stock available in 1981.

After J. S. Sharpe's death in 1957 his sons carried on the business until 1964, when the company was purchased by Charles Farlow and Company of Pall Mall, London.[8] Farlow continued making the popular rods designed by Alan Sharpe and Ritz, including the Farios, Featherweights, and Eightys. Considering for how long a period Sharpe rods were made, surprisingly few show up on American dealer lists today.

OTHER SCOTTISH MAKERS

Many other Scottish products found their way across the Atlantic, via both exports and, no doubt, accompanying immigrants. Many 19th-century salmon rods used by Canadian Scots were built in the old Homeland. To aid collectors, here's an abbreviated list of Scottish makers whose work tends to surface at swap meets and auctions. The street addresses were supplied by British tackle historian Graham Turner.[9]

Roderick Anderson showed up for Oliver's Fourth Annual Summer Auction, looking much like a staunch and elderly version of his 13-foot greenheart salmon rod built to the "old count" (one butt, two mids, three tips). The circa-1880–95 rod, with black wraps, brass hardware, and a wooden butt-cap button, came in a wood form and was marked "Andersons, Edinburgh." Later known as Anderson & Sons, Ltd., the rodmaker was active on Princess Street from 1880 to 1929, and at 49 & 51 Shandwick Place from 1930 to 1938.

Charles Ingram made his appearance at Oliver's Fifth Annual Summer Auction in the form of a nice 11-foot, three-piece, extra-tip, one-handed "light" grilse rod, looking very modern with stunning agate trimmings over split bamboo. The 1930s-era rod was marked "Ingram, Glasgow." In excellent condition in its original canvas case, it brought $100. Charles Ingram built rods from roughly 1895 until 1945. He

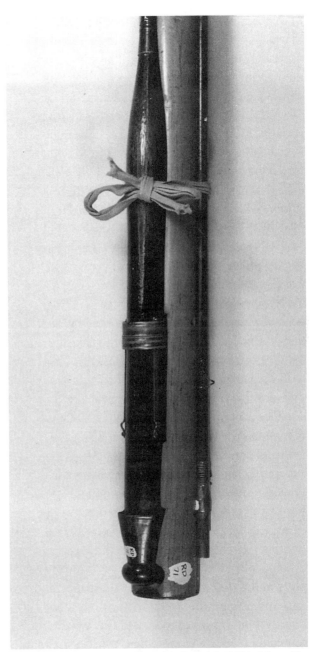

A Roderick Anderson trout fly rod in its original fitted form. The wooden butt-cap button was a fixture of many Scottish pieces, and was also used by the Maritime makers in Canada.

started out at 18b Renfield Street and took a second shop in Brothwell Street around 1920. By 1925 he'd moved to Waterloo Street.

Charles Playfair appeared at Oliver's Second Annual Summer Auction. He, too, was looking much like his two-handed salmon rod: 13½ feet, with "Weejer"-like reel fittings, spliced joints, and three sections, all in greenheart with a cork grip. This circa-1930 rod was marked "C. Playfair, Aberdeen, Scotland," where Playfair and his successors were active from 1820 on. The company became Chas. Playfair & Company in 1875 and survived until 1955. During this long tenure Playfair occupied various addresses on Union Street, evidently the tackle center of Aberdeen.

Robert Turnbull appeared in the July 1914 issue of *The Salmon & Trout*. He appears to have been another maker of the famous Scottish wooden rod; at least, there was no reference to split bamboo from this later craftsman. Robert billed himself as a "Maker of High Class Fishing Rods and Tackle." None o' that Low-Class stuff, Laddie, for the Champion Angler of 1906 and 1910. But where was he from 1907 through 1909?

Alex Martin arrived at Oliver's Third Annual Summer Auction, represented by two rods marked "Alex Martin, Glasgow Thistle." The first rod was a 9½-foot bethabara trout rod with snake guides and brass fittings, three pieces with two tips, and what appeared to be a sheet-cork grip. The second rod, a split-bamboo 9-footer, had three sections and two tips, with green wraps and intermediates. Alex Martin made his second U.S. appearance at Oliver's Fifth High Roller's Auction in the form of a 10-foot salmon rod marked "Alex Martin, Glasgow, Stirling, Edinburgh and Aberdeen." This was a 1915-era three-piece, two-tip, split-cane rod with Hardy-style ferrule plugs and tiptops. It had a two-hand-style cork grip and a large rubber butt cap. Both these rods were built by Alexander Martin's makers, originally established in 1838 at Argyle Street, Glasgow. Later Glasgow addresses included 18 Royal Exchange Square (1845–49) and 20 Royal Exchange Square (1850–1909). Martin then joined with Alex

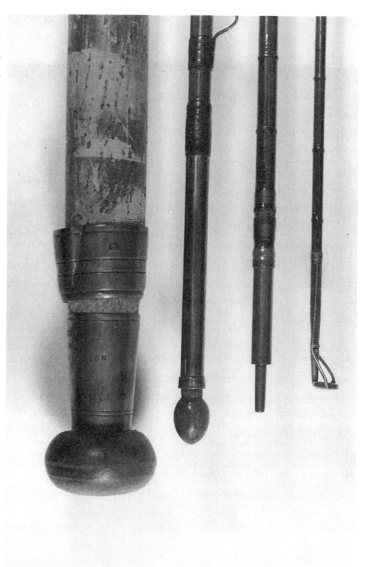

Marked "Alex Martin, Glasgow, Aberdeen & Edinburgh," this circa-1890–1910 wooden sea-trout rod has a rubber butt-cap button and classic ferrule plugs.

Henry & Company, Frederick Street, Edinburgh, and continued making rods there from 1910 to roughly 1920.

That wraps up our discussion of the Scots and Irish whose products are known to have landed in the New World. With the exception of a few bamboo models by makers like Muckle Alex Martin and Charles Ingram, the Celtic shaft continued to be made of traditional wood—greenheart or lancewood—often without ferrules. Throughout the 1800s the spliced rod

The Salmon & Trout, *July 1914*

Another successful day in the Empire. Equipped with two-handed rods, Victorian anglers toast a heap of salmon on the banks of a Celtic river. In their zeal, they've forgotten the gaff, now hiding in the grass to be found later. Salmon angling was a bit tougher in the New World.

never strayed far from the original Irish Castle Connell style, with a kick in the middle. In the early 1900s the model was revived as the "Grant Vibration Rod," testimony that good vibrations never go out of fashion. And Canadian makers were to produce their own variations on the two-piece Castle Connell theme.

Canada's Maritime Makers

The earliest recorded instance of fly fishing in the New World was in Maritime Canada in the late 1700s. By the mid-19th century the Maritime Provinces had become *The* Atlantic Salmon Hot Spot. Yet, for the most part, collectors of classic and antique tackle have neglected the Canadian scene. In the province of New Brunswick professional rod-makers were at work by the 1850s. For the next half century or longer these Maritime makers—only three of whom have been identified—built some of

the world's most fascinating examples of sporting apparati.

Our first glimpse of the Canadian rod comes through Frederic Tolfry, one of those early British tourists who visited the New World[10] in the wake of adventurer John Josselyn in the 1670s.[11] In Quebec Tolfry met a Major Browne, a provincial officer best described as a totally consumed salmon fanatic. Browne fashioned one or more salmon sticks prior to 1817 and displayed several in his garret. Tolfry described the Browne-built rod as being made in only "two joints," spliced in the middle. Said Browne, "Your dandy London rods, with their four or five joints and brass sockets and ferrules, are of no use here . . . you will find that they snap like a reed with a lively salmon at the end."

Another chapter in the life of the Dame Juliana Special, the two-piece spliced rod gained popularity on East Coast rivers—as mentioned by several American writers—between 1865 and 1892. And it influenced Bangor, Maine's first rodmaking community.

In the New World, Maine salmon were getting rare by the 1870s, and anglers had to venture east—to the rivers of the Bay of Chaleur. The style of fishing there contrasted sharply to that of Celtic waters, but Canadian tackle was amazingly similar.

DINGEE SCRIBNER'S LEGACY

Scion of one of the original Loyalist families removed from Connecticut, Dingee Scribner was Canada's first known professional rodmaker, establishing his tacklemaking shop around 1860. Area business records give the year as 1857, followed by six missing years. By 1863 Dingee was listed as a "fishing tackle maker." At this time the city was the last stop before the backcountry, and many a voyageur supplied his wanagan (canoeing essentials) from the bustling shops along Prince William, Sidney, Brussells, and Waterloo Streets. The growing numbers of American tourist-anglers stocked up in Saint John as well, buying flour and groceries from Alex M'Roberts at Prince William and tackle from Dingee Scribner at 5 Sidney.

By 1869 Scribner became a little better defined in the records. He was then listed as a "Manufacturer of Salmon and Trout Rods, Landing Nets and Flies," and had moved to 3 Waterloo. Scribner was the province's first and only full-time rodmaker. Other makers who followed augmented their incomes by selling products as diverse as cigars and picture frames; but in his long career, Scribner built and sold only fishing tackle. In 1879 young Frederick L. Scribner joined the firm,

The Maritime artisans congregated in the port of Saint John, which, with the arrival of 8,000 American loyalists in 1783, had become a city. The river Saint John rushed down from the border with northern Maine—a boundary finalized through the efforts of our good angler, Daniel Webster, and Britain's Lord Ashburton—through the middle of the province and to the Bay of Fundy. There, in Old Parr Town, a man with the fine Gaelic name of Dingee Scribner established a rod shop and left behind a few superb examples of a unique New World collectible.

1869 Saint John Business Register

which became "D. Scribner & Son," a stamping eventually found on several of their rods.

The Scribners remained on Waterloo until 1880, when they moved uptown to 59 King Square. The historic shop (now, alas, a parking lot) remained at this location for the next 40 years. In 1896 the partnership dissolved and Frederick Scribner moved to 15 King Square, opening a store under the heading of "tobacconist and fishing tackle." He built no rods, and during the next year was recorded as a "tobacconist" only. Dingee, it seems, died in 1898 or 1899. The 1900 *Saint John Business Directory* listed the company as "Dingee S. Scribner, fishing rods and gear" with no mention of manufacturing. In 1901 the business became "D. Scribner & Company," owned by Mrs. H. E. Scribner, apparently Dingee's widow. Scribner's remained a popular tackle shop throughout the early 1900s. In 1907 the store was managed by A. P. Watson, who became the sole proprietor in 1917.

This old Saint John maker was the most prolific of the Maritime group; several of his rods passed across various American auction blocks between 1985 and 1989. Dingee specialized in salmon and trout tackle built to the established Canadian style. The rods were planed from a combination of lancewood, greenheart, and/or "bethabara," which John Harrington Keene thought to be a "very unreliable" species of greenheart, and in fact may have been heartwood. A fourth material, "logwood," was probably the Mainer's "ironwood," also known as "hornbeam" in Connecticut. Given their construction period of 1860 to at least the late 1890s, a number of Scribners should appear in the future.

Maritime rods were signed with a gum-rubber ink stamp over wood in one of three places: on the reel-seat area, at the flats of the splices, or on the case.

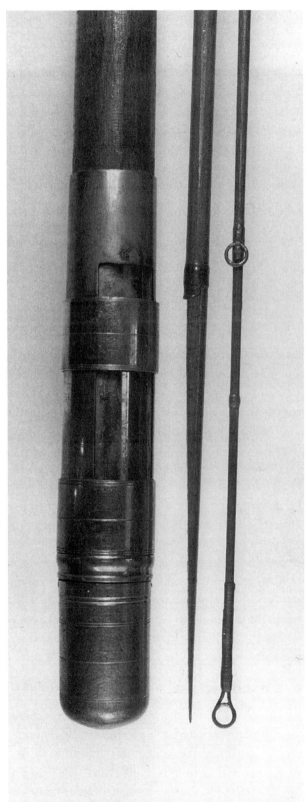

Details of a circa-1880 Dingee Scribner 11-foot, two-piece trout rod. Note the small metal "cup" wound to the splice area to serve as a receptacle for the opposite end. This "improved splice" is identical to one used by Scribner's contemporary, Hiram Leonard.

Known salmon rods were built in two styles, ferruled and spliced, and in lengths varying from 15 to 15½ feet. The Scribners also fashioned a number of "light" grilse models in these lengths. Standard later features included long rattan foregrips, seamed and soldered fittings, and two styles of butt cap. Some caps looked American, perhaps of Divine origin, while others had the British button. There were variations in reel seats as well; some were of the fixed and sliding-band styles, while others were full-length metal tubes.

Scribner salmon rods were built to the old count, two mids and three tips, and were put up in a pine form very similar to those built by Hiram Leonard. I think that this wooden form originated in Ireland or Scotland and was carried through to the Saint John design. Leonard had connections to the Maritimes and lived in New Brunswick for several years. His first personal fly rods probably had a Parr Town origin. The Maritime rod, as built by Scribner, influenced Leonard's early salmon designs, in particular those built with a splice. Additionally, Leonard became the only early American maker to specialize in two-piece models.

At least one of Dingee Scribner's trout rods has been found, a magnificently sculptured two-piece fly model 11 feet long. In its pine form, which reached almost 6 feet in length, two tips nestled on the opposite side of the butt section in a storage arrangement almost identical to that found in the wooden form built to hold a two-piece, two-tip Bangor Leonard rod. This Saint John–Bangor connection puts Scribner in the same class as Charles F. Murphy (see chapter 2) as an influence on the style of Leonard's early products.

C. BAILLIE, MAKER, SAINT JOHN, NEW BRUNSWICK

The second Saint John rodmaker to enter the trade, Charles Baillie, began his long career in 1874 in a shop at 136 Prince William. In 1877 he was upgraded from a "maker" to a "fishing tackle manufacturer," and listed at Duke, corner of Prince William. Competing against the strong lead established by Scribner, Charles had a secondary trade as a picture framer in 1879, when his shop and domicile had moved to Charlotte Street. Baillie remained

there until 1898, when he moved to the city's business center, with a storefront at 70 King Square.

After the death of Dingee Scribner, Charles Baillie's trade improved, and in 1901 he was recorded as a "Dealer and Manufacturer of Salmon and Trout tackle." As far as is known, he remained at King Square for the rest of his tenure. I followed Baillie until 1917–18, when his trade evolved into general fly-tackle sales as opposed to rodmaking. Baille's career lasted as long as Scribner's, but he seems never to have equaled his competitor in output.

Charles Baillie rods were typified by their large wooden butt-cap buttons and a plain grip, *sans* rattan. As did Scribner's, the reel seats varied from New World to British styles. Baillie's output included models built from either greenheart or lancewood as well as a combination of the two. Their wooden forms conform to the Saint John style; shorter rods were built in the two-piece configuration.

I've examined two Baillies. One was a diminutive 9-foot lancewood trout rod, packed in one of those great old wooden forms, built to the standard two-piece Maritime design but with only two tips. This delicate wand—surely no more than a 3-weight—had a semihollow taper where the shaft met the grip. The other Baillie was a 14-foot spliced Castle Connell–style salmon rod, which appeared to have been a light grilse model; it came in a Saint John form and had two mids and, again, only two tips. Both handsome fly rods were marked "C. Baillie, Maker, St. John, N.B.," and probably came from a single original owner. The rods were offered at Oliver's premier tackle auction in 1985 and sold for a mere $175 and $210, respectively. Great buys. If these two 1880-era products were typical, Baillie used brass hardware. Like all Canadian Maritime rods, regardless of maker, complete Baillies are fairly rare today because of the great length of their vulnerable tip sections.

JOE DALZELL'S CASTLE CONNELLS

Our last famous Saint John maker, Joseph "Joe" Dalzell, was the son of a cabinetmaker and furniture dealer. He first appeared in the business register as a "cabinetmaker" in 1871, boarding at 27 St. Patrick, the home of his father, Samuel B. Dalzell. A talented woodworker, young Joe was employed in related trades (including a stint as a cooper) until 1880, when

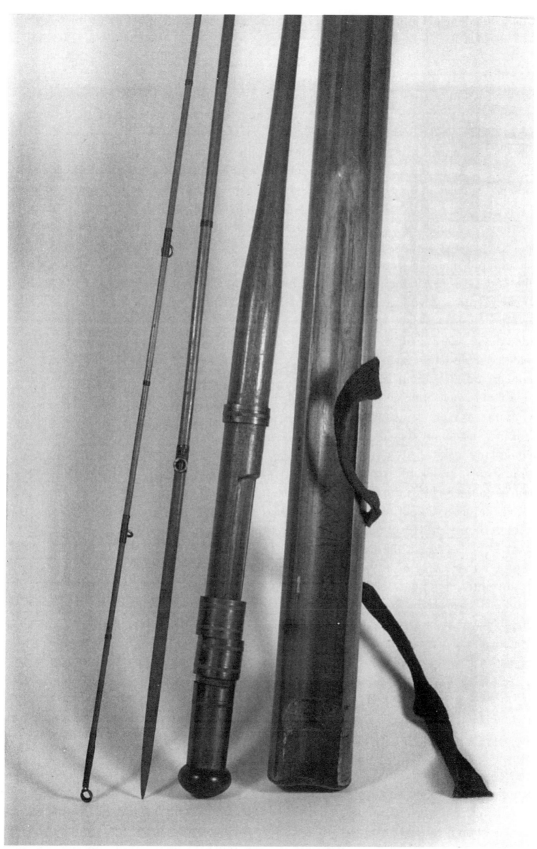

One of Charles Baillie's late trout models, this diminutive and priceless 8-foot Canadian wand was built in two sections of equal length and perpetuates the Celtic wooden-button butt cap. The woodworkers of Saint John, New Brunswick, were the last makers to produce the ancient spliced or "lashed" rod.

he was finally listed under "fishing tackle." He made the first of his rods at 85 Germain, boarding at the same address.

In 1883 Dalzell occupied a short-lived shop on King Street, placed the first of several display advertisements, and was finally recorded as a "fishing rod manufacturer." He would build many greenheart rods in the Castle Connell style. In 1884, Joe moved his business to 57 Germain and stayed there until 1917-18 or later. As noted in his early ad, Dalzell also pushed "choice Havana cigars" as a sideline. As with Baillie, Scribner's passing brought Dalzell increased business. In 1902 he was specifically recorded as a "Manufacturer of Salmon & Trout tackle."

At the very peak of his rodmaking career Joe Dalzell was mentioned and quoted in Shields's *American Game Fishes*[12] by J. H. Keene. In regard to design, Keene wrote:

> I prefer a double-action, solid-wood greenheart rod, for Salmon, built on the Castle Connell principle . . . a rod with rather exaggerated resiliency, insomuch that its tip, when striking a fish, first goes forward and *then* backward—i.e., its action is double. The Castle-Connell rods are also without ferrules, the joints being put together by splicing . . . elasticity, is preserved along the *entire length* of the rod, and not interfered with by the unyielding ferrule. There is . . . considerable play in the butt-joint, which is not the case with the ordinary make.

Keene also gave Dalzell credit as "the best Salmon-rod maker I know of." Championing his Castle Connell–style rods in a letter to Keene, Dalzell wrote:

> I think you will find every part of my rods does its share of the work. *I make it work from handle to tip.* Most rods are made with [a] stiff butt. This is good enough for its purpose—giving length. In a rod of that kind the fish is killed on the weakest part; in my rod you kill him on the strongest. Also, in casting a line—if you get impetus from the resilient butt you can cast with less exertion, and a

smaller rod of this kind will do the work of a much larger one of the ordinary build.

Basically, this was the parabolic idea used by later rod designers, including Charles Ritz, Paul Young, and Jim Payne, although its first American advocate was Hiram Leonard. Such a rod, whether sized for salmon or trout, had an extremely slow casting action that, as I've mentioned, required little effort to set in motion. Years later the notion was seen again, in fiberglass, in the Berkley Parametric and the rods of Russ Peak. Like Dingee Scribner, Joe Dalzell had an American following, and angler-writers such as John Harrington Keene and A. G. Wilkinson owned and used his Castle Connell models. Dalzell also made a number of rods with standard ferrules because, as he put it, "I have to." The average salmon

ST. JOHN. 83

JOSEPH DALZELL,
Manufacturer of
FISHING TACKLE, RODS, FLIES, &c,
CHOICE HAVANA CIGARS AND TOBACCOS.
55 KING STREET.

1883 Saint John Business Register

angler still preferred ferrules, which made putting a stick together or breaking it down a lot easier.

The Dalzell-marked rod, whether trout or salmon model, is an extremely tough bugger to find today and in good condition commands a healthy figure. The Maritime makers as a group produced some of the most interesting of northern collectibles. The total output of Scribner, Baillie, Dalzell, and others still unknown was relatively small, and this scarcity will ensure that the value of the Parr Town rods increases. Of course, spliced wooden rods present problems for collectors: Their extreme length makes it tough to find a wall large enough to hang one intact. Nevertheless, the two-piece Maritime trout rods, in particular, are unique historic pieces. The only other original North American models built according to this very old idea were made in Bangor, Maine, by a man named Leonard.

So ends our look at the handsome Canadian rods, symbols of an era when American sportsmen first began venturing into Canada's near-virgin waters. Like the Celtic fly, these rods are rare and extremely collectible. The makers were influenced by a river of tradition—no gimmicky claims, silly patents, or fancy engraving. These were honest tools made by humble craftsmen and built with "beauty without alloy," as Genio C. Scott once put it. They may not be fully appreciated now, but they will never fade. Good things never do.

LATE TO DINNER.

Notes

[1]Anonymous Gospels. *Book of Kells.* Writing commenced at Iona, c. 760. After the First Viking Invasion (806), monks took the book to Kells, where it was finished decades later. At least three different scribes and painters produced this most artful volume—680 pages, all but 2 in color. The book is now at the Library of Trinity College, Dublin.

[2]Stoddart, Thomas Tod. *The Art of Angling, As Practised in Scotland.* Edinburgh: W&R Chambers, 1836.

[3]Scott, Genio C. *Fishing in American Waters.* New York: Harper & Brothers, 1869.

[4]Fitzgibbon, Edward "Ephemera." *A Handbook of Angling,* Third Edition. London: Longman, Brown, Green, and Longmans, 1853.

[5]Kelson, George. *The Salmon Fly.* Edinburgh: Privately printed, 1895.

[6,7,8,9]Turner, Graham. *Fishing Tackle: A Collector's Guide.* London: Ward Lock, 1989.

[10]Tolfry, Frederic. *The Sportsman in Canada.* London: Newby, 1845.

[11]Josselyn, John. *Account of Two Voyages to New England.* London: Roger L'estrange, 1673.

[12]Shields, George O. (ed.). *American Game Fishes.* Chicago: Rand McNally, 1892.

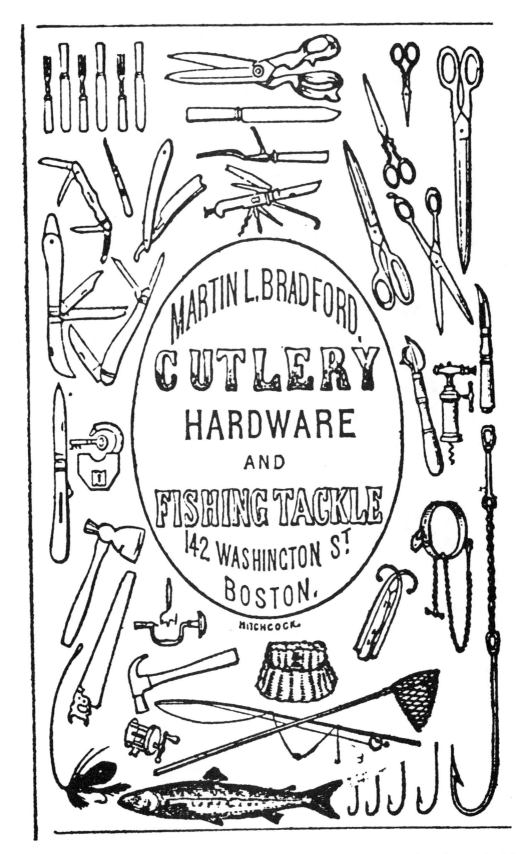

Martin L. Bradford's 1855 business directory advertisement—one of the earliest in the American "display" genre. Doubtless the multi-pronged knife, just above the oval, inspired the Swiss Army. Notice the bobber and the little squiggly things just below the rod-'n'-reel combo. These must be worms.

13 Great Fly-Tackle Shops

Collectors often discover fly tackle marked with a name other than that of the maker—usually a famous old tackle shop. Heddon rods, for instance, are often found marked Lyon & Coulson, a tackle seller in Buffalo. Thomas & Edwards wands may be found stamped Empire City, one of several brands used by the great old New York firm of Abbey & Imbrie. The tradition extends back at least to the 1850s, when we see Conroy reels marked "Bradford, Boston." Because many makers had links to specific tackle houses, it's often possible to identify mysterious items by dealer markings, whether they be indelible die stampings or more emphemeral acetate labels.

American tackle shops date back to the early 18th century. In those days, a typical small retailer served as an intermediary between customer and maker, typically conducting his business from a semipermanent table or stand and carrying a variety of goods appealing to housewives, farmers, woodworkers, and sportsmen. At the end of the day the vendor packed his "Hard-Wares" into a double sack, tossed it over his horse like a giant saddlebag, and went home. The term "stand" remained in use long after merchants were finally quartered indoors.

By the 1770s quantities of English fly tackle began arriving in Boston, exported by John Souch of London and imported by Jeremiah Allen, a hardware merchant. That same decade, Philadelphia tavernkeeper Edward Pole (a great name for a tackle dealer) began selling fishing tackle. Until the 1840s aristocratic fly tackle was advertised as "sporting apparatus." Around 1850 the term subdivided into "shooting tackle" and "fishing tackle." At this time major New York wholesalers, such as John Conroy, T. H. Bate, and Andrew Clerk had woven a sales base extending to Bradford's in Boston and up to Bangor, Maine.

J. HAPGOOD,

Manufacturer, Importer and Dealer in

GUNS,

Rifles, Revolvers, Pistols,

AND GUN MAKERS' GOODS.

ERICAN AND ENGLISH SPORTING POWDER,

In Kegs and Canisters, and all kinds of

SPORTING APPARATUS,

At Wholesale and Retail,

a. 12 Washington Street, Boston.

Gunmaker Joab Hapgood, active from the 1840s until after the Civil War, was the first recorded Boston dealer to advertise "sporting apparatus." By 1865, Hapgood had moved to 15 Washington and was pushing "fishing tackle." From the 1855 Maine Business Directory.

The Tackle Houses of Boston

The Bradford family, Hard-Ware dealers almost from their landing at Plymouth Rock, carried on a lively Boston trade through several generations. From the late 1700s to the 1840s, William B. Bradford and a son, John, sold "hardware and carpenter's planes." William's brother, Samuel Bradford, was probably a hardware merchant and auctioneer. The Bradfords were contemporaries of gun- and tacklesmiths John Lovell of Dock Square and Joab Hapgood of Washington Street, the area of the city that eventually became home to many tackle establishments. From the 1840s on good records of the fly-tackle trade exist, fixing Boston as a honey hole for Conroy and vom Hofe reels and rods built by Chubb, the Horrocks-Ibbotson Company, Thomas, and Edwards.

SAMUEL BRADFORD TO STODDARD'S

Samuel Bradford founded his business in 1800, but it was his son, Martin, who became America's first great tackle dealer. Although he took the reins in 1845, Martin L. Bradford wasn't listed in the *Boston Business Directory*, under "hardware," until 1846. With a shop at 142 Washington Street, a stone's throw from the historic Old South Church, Bradford advertised hardware and "gun cotton" in 1848. In 1850 he was the first Boston merchant to be listed in the new category of "Fishing Tackle." An 1855 advertisement also listed Bradford as a dealer in cutlery, an item the company and its successors would offer throughout their history. In 1856 the business became Martin L. Bradford & Company, and in 1861 it was listed as

Forest and Stream, *July 1, 1875*

"Manufacturers of Fishing-Rods." Just who built these early rods remains a mystery. In February of that same year the business moved to a "new granite store" at 178 Washington Street.

Bradford's grew as a wholesaler, sending salesmen along the post roads into rural Massachusetts, Vermont, New Hampshire, and, especially, Maine, where local smiths ordered sporting apparati shipped from Boston via Concord stagecoach. Bradford's lines included English wares and Conroy rods and reels. In the 1850s Martin was an agent for William Henry Alfred & Son of 54 Moorgate Street, London, which made split-bamboo rods with the enamel on the inside. Many tackle items marked "Bradford, Boston"—usually reels—appear to have been J. C. Conroy products.

In 1855 Bradford hired young Lorenzo Prouty as a salesman in the tackle department. An avid fly fisher,

Boston Gazette, *May 7, 1805*

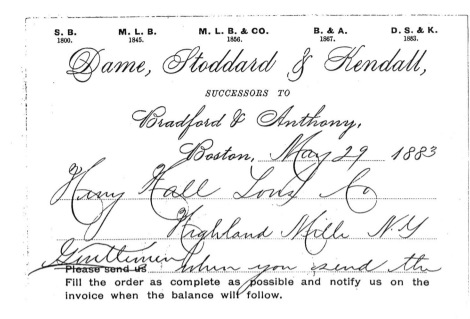

Around the turn of the century the company became known as the Dame, Stoddard Company. One of its rod brands, the Never-break, appears to have been the work of two makers—late Chubb or high-quality Varney-Montague, and very early F. E. Thomas. A little later, Dame, Stoddard changed its rod line to the F. E. Thomas Special grade, marked only as the Stoddard's Special. In 1920 the partnership ended, and the new business became known as Stoddard's. In the early to mid-1940s the house

Prouty had a keen knowledge of angling and a gift of gab without being obtrusive. He rose to department head and remained with Bradford's for 27 years. Called the "genial Prouty" by Doctor A. G. Wilkinson, Lorenzo was largely responsible for the great success of Bradford's tackle business.

Starting in 1864 the Boston store was listed as M. L. Bradford & Company. In 1867 Martin Bradford took on Nathan Anthony as a full partner and the company became known as Bradford & Anthony. Subsequent tackle was so stamped, along with the "Boston" address. Today, the wonderful antique reels marked to this prestigious New England company have a keen following. And collectors anxiously await the discovery of the original H. L. Leonard rods built for the historic Hub partnership.

In 1883 Bradford & Anthony was sold to the partnership of Dame, Stoddard & Kendall, "importers and dealers," which appears to have abandoned Bradford's wholesaling trade. The company continued to sell British flies, sold rods built by a variety of makers—including Nichols and Chubb—and, increasingly, bought wares from the vom Hofe family. The business moved several times along Washington Street; at one point it had a triple address of storefronts. Product trade names included Optimus, HUB, DS&K, and Bray brand tackle.

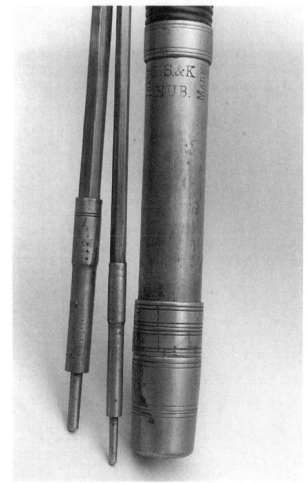

The "DS & K, HUB, TRADE MARK," a Dame, Stoddard & Kendall rod attributed to B. F. Nichols, Boston.

continued the Stoddard's Special, this time built by Gene Edwards. By the late 1940s Stoddard's carried Duckie Corkran's new Orvis rods, including a selection of early impregnated models. Still in the trade at Temple Square today, Stoddard's is perhaps the oldest continuously operated cutlery/tackle store in the United States. Collectors will find a good amount of tackle made by Julius vom Hofe, B. F. Nichols, Edwards, and Thomas, all marked with the name of this fine old company.

JOHN P. LOVELL TO THE IVER JOHNSON SPORTING GOODS COMPANY

According to the date given by Iver Johnson's in 1921, gunsmith John Lovell established his trade at Dock Square in 1840. He didn't actively advertise "fishing tackle" until after the Civil War, however. In 1867 the partnership of John P. Lovell & Sons began catering to anglers.[1] Marked tackle from this two-decade partnership has been found and at least one reel appears to have been of New York origin, made by Conroy or Fritz Vom Hofe.

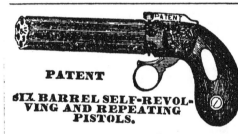

JOHN P. LOVELL,

Importer, Manufacturer and Dealer in

GUNS AND GUN MATERIAL

OF EVERY DESCRIPTION,

PATENT

SIX BARREL SELF-REVOLVING AND REPEATING PISTOLS.

No. 27 Dock Square, (up stairs,) Boston.

Four doors East of Exchange Street.

AGENT FOR HITCHCOCK & MUZZY'S RIFLE BARRELS.

1855 Maine Business Directory

The second generation of Lovells sold not only guns and fishing tackle but also a line of "Athletic and Sporting Goods of every description." Business grew, and by 1883 the Lovells occupied expanded new quarters at 147 Washington Street. The Lovell Brothers helped pioneer the idea of the big sporting goods store. Others, such as A. G. Spalding & Brothers, would do likewise, selling every conceivable type of outdoor equipment.

By 1886 the business was restructured as the John P. Lovell Arms Company, selling at wholesale and retail. At this time the Lovells also opened a branch store in Portland, Maine. Managed by Warren H. Chase and Albert E. Hall, the new Portland store offered a complete array of fishing tackle as well as accessories for the new sports of baseball and football. The tackle guru at Lovell's Portland store was noted Maine angler John W. Lewis.

Upon the death in 1900 of Colonel Ben Lovell, the last of the family, Lovell Arms became a subsidiary of Iver Johnson Arms & Cycle Works. The huge Boston retail store was renamed the Iver Johnson Sporting Goods Company. Iver Johnson's offered a complete stock of H. L. Leonard and Mills Standard rods, Julius vom Hofe and Montague City reels, Harrimac nets, and thousands of smaller tackle items, all sold under the makers' names.

Under Iver Johnson's house brands, classic Thomas rods of Dirigo grade can now be found marked as upper-grade Sagamores. Some of the early Thomas creations were built from Calcutta cane. The midpriced Johnson Radio rods appear to have been of Horrocks-Ibbotson origin. In 1921 Iver Johnson Sporting Goods was still a heavyweight in the trade at 150 Washington Street. The business continued until 1927 or later.

LANE & READ TO WILLIAM READ & SONS, INC.

The Read dynasty began in 1848 as the partnership of Lane & Read, located at 6 South Market Square and selling "Guns and Sporting Apparatus." Two years later the firm was renamed simply William Read. In an 1855 "Fowling Pieces" advertisement, Read also listed "Sportsmen and Travelers' Liquor Flasks, Rods, and Worms." That just about covers all an angler's needs: Thy rod, Thy worm, and Thy old Mr. Boston. By 1857 the business had grown into William Read & Son and moved to 13 Faneuil Hall Square; a decade later, Read took a second son into the partnership. At this time the Reads sold a wide variety of sporting paraphernalia, including tackle marked with their name.

In 1885 William and his sons moved to 107 Washington Street; collectors will find this address on most stamped Read products. Many marked William Read rods dated to this era feature excellent Calcutta

cane with a swelled butt above a sheet-cork grip, possibly the work of Charles E. Wheeler. The hardware on these 10- to 10½-footers was usually second class, made from nickel-plated brass, but the multitude of beautiful Chinese red wraps and German silver twist guides made for stunning rods.

The business was finally incorporated and survived until after 1900. Some high-quality rods from this era have been found with the marking "William Read & Sons, Inc., Boston." This is also the period of the "Read's Special"; the word "special" may signify a liaison with Fred Thomas. Thomas did make the suptuous circa-1899–1900 Imperial, an Izaak Walton–grade fly rod with the "new" classic look and a split-bamboo shaft made from Calcutta cane. These models had quality German silver fittings and ringed cork or rattan grips. No doubt reels were also stamped with the Read name; collectors should be on the lookout.

J. S. TROWBRIDGE & COMPANY

The tackle house of J. S. Trowbridge & Company was established in 1873 at 6 Faneuil Hall Square. Trowbridge's marked a great number of modestly priced rods and reels "Trowbridge, Boston." Most reels were small brass winches, often with rather robust side plates; some had John Kopf's patent reel foot. The rods were often huge 10½-foot Calcuttas with cheap fittings; many were designed in the all-purpose style.

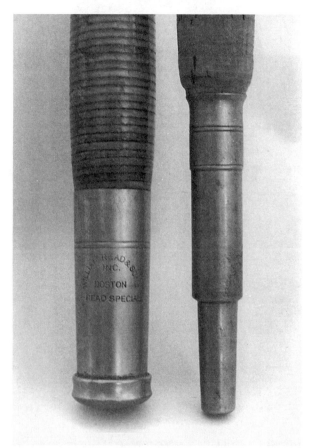

Not really a fly rod, but marked "Wm. Read & Sons, Inc. Boston, Read Special." "Special" may point toward Fred Thomas as the maker.

FOWLING PIECES
OF EVERY DESCRIPTION.

Rifles, Pistols, Powder, Shot, Balls, Percussion Caps, Copper and Leather Powder Flasks, Leather Shot Belts and Pouches, Game Bags, Gun Making Materials, Sportsmen and Travellers' Liquor Flasks, Double and Single Gun Barrels, Locks, Stocks, Rods, Worms, Gun Cases, Cylinders, Parts of Locks, Brass and Steel Mountings, constantly for sale by

WILLIAM READ,
NO. 6 MARKET SQUARE, BOSTON.

1855 Maine Business Directory

In 1885 J. S. Trowbridge & Company moved up to 88 Washington, and three years later the company opened a second retail store at 40 Friend. Both storefronts sold retail tackle until 1889 or later. Trowbridge's remained in business until around 1900, stamping its goods for well over a quarter century. Some of this company's marked reels were diminutive and charming.

PROUTY & APPLETON TO APPLETON & BASSETT

This historic line of partnerships was an offshoot of Bradford & Anthony, who were scheduled to retire and sell their business at the end of 1882. Upon learning that he wouldn't be their successor, a disappointed Lorenzo Prouty left Bradford & Anthony in December, taking long-time friend and associate George B. Appleton with him. Two weeks later Lorenzo Prouty was dead at the prime age of 43—the tackle house of Prouty & Appleton being the shortest-lived enterprise in the biz.[2]

George Appleton took Prouty's sales philosophy—high quality and personal attention—and continued the type of work he and Lorenzo had done at Bradford's. In January 1883 Henry Litchfield became Appleton's second partner, with a storefront at 304 Washington Street. At least one superb reel has been found marked "Appleton & Litchfield, Boston, Mass." as mentioned in chapter 3 in reference to Frank Philbrook.

Throughout the history of the company and of George's various partnerships, Appleton's always carried fly tackle of very high quality. Rodmakers Thomas, Edwards and Payne were also connected to this prestigious outfit. Some rods marked "APPLETON & BASSETT, BOSTON," were Isaak Walton–grade models built from Tonkin cane, not the usual Calcutta. These rods had ⅛-inch cork ring handles and early Kosmic-Walton grip checks; they were stamped with Fred Thomas's upper-case die set.

Appleton & Litchfield broke up in 1888, and 304 Washington reverted to Geo. B. Appleton & Com-

pany. George continued alone until 1893, when he took on Charles Bassett as his final partner. Reels from this association were marked "Appleton & Bassett, Boston," and known examples may also have been stamped by their maker, Julius vom Hofe. Appleton & Bassett remained in business until 1901, according to reel historian Jim Brown.

APPLETON & LITCHFIELD,

SUCCESSORS TO PROUTY & APPLETON,

IMPORTERS AND RETAILERS OF

FINE CUTLERY AND FISHING TACKLE.

From Fish: Their Habits and Haunts, *L. Prouty, 1883*

HENRY C. LITCHFIELD & COMPANY

After splitting with Appleton, Litchfield moved next door to 302 Washington, and in 1888 started his own retail tackle business as Henry C. Litchfield & Company. Litchfield also had an eye for quality. His rods were built by Benjamin Nichols, who joined him in the same shop space a few months later. These rods were marked "The B. F. Nichols, H. C. Litchfield & Co., Boston, Mass." At least one split-bamboo rod sold through this association was built to very old standards—two mids and four tips. They just don't make 'em like that anymore.

1923 advertisement, J. B. Hunter sold "high grade" tackle, and it was making enough money from this business to send out catalogs containing three color pages of flies. The best of Hunter's offerings were house-labeled Thomas fly rods, apparently Dirigo grade, marked only by Hunter.

BOB SMITH, SPORTING GOODS

Whoever Bob was he had great taste in tackle, and his successor is still in business and still called Bob Smith, Sporting Goods—a tribute to a great name. In Bob's "decal" era he sold a variety of rods—

Displayed in the April 1923 issue of National Sportsman, this Hunter ad accurately claims that its "Hand Made Split Bamboo Rods are unsurpassed for Quality." The fly models described were built by Fred Thomas.

H. A. WHITTEMORE & COMPANY

Whittemore's sold quality tackle. A number of Kosmic-marked rods have surfaced with the Whittemore company's stamping and address, giving us a link to the U.S. Net & Twine Company, probably through the great salesmanship of Herb Frost. The H. A. Whittemore store was active during the 1894–99 period, when Herb traveled from New York to Boston by train, selling his wares along the way. Whittemore's also sold a number of Kelso and Otter brand products.

THE J. B. HUNTER COMPANY

Hunter's was active from the early 1900s until the 1930s or later, selling a quantity of Montague rods with the company's arrowhead decal. According to its

Montagues and an early line of F. E. Thomas models. Then came some nice later W. E. Edwards models, and finally some Gene Edwards classics; the latter two offerings had "Bob Smith" scripted on their shafts.

New York's Tackle Trade

Although I know of no surviving advertisements from the era, New York tackle sellers probably date back to the early 18th century. By the late 1820s a number of hook, hardware, and sporting apparatus dealers were supplying Knickerbocker gentlemen with British fly-fishing accoutrements and with locally produced tackle as well—a large salmon or

National Sportsman, *April 1916*

saltwater reel marked by G. C. Furman in 1826, for instance.

After 1830 a small number of sellers began to appear in the *New York City Business Directory.* As with Boston firms, the majority of these early houses were direct importers of hardware, guns, and cutlery. Between 1830 and 1832 we find tackle listings for Thomas W. Horsfield on Fulton Street, Chas. R. Taylor at 1¼ Maiden Lane, John Lentner at 39 Nassau Street, and an obscure chap named Lewis at 3 Wall Street. The tackle house of Abraham Brower, at 230 Water Street, touted "Grass, Gut silk, and (horse) Hair Lines of English and American manufacture," as well as Limerick and steel fishhooks and "A large assortment of Canton and Calcutta, [solid] Bamboo Fishing rods with or without joints, for Fly and other fishing." Two dealers destined to became giants in the trade, T. & J. Bate and Andrew Clerk, were in business at this time, but don't seem to have advertised.

ANDREW CLERK & COMPANY TO ABBEY & IMBRIE

One of New York's Big Three, Andrew Clerk's was established in 1820, according to its extensive advertising. Obscure for its first two decades, Clerk's ultimately became a large and well-frequented tackle house, venerated by collectors for the early and steadfast patronage given to emerging split-bamboo rodmakers by "Dr. Clerk" and his successor, Charles Imbrie. Clerk carried the handcrafted rods of Samuel Phillippe and Charles F. Murphy until 1875 and was probably the first tackle house to export large quantities of American tackle, especially split-bamboo rods, to the British Isles.

Andrew Clerk's also carried an extensive selection of New York–built reels, presumably by the Conroys and Jabez Crook, and flies of all sizes and descriptions. The rapid growth of this early tackle house can be credited to the keen business sense of Andrew Clerk, who was not afraid to invest in new and sometimes totally odd products. This form of trend setting would not be seen again until T. J. Conroy arrived on the New York scene some decades later.

Forest and Stream, *May 11, 1875*

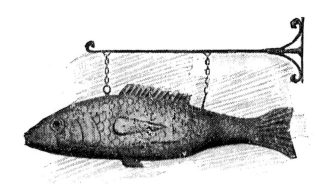

This is the most famous gold fish in the world. It hung in front of Abbey & Imbrie's door for forty years and was honorably retired in 1907.

Field & Stream, May 1920

In 1875 Abbey & Imbrie succeeded Andrew Clerk & Company. A&I's driving force, Charles F. Imbrie, who'd been with the original Clerk's staff for a number of years, decided to drop Murphy and gamble on a promising new split-bamboo rodmaker from Maine named Hiram Leonard. Leonard proved to be the independent type, however, and after less than two years he stopped supplying A&I exclusively. Abbey & Imbrie's later rods included the inventive products of the Pritchard Brothers and quality pieces by Horrocks-Ibbotson and Thomas & Edwards.

A&I continued the Clerk's tradition of importing British wares, in particular reels made by Samuel All-cock & Company, whose brass winches could be ordered in a fine nickel plate for 25 cents extra. Reels sold from 1878 through 1882 had the original A&I logo—crossed fishhooks with an A and an I located *inside*. In 1883 the initials extended *outside* the hooks. This logo, used for the greater part of the late 19th century, also appeared within the center of the company's last logo, the dolphin trademark of the 1900s. Careful examination of these markings will aid you in aging Abbey & Imbrie reels.

Many fly reels sold after 1883 were marked "Abbey & Imbrie, N.Y.," and the best of these were built by Julius vom Hofe. They ranged from "squared-off" raised-pillar models to protected-rim reels, usually constructed from hard rubber and German silver, although nickel-plated brass may be seen in lesser grades.

The company sold an immense volume of tackle during its century-long tenure, wholesaling to thousands of dealers from Maine to Chicago and at one time cataloging over 16,000 items. Originally located at 18 Vesey Street, A&I eventually moved to Chambers, and in the 1920s became a division of Baker, Murray & Imbrie, Inc. In this high-classic period it carried a series of E. W. Edwards Deluxe fly rods that also sported an A&I decal on the lower shaft. The contemporaneous Duplex, a bait-caster-fly-rod combo, appears to have had Thomas origins. In 1930, the corporation sold A&I to Horrocks-Ibbotson of Utica.

T. & J. BATE TO WILLIAM MILLS & SON

The old firm of Thomas & James Bate was started by two needle- and hookmaking brothers from Redditch, England, who sent a representative to New York in 1822. In 1828 Thomas Bate arrived from the Isles and remained. The earliest advertising

THOMAS H. BATES, MANUFACTURER AND IMPORTER of Needles, Fish-Hooks, and Fishing Tackle, has constantly on hand a large assortment of Fishing-Rods, Reels, Silk, Linen, and Cotton Lines, Chinese Grass Line, and Silkworm Gut, Bamboo and Reed Poles.
    Patentee of the Improved Serpentine Spinner, the best Trolling Bait in use.   [81-132]   85 Maiden Lane, New York.

Note the typo: "Thomas H. Bates." The real T. H. Bate was not related to noted Hollywood hotelier and taxidermist Norman Bates. From Porter's Spirit of the Times, January 20, 1859.

for the company appeared in 1832, but it offered needles and hooks only—no general or fly tackle. James's son, Thomas H., came to New York 11 years later, and the partnership was restructured as T. & T. H. Bate. In 1853 son-in-law William Mills moved from England to the United States and entered the partnership. Before 1859 the Maiden Lane company became Thomas H. Bate, engaged in the general tackle trade.

By 1867 the company was known as Thomas H. Bate & Company and had moved to No. 7 Warren Street. Its catalog that year offered a variety of tackle—English (Allcock's) and New York plain and click reels, Hartill's patent reels, plus rods, lines, gut, flies, and even rodmaking supplies for the home builder. Collectors can today find some beautiful little brass winches marked by this last of the structured Bate companies. The rods are much rarer; some were evidently built in a plant owned by the Bates in the small town of Central Valley, just north of the City. These wooden rods have escaped collectors to date, but we should be on the lookout. It must have been a good factory; Hiram Leonard occupied it in 1881.

In 1873 the last Bate family member "fished 'round the bend," and the company became William Mills & Son, which grew into a most prestigious New York tackle house. One of its greatest coups was the purchase of James Kidder's holdings in H. L. Leonard's rodmaking business. After the death of his father in 1883, Thomas Bate Mills took over the company but never changed its name. By 1905 William Mills & Son's retail section moved to its famous store at 19–21 Park Place—a dimly lit emporium that served

the wants of fly fishers for 70 years.

The company continued to operate its factories in Redditch and Central Valley, and created an extensive jobber-wholesale base across the United States. The sons of Thomas Bate Mills—Eddie the tournament fly caster, Old Arthur, and Chester the fly tyer—cut their teeth on the salesman's circuit, stopping at Mom-'n'-Pop general stores and local hardware shops to convert another rural business into a William Mills & Son retailer.

The Mills products are legendary—the Leonard rod and reel, the Mills fly reels and Fairy models, the Standard rods, and on and on. Most of the tackle sold through this great house was marked

The Outdoorsman's Handbook, W. H. Miller, 1916

by Mills, and is easily identified. When Old Arthur died, in 1942, his son Arthur C. Mills—the last surviving member of the family—became the company president. In the early 1970s, after the passing of "Young Arthur," the great old firm of William Mills & Son became history.

JOHN P. MOORE TO SCHOVERLING, DALY & GALES

One of the city's oldest sporting goods houses was established in 1823 by gunsmith John P. Moore. After the Civil War the store became John P. Moore's Sons and offered an extensive line of quality fly tackle, including the rods of Mainer Charles E. Wheeler. At the end of 1887 the last active son, George G. Moore, sold the business to famous gun dealer Shoverling, Daly & Gales.

Maker of the superb Charles Daly shotgun, SD&G relinquished its original Chambers Street shop, took over the old Moore addresses at 302 Broadway and 84 Duane, and continued to carry Wheeler fly rods—along with a quantity of tackle—at the retail and wholesale levels. A later fly rod of unknown make, the Victory, appeared after 1900. The

Bass Flies on Edgar's Barbless Hooks.
(Wm. Mills & Son.)

German Silver Click-Reel.—40 yards.
(Wm. Mills & Son.)

William Mills & Son Tackle. Three very collectible Mills fly-tackle items include Lake flies tied on Edgar's barbless hooks, plus the extremely rare all–German silver Click Reel and a unique raised-pillar revolving-plate version of the Mills CROWN. Examples like these, along with Mills's renowned H. L. Leonard fly rods, could form an entire collection.

Forest and Stream, May 26, 1887

SD&G Perfection series, which appeared at about the same time as the Victory, was built by Fred Divine. Both models can now be found marked by the seller. The well-heeled emporium sold fishing tackle until the 1920s or later. The company, perhaps restructured, continued production of the Charles Daly shotgun until the 1960s. The last models were built in Japan by B. C. Miroku.

JOHN CONROY TO T. J. CONROY

John Conroy, a machinist, probably began making reels in the mid-1820s. The company claimed it began selling fishing equipment in 1830, but the first general tackle ads didn't show up until around 1837. Along with Mills and Abbey & Imbrie, the Conroy dynasty was an important firm, stamping a great number of fine collectibles with its name. This large jobber, wholesaler, and retailer was covered in chapter 6 under Brooklyn tacklemakers, but note that Conroy's was perhaps the first New York outfit to wholesale tackle to the interior rural states, sometime during the 1840s.

Starting in 1875, at the beginning of the Thomas J. Conroy era, the house had grown to the point that it opened a branch store in London. T. J. had a keen eye for innovation and actively sought new and different fly-fishing items, including the rods of Malleson, Landman, and Divine, and a line of the best Malleson and Meisselbach reels.

J. B. CROOK & COMPANY

J. B. Crook & Company was established in 1837 at 50 Fulton at roughly the same time as Conroy. Jabez Crook, who may have started his career as a machinist for Conroy, became a well-known maker of reels and split-bamboo rods, and also imported British guns and tackle. Starting in late 1883 or early 1884 the shop address also included 52 Fulton, Conroy's old space. During this decade Crook's wares expanded to include a general line of sporting goods, archery equipment, lawn tennis stuff, and all that rot. Many ball-handled reels, a very few being small fly-sized cuties, were marked "J. B. Crook, Maker, N.Y." You *should* find rods marked by Crook as well, but for reasons unknown, J. B. Crook–marked fly tackle is scarce. What happened to Crook? He was advertising in 1884, but by 1887 the business had disappeared.

1919 SD&G letterhead

J. F. MARSTERS

In 1861 James Marsters had a stand at 578 Atlantic Avenue in Brooklyn. Marsters was then a small-time reelmaker but he later became a major fly dealer, with a large crew of New York women fashioning the fancy patterns so popular throughout the latter part of the 1800s. By 1894 his store had moved to 55 Court Street, Brooklyn, where it remained as a popular retail spot. His flies were advertised nationally and

CONROY, BISSETT & MALLESON,

(Successors to J. C. Conroy & Co.)

65 Fulton St., N. Y.,

TRADE MARK.

Would invite the attention of amateurs to the additions they have made to their regular stock of Fine Goods, comprising in part Six Strip Split Bamboo Rods for trout and salmon. The famed "McGinnis" Black Bass Rods, made of Split Bamboo (new this season). Vom Hofe's Rubber steel Pivot, Click and Salmon Reels. The celebrated "Frankfort" Reel, &c., &c. Sole manufacturers of "Mullaly's" Patent Flies. The New style Cuttyhunk Bass Lines of Extra quality.
  Parties fitted out for the Adirondacks, the Maine Woods, Lake Superior, Newport, Cuttyhunk, Pasque Island, West Island, Barnegat, &c.

ORDERS BY MAIL WILL RECEIVE CAREFUL AND PROMPT ATTENTION.

J. B. CROOK & CO.,

IMPORTERS AND MANUFACTURERS OF

FISHING TACKLE,

50 FULTON STREET, N. Y.

Sole manufacturers of the celebrated "GREENHART AND SPLIT BAMBOO RODS," for salmon and trout fishing.

wholesaled to dealers in the field; the fly rods appear to have been made by Chubb and, later, Montague. A 1916 ad showed James F. Marsters still in business at 52 Court and offering tackle through retail and catalog sales.

No hard feelings here. A rear view of an aluminum and hard-rubber Fred Malleson reel, marked to his former business partners—the logo of Conroy & Bissett.

HAWKS & OGILVY TO THE ROBERT OGILVY COMPANY

The fine old company of Hawks & Ogilvy, established in 1871, became the largest New York seller of Fritz Vom Hofe & Son reels. Of course, it also stamped its name on a large quantity of fishing tackle built by other makers. At some point around 1900, the business became the Robert Ogilvy Company. In 1910, as seen in its 1923 ad, the outfit was incorporated. Most of the marked Ogilvy tackle came from the early Hawks partnership years;

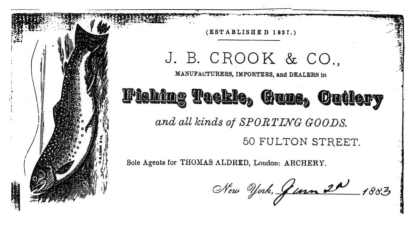

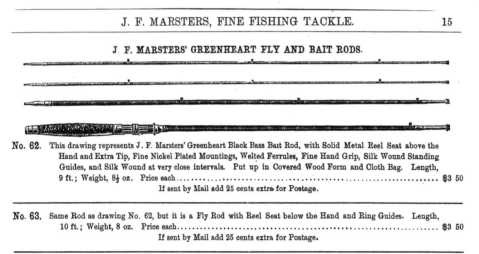

J. F. MARSTERS, FINE FISHING TACKLE.   15

J. F. MARSTERS' GREENHEART FLY AND BAIT RODS.

No. 62. This drawing represents J. F. Marsters' Greenheart Black Bass Bait Rod, with Solid Metal Reel Seat above the Hand and Extra Tip, Fine Nickel Plated Mountings, Welted Ferrules, Fine Hand Grip, Silk Wound Standing Guides, and Silk Wound at very close intervals. Put up in Covered Wood Form and Cloth Bag. Length, 9 ft.; Weight, 8½ oz. Price each...$3 50
If sent by Mail add 25 cents extra for Postage.

No. 63. Same Rod as drawing No. 62, but it is a Fly Rod with Reel Seat below the Hand and Ring Guides. Length, 10 ft.; Weight, 8 oz. Price each...$3 50
If sent by Mail add 25 cents extra for Postage.

American Game Fishes, *Shields, 1892*

National Sportsman, *April 1916*

these items bring excellent prices today.

E. G. KOENIG COMPANY

A fine old Prussian name in tackle, E. G. Koenig started retailing in the early 1880s in the Benedict Building on Court Street and remained in business until 1926 or later. By 1884 Koenig had established a branch salesroom at 875 Broad Street, Newark. In the 1890s Koenig briefly entered a rodmaking partnership. The only model I've found marked "Nichols-Koenig Rod Company, Newark, NJ" had a quality George Varney reel seat, Montague-Chubb German silver ferrules, and a grip built from ⅛-inch cork rings. These features indicate an 1895–98-era origin; the rod could be mistaken for a Thomas & Edwards. It may have been built by Benjamin F. Nichols in Boston and labeled by the New Jersey store.

The E. G. Koenig Company was a major fly-tackle house, selling the early Kosmic line and Julius vom Hofe's fly reels. In his early years Koenig seems to have suffered from financial anorexia. Julius vom Hofe sent a plea for payment in July 1886, telling Koenig that "I am losing my patience, you

E. G. KOENIG,

No. 1 Courtland street, New York.
BENEDICT BUILDING.

GUNS,

RIFLES,

Ammunition.

FINE

Fishing
TACKLE,

Boats and Canoes.

Sportsmen's supplies of every description at lowest prices. Shells carefully loaded.

Branch Store, 875 Broad Street, Newark, N. J.

Book on Fishing with Hook and Line, by Frank Forrester, mailed to any address on receipt of 25 cents.

The American Angler, September 27, 1884

owe me the money $104.88 long enough now, almost a year. I should think you ought to pay, let me know its cause, and send me money as soon as possible, as I am in need of it."[3] Julius's postscript read—one Prussian to another—"This is no business way of doing."

VON LENGERKE & DETMOLD

VL&D was born in 1882 and gradually built up a solid clientele by selling high-quality goods. It didn't just carry the high-priced spread, however; throughout its history Von Lengerke & Detmold carried fly tackle made by the likes of Gus Meisselbach, Yawman & Erbe, Horrocks-Ibbotson, Martin, and Pflueger. In 1910 the company was located at 700 Fifth Avenue, but sometime between then and 1923 it moved to 349 Madison. During the early 1920s VL&D incorporated, with F. H. Schauffler as president.

Von Lengerke & Detmold sold a fine selection of hand-tied dry flies by Jean Erskine as well as bass bugs by Heddon, Hayes, Tuttle, and Callmac. For a few years prior to 1920 it carried the Favorite, an excellent fly rod built by E. W. Edwards in Brewer, Maine. VL&D was the second largest New York outlet for Fred and Leon Thomas, and a prolific retailer of early Jordan-built Orvis rods. The company also

imported British tackle, including Halford silk fly lines and Hardy Brothers reels.

U.S. NET & TWINE COMPANY

U.S. Net & Twine, located at 316 Broadway, entered the retail-wholesale trade in 1887, at first as a reelmaker; in 1894 it became a rod manufacturer although it supposedly bowed out of manufacturing in 1899. Charles Pratt, of Standard Oil and Pratt Institute fame, was the company's president and driving force. U.S. Net's most famous line included Kosmic rods and reels, and Walton rods. Lesser tackle encompassed a line of Climax reels, which were wholesaled throughout the country. After 1897 it carried Thomas & Edwards rods. Then for a couple years the company sold early F. E. Thomas fly rods, often marked with the old Isaak Walton die or stamping. Recently I acquired a charming little 7½-foot Isaak Walton that nicely handles a delicate 4-weight line. Today, U.S. Net & Twine rods are held in high regard both as valuable early classics and as dependable working tools.

H. J. FROST & COMPANY

In 1896 U.S. Net & Twine's principal drummer, Herbert J. Frost, started his own company. Located

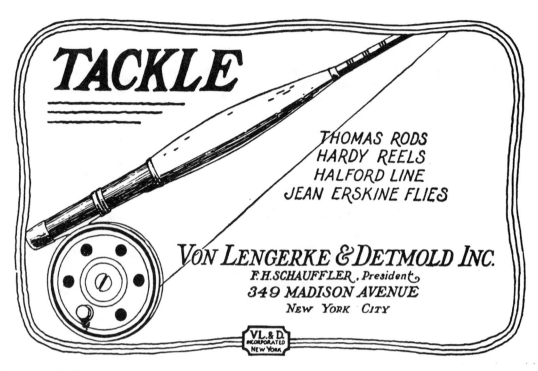

TACKLE

THOMAS RODS
HARDY REELS
HALFORD LINE
JEAN ERSKINE FLIES

VON LENGERKE & DETMOLD INC.
F.H.SCHAUFFLER , President,
349 MADISON AVENUE
NEW YORK CITY

VL.& D.
INCORPORATED
NEW YORK

National Sportsman, *April 1923*

at 90 Chambers Street, Frost's offered an extensive line of house-labeled tackle, including Senate Steel Vine rods, Otter brand rods (made by Divine), the Kelso Automatic reel, and the Gem and Ideal single-action fly reels. The higher-quality merchandise was Frost's Kelso brand—not to be confused with Forrest of Kelso tackle from Scotland, of course. Herb was a salesman, and went to whatever lengths it took to sell tackle. Some Frost's Kelso rods appear to have been house-marked Kosmics and carried the celluloid reel seat. The company also wholesaled, and a number of Frost-labeled products were retailed through Boston stores, particularly the Senate brand, which was sold through Iver Johnson's.

The rods were built by Horrocks-Ibbotson and had a Senate Steel Vine decal at the junction of the grip and shaft. Two models were sold; a nickel-plated version at $6.65, retail, and a German silver–appointed line at $10. All the rods were built round, and they were available in lengths of from 8 to 10 feet. Otter brand rods were stamped on the reel seat and the entire round split-bamboo blank was wrapped with clear (white) silk thread, making them strong and heavy enough to dislodge mud-dauber nests 'neath the eaves of many a turn-of-the-century angler's home. Surprisingly, the most collectible

Frost products today are the scarce German silver versions of his Gem and Ideal fly reels, which appear to have been contract items. I believe that Herb remained in business until the 1920s.

ABERCROMBIE & FITCH

The tackle house founded by David T. Abercrombie and Ezra Fitch was a famous one, lasting from the turn of the century until the 1970s, although the partnership was over by May 1908. The original store, located at 57 Reade Street, remained in operation under the same name. Abercrombie went on to form the David T. Abercrombie Company and established "Abercrombie's Camp" at 311 Broadway. This store was a tad smaller than the original, but still had room for demonstration fly casters and shootists such as "Flyrod" Crosby and Adolph Topperwein. Before 1920 Abercrombie sold his business to Baker, Murray & Imbrie, Inc., which still had showrooms at the above address in April 1942.

Abercrombie's Camp would eventually pass by the wayside, but Ezra Fitch's place remained high profile. As early offerings he carried a complete line of Divine rods and reels made by Orvis and Frost. Around 1915 the original Abercrombie & Fitch moved to a new store—the largest retail angler-

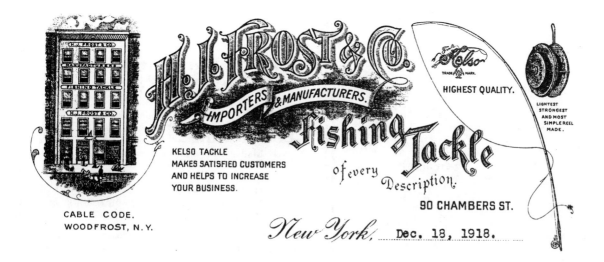

CABLE CODE.
WOODFROST, N.Y.

New York, Dec. 18, 1918.

Mr. Kaufman,
C/o., Pequea Works,
Strasburg,Pa.

Dear Mr. Kaufman:

I have been going over the shipments and I find that we have got about 35 shipments partly packed waiting for your goods, and in looking over the letters we have written you about this I find that they are very numerous but fail to find any answer from you as to when you will ship.

Thought you might not have received these letters so I am writing you personally. If you can't ship all of our orders, send us some of each so we can take care of our customers and get some of the cases out of the place that we have packed up waiting for your goods.

I have one order before me #6456, which was dated last June. I find there is only one item that has been shipped on it, and that only part, 5 gross out of 50 BB Split Shot. Now this won't do. We must have some goods. I want you to let me hear from you by return mail. Also please let us know if the Ball Twine that we sent you is all right and if you want the balance of the goods shipped, as I thought it best to hear from you after you received this lot to be sure it was right. At the time we sent you a sample we sent one to Chester and he has ordered the goods subject to prior sale, so if you take these we will not have it for him.

Very truly yours,

H. J. Frost

HJF/M

hunter establishment in New York—at the corner of Madison Avenue and 45th Street.

During the Depression Ezra Fitch held the distinction of almost single-handedly saving the quality American fly rod. William Mills & Son pushed Leonard rods almost exclusively; rods below Leonard-quality were very basic factory-built production items. When times were tough, however, the old Leonard gang—Jim Payne, Hiram Hawes, Eustis William Edwards and sons, and Fred and Leon Thomas—found financial salvation in the patronage of Ezra Fitch. Fitch also imported Hardy rods and reels, and in the late 1930s contracted with Hardy to produce the little Flyweight Silent-Check model. The Silent-Check, one of the nicest of tiny modern midge reels, avoided all that streamside clicking that drives some of us up a tree.

Chances are, if you couldn't find it at Abercrombie & Fitch, it wasn't worth buying. The store carried the best and the second best, from the Yellowstone Special rod by Thomas to the Folding Trout Knife by Marble. After the Thomas Yellowstones came those built by Edwards and, later, Bill Phillipson. The Favorite, made by E. W. Edwards & Sons, was indeed a favorite, especially the classic two-piece 7½-foot model that handled a 5-weight line. A&F house-stamped items were sold for well over 50 years and appear often today. In the late 1960s, after Jim Payne's death, the store was overrun by speculators in Payne rods. A few years later Abercrombie & Fitch closed its doors and the contents were auctioned off. With the 1970s demise of A&F and the old William Mills & Son store, New York's classic tackle era came to an end.

Philadelphia Tackle Shops

Philadelphia is steeped in fly-fishing tradition. America's earliest recorded fly tyer, David Hugh Davis, was a Philadelphia innkeeper around 1795. A Quaker, Davis also fashioned "fishing tackle deepseas."[4] No doubt his freshwater "fly feathers" were somber, probably Cow Dungs and Alders, with per-

"Where the Blazed Trail Crosses the Boulevard"

The new building of

ABERCROMBIE & FITCH CO.
Madison Ave. and Forty-fifth Street
New York

Is the sporting headquarters of the Western Hemisphere, unique in its appointments and unequalled in size.

Its fishing and hunting departments, its rifle ranges, trout pool, casting tank, guides' room, proving ground for tents and all camp equipment —its general and practical completeness, give it a character presented by no other establishment on earth.

To the Outdoorsman it is the most interesting spot in New York—the place where his every thought has been correctly anticipated.

"The Greatest Sporting Goods Store in the World"

The Outdoorsman's Handbook, W. H. Miller, 1916

haps an Olive Dunfly for special occasions. A contemporaneous tackle dealer, Edward Pole, had a tavern on the outskirts of town called the Wigwam where patrons not only could buy tackle but also have a snort of brandy and tell lies into the night. Pole sold his business to George R. Lawton, who moved it to Great Dock Street around 1800. Lawton's most notable tackle sale was to noted worm-dunker Merriwether Lewis, who was mounting an expedition into the puckerbrush of Big Fish Country.[5]

A. B. SHIPLEY TO M. A. SHIPLEY SPORTING GOODS

Shipley's was an old and long-lived company. Augustus B. Shipley started the business in 1793, operating one of the original stands on Commerce Street, in the heart of the city. Later he constructed a building at more or less the same spot, 503 Com-

National Sportsman, *April 1916*

merce Street. Shipley was a dedicated angler and disciple of the long wand; he collaborated with Edward Fitzgibbon on a fly-fishing volume in 1839.

Shipley always claimed to have made his own rods and reels, and many collectors take his word at face value. I believe, however, that Augustus used the term "make" rather loosely; the reels of his "own pattern" were either directly influenced or made by New York makers, differing primarily in finish details. The early reels looked like Conroys, and later models, sometimes utilizing hard rubber in their construction, looked like vom Hofes. The Shipley "pattern" included a straight, stylized crank knob, with a deep chamfer at the end adjoining the handle.

Fly tackle marked by A. B. Shipley is rare, high quality, and quite collectible. Shipley's son, Malcolm A., ran the business during the second half of the 19th century, and in 1895 changed the firm's name to Malcolm A. Shipley. At this time the Shipley-marked

Meisselbach reels first appeared. Around 1905 the company became M. A. Shipley Sporting Goods. It remained in the business until approximately 1920.

GEORGE W. TRYON TO THE EDWARD K. TRYON COMPANY

Established in 1811, Tryon went through more name changes than any other sporting goods house. The business went by the founder's name until 1836, when "& Company" was added to recognize George's son as the new junior member. Edward controlled the house after 1841, making room for a brother in the 1860s. Tackle markings followed the changes in hierarchy: from 1841 to 1843, "Edw. K. Tryon & Co."; 1843–59, "Edw. K. Tryon"; 1859–63, "Edw. K. Tryon & Co." again; 1863–66, "Tryon & Bro."; 1866–68, "Tryon Bros. & Co."; 1868–1905, "Edw. K. Tryon Jr. & Co."; 1905 to approximately 1925, "Edw. K. Tryon Co." Whew! Throughout its history, Tryon imported, wholesaled, and retailed a huge selection of quality rifles, pistols, cutlery, and fly tackle.

The Tryons imported H. C. Pennell's hooks and were evidently personal acquaintances of this respected British maker-writer; the Pennell name is stamped or labeled on a variety of Tryon's tackle. The reels were probably first built by U.S. Net & Twine; later reels were usually from the Montague factory in Brooklyn, and the rods also seem to have been of Montague lineage. Some collectors have tried to link the Pennell name directly to a Philadelphia manufacturer. Based on the features, including handles, chamfered screws, and bearing caps, the Pennell-marked reels were quality Montagues. Sears, Roebuck also used the Pennell name, but these models, also made by Montague, were not so stamped. Tryon elevated the Pennell name to a buyer's buzzword, and Sears simply took advantage of it.

The Tryons also used the Kingfisher trademark, which was pasted on countless Montague rods during the first two decades of the 1900s. Other companies used the name as well, including the E. J. Martin Line Company. During this period, the E. K. Tryon

Forest and Stream, *September 9, 1880*

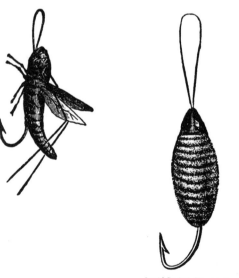

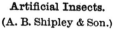

Artificial Insects.
(A. B. Shipley & Son.)

toric old sporting goods business continued into the 1930s. An early Heddon-built fly rod, an 8-footer marked "#103 Kingfisher," has been found. Later rods were marked by Jay Harvey, perhaps a cousin of J. C. Higgins. These fly rods were built by Heddon in the higher grades, and have "Jay Harvey" engraved on the reel seat.[6] Tryon also carried a lesser Jay Harvey grade built by Montague.

JOHN KRIDER TO THE L. C. SIMER COMPANY

John Krider was an early and important figure in Philadelphia sporting goods—an importer, wholesaler, and retailer active from the 1820s until around 1880 (his rodmaking career is discussed in chapter 2). From Krider's 1853 catalog (illustrated) a collector can scope the earliest detailed listing of American fly tackle yet found. Krider had much in common with Thaddeus Norris, who moved to Philadelphia to start his family's branch store. Any item of fly tackle marked by either Krider or Norris is a prime collectible. At some point after 1878 John Krider's famous shop, at the northeast corner of Second and Walnut Streets, was occupied by the L. C. Simer Company. Krider's probable successors, L. C. Simer continued to sell sporting goods and repair rods.

M&H SPORTING GOODS

M&H Sporting Goods was an industry giant. Owner of the Ocean City Reel Company, M&H also purchased Montague's rod and reel business in the 1930s, and later the Edward vom Hofe Company. M&H also sold classic Heddon rods marked "M & H Special by Heddon" that featured an ivory-colored down-locking reel seat and blued ferrules.

M&H carried classic fly tackle for over four decades. The company's principal stockholders, Lou Moskowitz and Paul Johnson, merged Montague with Ocean City in 1953. M&H had the best retail staff obtainable, including a number of outdoor writers. Collector

Company was located at 817–819 Arch Street, with Charles Z. Tryon as president—the last of the Tryons and the major holder since 1905. After the death or retirement of Charles, sometime after 1921, the his-

Albert Chance, writing to me in 1981, described his first tackle purchases as a young man:

By 1931, I had a job and had saved sufficient money to invest in a split bamboo rod together with an Ocean City reel. These were purchased from the M&H Sporting Goods store in Philadelphia. The salesman was Bob Hall who was the Salt Water Editor for, I think, *Outdoor Life*. Bob was one of the few peerless salesmen who knew his products, how to use them, and would give honest and knowledgeable advice to his customers. I also have both a split bamboo casting rod and a bamboo fly rod purchased from him.

 ## Tackle Houses in Chicago

Chicago may have been late to join the tackle trade, but several houses, notably Spalding and Wilkinson, managed to grow into substantial businesses. So, too, did a couple of local companies that sold affordable tackle through the mail: Sears, Roebuck and Montgomery-Ward. Doubltless many young rural anglers got hooked on the sport while sitting on their two-holers and dreaming of saving up $1.05 for a St. Croix River Lancewood Fly Rod.

Other Chicago tackle merchants offered high-quality items built by Thomas Chubb, Fred Divine, and the former Leonard crew. Utica-built tackle probably arrived in Chicago first, hauled by oxen and mules over the Erie Canal and then shipped—literally—across the Great Lakes to the Windy City. D. H. Lamberson, one of the city's first large fly-tackle dealers, sold New York State products—rods by Divine and guns by Philo Remington—from its store on State Street, where it remained in business until 1881 or later.

E. E. EATON'S

From the 1870s on the quality tackle house of E. E. Eaton's carried a large

National Sportsman, *April 1916*

JOHN KRIDER,

MANUFACTURER OF

SHOT GUNS, RIFLES AND PISTOLS,

ALSO, IMPORTER OF

GUNS AND ALL SPORTING APPARATUS,

FISHING TACKLE AND FINE CUTLERY,

N. E. CORNER OF SECOND AND WALNUT STS.,

PHILADELPHIA.

FISHING TACKLE.

HOOKS.—Genuine Limerick Salmon Hooks; best Limerick Trout Hooks; best Limerick Salmon Hooks, flatted; best Limerick River Hooks, flatted; best Limerick Hooks, bowed; genuine Virginia Hooks, all sizes; Kirby Black Fish Hooks, all sizes; Kirby Salmon Hooks; Chestertown Hooks; best Kirby Hooks, bowed.

LINES.—Plaited Silk Lines, Twisted Silk Lines, Silk and Hair Fly Lines, Twisted Hair Lines, China Grass Lines, and also a large assortment of Cotton and Linen Lines.

FISHING RODS.—Walking Cane Rods, three and four joint, plug end; Walking Cane Rods, three and four joint, screw ferrule; Walking Cane Rods, three and four joint, ash butts; Hazel Rods, three and four joint, brass ferrule, whalebone tips; Bamboo Rods, four joint, ringed; finely mounted Trout Rods, three and four joints; Trunk Rods, five and six joints. Also, a large assortment of common Rods always on hand.

BRASS FISHING REELS, multiplying and plain; Fly Tackle Books; Trout Baskets, best white gimp, all sizes; Bait Boxes; Ferrules, Tips and Rings, for Rods; best quill Floats, bound and unbound; Egg-shape Cork and Wood Floats; large bound Floats, assorted; Swivel and Lead Sinkers; Limerick and Kirby Hooks on gimp; Limerick Trout Hooks on single gut; Limerick Salmon Hooks on twisted gut; superfine Kirby Hooks on gut; Virginia Hooks on gimp; Limerick Hooks on bristles; Kirby Hooks on hair; fine Artificial Salmon and Trout Flies; Black Fish Snoods, single and double; Artificial Minnows, of leather, tinsel and pearl; Artificial Grasshoppers, Frogs, Shrimps and Caterpillars; Spoon Bait, for bay fishing; one, two, three and four hook Gut, Grass and Hair Snoods; Float and Deep-sea Lines, assorted; Jointed and Common Bows. Besides many articles too numerous to mention, sold wholesale and retail, on the lowest terms.

From Krider's Sporting Notes, J. Krider, 1853

Forest and Stream, *November 8, 1877*

selection of "Fishing Rods, Guns, Rifles." Eaton's catalogs devoted many pages to the angler. A 40-page Eaton's catalog published in 1881 sold at auction in 1987 for $180 plus the buyer's premium—almost $5 a page! At these prices, we should start saving all those Orvis catalogs.

Eaton's sold a striking array of high-quality merchandise, including custom rods by Malleson and superior Milam reels. I've seen a matched set of 7½-foot E. E. Eaton's-marked Malleson rods that simply exuded the rich aura of the Gaslight Era. The smaller model was a light reel-above-the-hand rod of excellent workmanship, including hand-fashioned tunnel guides, designed for light bait fishing for trout. Eaton's remained a large tackle house through the last quarter of the century, and sold wholesale as well as retail.

JOHN WILKINSON TO THE WILKINSON COMPANY

John Wilkinson began business in 1872 as a gunsmith. By 1877 he was retailing a huge line of sporting goods that included "fishing tackle," and had a wide-ranging mail-order clientele. The retail storefront was at 75–77 State Street. As sales grew, so did the company; by 1884 it was known as The John Wilkinson Company and offered an entire mail-order catalog devoted to angling. Along this same time the business moved to a double storefront on Wabash Avenue.

In 1892 the firm was incorporated as The Wilkinson Company, "Manufacturers, Importers, and Dealers" in fishing tackle. Four years later, Wilkinson moved to 83 Randolph, it had become one of the largest retailers of Kosmic rods, reels, and fly lines. Wilkinson was the original Chicago outlet for Abbey & Imbrie flies as well as early Pflueger luminous flies and Meisselbach and Hendryx reels. The Kosmic reel sold for $7, but entry-level anglers could buy a Hendryx model for 10 cents. Later, this old tackle house evolved into something or other to do with razor blades.

Congress Tackle's impaled carp logo, in a 1916 full-page ad placed in Miller's Sportsman's Handbook. *Between 1903 and 1926, Supplee-Biddle sold a bunch of German silver–appointed Congress-brand rods finished for it at the Hall Rod Works (wrappers and finishers) of Highland Mills, New York. The blanks were made by Montague. Doubtless this company was in business well on either side of those dates.*

A. G. SPALDING & BROTHERS

Established in the centennial year of 1876, A. G. Spalding & Brothers set out to make an impact on the angling business and within a decade had opened branch outlets on New York's Broadway and in Philadelphia. The Spalding Brothers were the first retailers to carry the rods of Thomas, Edwards & Hawes, and around 1892 bought the Highland Mills company from the remaining original partners and Edward Payne.

The company's Kosmic and Isaak Walton–brand tackle were marked with a number of nearly identical stamping dies, including "A. G. Spalding & Bros., Chicago," "Spalding & Bros., Chicago," and "Spalding & Bros." with no address—this being the New York–Philadelphia marking. Spalding also carried rods built by Fred Divine and others, so these stampings didn't necessarily indicate that all such rods were made by the Kosmic crew. After three to four years in rod manufacturing, Spalding decided to get out of the tacklemaking business and sold its Kosmic label and inventory to U.S. Net & Twine. After 1900, the Spaldings carried an extensive line of Abbey & Imbrie products at the retail level. And in due time, old A. G.'s successors began to focus on such mundane products as golf balls and tennis rackets.

VON LENGERKE & ANTOINE

The VL&A partnership, formed in 1891, quickly became prominent, then seemed to fade slowly into obscurity. By 1927 it had become a retail store at 33 South Wabash Avenue. Early on, however—from 1891 to 1897—Von Lengerke & Antoine sold a selection of rods built by John Landman, usually marked only by the seller. From 1897–98 into the early 1900s VL&A carried the products of Thomas & Edwards, then Thomas alone. These rods were all stamped "VL&A, Chicago," or "VL&A, Special Grade"; the former indicated models by Landman and Thomas & Edwards, and the latter rods built in Brewer, Maine. Von Lengerke & Antoine rods made by these three prestigious makers evoke a keen interest among collectors today, and figuring out just who made what is one of the great fascinations of the hobby.

THE WISH BOOK TACKLE MERCHANTS

Legend claims that drummer Richard W. Sears began his career selling watches to midwestern farmers who needed to know when the next train rolled

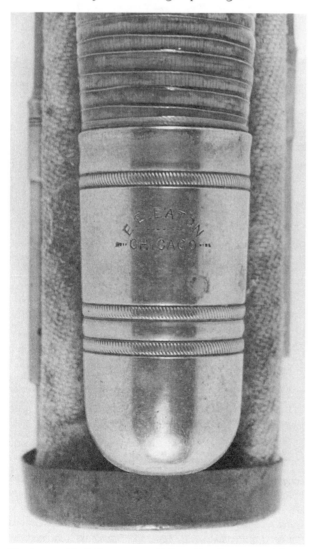

The impeccable butt cap on a high-quality circa-1885 Malleson trunk rod, marked "E. E. Eaton, Chicago."

Chicago Field, *June 11, 1881*

by, perhaps so they could move their cows off the tracks. By 1900 the farmers had a 2-inch-thick catalog from Sears, Roebuck & Company of Chicago, and it featured an array of fishing tackle that found its way across the entire country, especially to anglers living

Two early Kosmic-Walton models. At left, a 10-foot trout fly rod marked "THE ISAAK WALTON" and "A. G. Spalding & Bros." The light grilse rod at right bears the stamp "Spalding & Bros., Chicago," along with the 1890 Edwards and Hawes patent dates on the lower female ferrule.

in the bass belt. Sears, Roebuck carried a limited selection of fly gear made by Hendryx, Horrocks-Ibbotson, Meisselbach, and Montague City.

In 1902, $4.25 would buy the Beaverkill, a top-of-the-line lancewood fly rod made by Horrocks-Ibbotson, 8½ feet long and 8 ounces in weight; it had German silver hardware and came packed in a velvet-covered wooden form. Sears's best split-bamboo rod, made by the Montague City Rod & Reel Company, was the Rio Grande, which featured a nebulous length of "about 9 to 10" feet. Quite a difference in handling characteristics! Despite its cheap nickel-plated furnishings, the rod had nice mortised cedar inlays above the handle. As an option, the customer could order the rod with a black-and-white celluloid grip.

Early "quality" reels included the "Genuine Feather Light Reel" by A. F. Meisselbach and the "Acme Trout Reel" by Montague. The cheapest reel was a Hendryx, of course. To compete with or undersell its competition, Sears asked Montague to cut production costs by using only a single dip in the nickel bath. The Meisselbach Feather Light came in a "fine oxidized finish," eliminating the standard double nickel plating that the Newark company used on the reels it supplied to Abbey & Imbrie and Edward vom Hofe. With a little creativity, Sears managed to sell the same product, or nearly so, for less.

Over the years Sears synthesized its own brand names, such as David Bradley farm implements and J. C. Higgins—sounding like everyone's favorite uncle—sporting equipment. The J. C. Higgins marking can now be found on a number of tackle items sold by Sears in the 1940s and '50s. Ultimately, Sears managed to secure the right to apply the name of Ted Williams to practically everything. This is Fenway Park's Ted Williams, not *Fly Rod & Reel's* and *Audubon* magazine's environmental gadfly—long may he rave.

Sears continued to sell Horrocks-Ibbotson products well into the 20th century; I've seen split-bamboo "Beaver" models obviously made by H-I. After Montague's demise, Sears sold Ocean City and Shakespeare reels under the J. C. Higgins label. Some are quite nice in a still-usable sort of way.

Forest and Stream, July 28, 1900

Illustrated here is a J. C. Higgins fly reel from the 1940s, equipped with the handy line guard. Sears carried higher-quality tackle as well. High-end rods were by Heddon—notably the Premier 115 and the Expert 125, both fitted with distinctive gold-colored ferrules.

Failing to change with the times, Sears finally bowed out of the mail-order business and closed all those little Mom-'n'-Pop branch stores with their small counters topped with order catalogs. This nice bit of Americana brought me my first new, non-hand-me-down fly reel in my early teens.

Our other big-time Chicago mail-order company was Monkey Ward's, officially titled Montgomery Ward & Company and actually based in Chicago, Kansas City, and St. Paul. Also big on sporting goods, Montgomery Ward offered the same stuff as Sears but tried to undersell its competitor. Best described as "usable stuff," classic Monkey Ward artifacts worth more than $25 have yet to be found.

DENVER'S CLASSIC TACKLE TRADE

Chicago was just one gateway to the west. In St. Louis, turn-of-the-century dealers such as the Simmons Hardware Company retailed and wholesaled privately labeled fly tackle. Denver is a true collector's hot spot; a number of big-time tackle outlets there were active into the 20th century. C. B. Whit-

An offering in Sears, Roebuck's 1902 catalog, the Rio Grande appears to be a mortised Montague fly rod. It was one of Sears's two "top-of-the-line" models, and seems to have Richardson guides.

ney & Company, perhaps Denver's largest wholesaler, sold tackle to dealers in every state west of the Missouri River. The A. W. Colburn Mercantile Company was another active Denver wholesaler, selling its Silkaline- and Eureka-brand goods to department and Mom-'n'-Pop stores throughout the West.

Denver's retailers also marketed a quantity of classic fly tackle starting around 1890. Dating back to at least 1891, H. M. Bostwick had a walk-in trade at 1225 16th Street and sold rods obtained from eastern wholesaler Fred Malleson. Bostwick continued in business for a number of years. The Mayer brothers, known as The Geo. Mayer Hardware Company, retailed at 1520 Arapahoe Street in 1905–07. Well into the late 1920s anglers could buy tackle at The Broadway Department Store Company on the corner of Broadway and Archer Place. Another old Denver outfit, Dave Cook Sporting Goods, is still active. In the mid-1940s Cook sold a line of fly rods built from Phillipson and Wright & McGill parts. The models

Sears sold wish book fly-fishing tackle from the 1890s until the recent demise of its catalog. This "J. C. Higgins"–brand fly reel, with a nice little circular line guard, was made by Ocean City. It was substantial, pragmatic, and looked "right" with the 1940s classic production rod. Perhaps not all that collectible, the reel can still serve as a functional piece in the field.

were marked "Courtney Riley Cooper" and were built by Denver's Fred DeBell.[7]

UPCOUNTRY TACKLE DEALERS

Great tackle also came from some out-of-the-way places, usually through catalog sales. At Stevens Point, Wisconsin, O. L. Weber conducted a giant business in flies, spinners, and snelled hooks. The Weber Lifelike Fly Company was an important outlet for W. E. Edwards fly rods from 1920 to 1929. From 1930 until the end of the split-cane era, Weber marketed a better grade of rod built by Heddon.

Many collectors have discovered the excellent Heddon fly rods marked "L&C," for Lyon & Coulson. The twin lion logo of this Buffalo-based outfit appeared on a lot of classic-era equipment. Throughout the era of the 1920s to '40s Lyon & Coulson conducted one of the largest mail-order and retail tackle business of the day. Its popular Tiger-pattern streamers (illustrated) were all the rage during the classic streamer era. Lyon & Coulson had stiff competition from other catalog dealers, such as the Folsom Arms Company, which also carried classic Heddon rods marked by its house. For more on private-label Heddon rods, see chapter 10.

The famous Maine firm of L. L. Bean is synonymous with quality fly tackle. In 1912 Leon Bean began making woodsman's boots with rubber feet and leather uppers, and soon expanded into an ever-growing line of hunting and angling equipment, all sold with a money-back guarantee. By the 1920s Bean was selling a variety of tackle stamped with his own name—for example, a Montague City reel marked "Bean's Special" on the face plate in fancy script. Montague City tackle was carried into the late 1930s; I've seen Montague models sporting Chubb's reinforced ferrules carrying the name "Bean's Light Salmon Rod."

Around 1940 Bean introduced his Double-L series of quality fly rods and reels. The first examples were built by Bristol's Gene Edwards, perhaps in conjunction with his brother, Bill. After the war they were made by Bill Phillipson. Double-L rods are collectible and practical fishing tools, but if you wish to shoot line for distance, don't rig the rod as pictured in the accompanying 1940 Bean advertisement—with the line passed through the keeper ring. Bean

Denver, Colo. 10/27 1891

"L & C" FLY RODS
BRING ADDED PLEASURE
TO EVERY CAST AND
EVERY CATCH

YES!

"L & C" rods will help you catch better
fish—their perfect balance and action
will give you greater casting pleasure
and longer service!

Our new catalog illustrates these rods
and other fine "L & C" angling equipment.

"L & C's" 88 Page CATALOG
HANDBOOK OF FLY FISHING
By Breems Forrest

Send 25c for famous YELLOW TIGER streamer or 30c for sensational new IRIS No. 1
streamer and we will include without charge

IRIS No.1   Breems Forrest's
Handbook and
"L & C" Catalog.
Write Dept. A,
1280 MAIN ST.,
YELLOW TIGER   BUFFALO, N.Y.

Dealers Everywhere Sell "L & C" Tackle

Hunting & Fishing, March 1941

continued to offer Phillipson's Double-L rods into the fiberglass era and up until around 1970. Today, L. L. Bean is run by folks who wear business suits and conduct team-spirit seminars, but the quality and selection in its fly fishing department is still top drawer. And there's a whole new generation of graphite Double-L rods.

Bean's 1940 "Double L" Fly Rod

$13.50
Postpaid

Made of the highest grade bamboo, cured by the new heat process which brings out the natural, permanent, light brown color and shows up to advantage the professional silk windings. We are ready to put it up against any Rod on the market that sells up to $25. The case is genuine black calfskin with zipper opening. Our own manufacture. Lengths, 8½ and 9 feet. Weights 5 and 5½ ounces. Price, Rod and Case, $13.50 postpaid.

Write for Spring Catalog—Just off the Press

L. L. Bean, Inc., 220 Main Street, Freeport, Maine
Mfrs. Fishing and Camping Specialties

Hunting & Fishing, *April 1940*

Notes

[1]Brown, Jim. *A Treasury of Reels.* Manchester, Vt.: American Museum of Fly Fishing, 1990.

[2]Prouty, Lorenzo. *Fish: Their Habits and Haunts.* Boston: Cupples, Upham and Company, 1883.

[3]Reback, Ken. Photostat of original J. vom Hofe letter.

[4, 5]Schullery, Paul. *American Fly Fishing: A History.* New York: Lyons & Burford, 1987.

[6]Corsetti, Bob. Phone conversation, 1994.

[7]Sinclair, Mike. *Bamboo Rod Restoration Handbook.* Grand Junction, Colo.: Centennial, 1994.

Antique trout flies in a Bray fly book, made originally by New York's Schoverling, Daly & Gales and continued by Dame, Stoddard & Kendall of Boston. Dame, Stoddard used the Optimus brand name on many items, from rods to the leader envelope shown.

14 Flies and Accessories

Tackle buffs may specialize in either rods or reels, but all seem to share a fascination with the wonderful accoutrements of old-time fly fishers. From antique patterns by the likes of Theodore Gordon and Carrie Stevens to the modern dressings of Vince Marinaro, Mike Martinek, and Bill Catherwood, flies make lovely groupings under glass in shadow boxes. Fish carvings and landing nets are impressive wall decorations in the tackle room, as are old piscatorial prints and, if your budget allows, original paintings.

Much desirable stuff is practically new—fashioned within the few past decades. Some items actually are new, but poised to become future collectibles. Other accessories may date back to the 19th century or even earlier. I'll try to give an overview of this piscatoriana, mentioning what might increase in value or popularity and warning of pitfalls to avoid.

THE ANTIQUE FLY

The heavy bidding on classic flies at any tackle auction is testament to the value anglers place on these minuscule links to our fly-fishing heritage. Today's Olive Dun dates back at least to the 15th century, one of Dame Juliana's original Jury of Twelve. Walton mentioned the palmered fly and that method of tying. The Gnat, Stone Fly, and Alder all came to us from Britain, but the Royal Coachman, contrary to popular belief, is American, conceived in 1878 by New York tyer John Haily and named by L. C. Orvis.

Trout, lake, and bass flies from the antique era were tied on short silkworm gut leaders; salmon flies had eyes made from twisted gut. This old natural

Brand-name flies can bring good prices, even when one is missing from the package. This celluloid-windowed Kosmic pack, the only one known, sold at auction for $33 per fly—a bit pricey for fishing.

leader material is very brittle and often breaks at the hook; you must take extreme care when handling these artifacts. Gut-snelled flies were commonplace long before 1869, when Genio C. Scott listed their widespread availability—from major cities such as New York, Boston, Montreal, and Quebec, to modest burgs like "Rome, Rochester, or Mumford, New York." The flies in Mumford, we know, were the products of John McBride, one of our earliest known fly tyers.

Very few antique flies were carded by the actual maker; most were sold under the names of retailers such as Orvis, Stoddard's, and Abbey & Imbrie. Any 19th-century fly in its retailer's envelope or on the original seller's card is more collectible than a similar "loose" fly. Orvis Superfine trout and bass flies of the 1880s and '90s, probably tied under the direction of author Mary Orvis Marbury, now bring a premium. Pflueger's Luminous trout flies from the mid-1890s, with bodies that glowed after exposure to light,

Famous Lady Fly Tyers

Many of our great early fly tyers were women. The first famous female tyer was Sara Jane McBride, of Mumford, New York, who was active in the 1870s and '80s; she was the daughter of John McBride. Sara McBride is considered one of the founders of the imitative school. An expert angler, she penned erudite articles on stream insects, one series published in 1876 through *Forest and Stream*. Mary Orvis Marbury, daughter of Charles F. Orvis, was an exponent of fancy flies and collected hundreds of patterns through personal correspondence; these became the core of *Famous Flies and Their Histories*, published in 1892. In Rangeley, Maine, Mrs. Lydia Jacobs produced colorful trout and land-locked salmon patterns in the 1890s. Mrs. Jacobs also started one of the region's first tackle shops. Another early Rangeley tyer, Mrs. Etta Dill, tied great flies when not canning pickles. (My apologies to Mrs. Dill.)

McBride Flies.

Medal and Diploma from International Exhibition Philadelphia, 1876, decreed for exceedingly neat work with solidity of construction. A fine assortment of flies for trout, salmon and bass constantly on hand. Also casting-lines. Flies carefully selected for any locality. Goods promptly made to order from any pattern desired. Address SARA J. McBRIDE, Mumford, Monroe co., N. Y. W. Holberton, Sole Agent for New York and vicinity. feb15—4m.

Forest and Stream, May 17, 1877

Rangeley Flies

— MANUFACTURED BY —

MRS. ETTA DILL,

Rangeley, Maine.

Every Variety of Trout and Salmon Flies.

1898–99 Franklin County Business Directory

Miss C. J. FROST

MANUFACTURER OF

Trout and Bass Flies

TROUT SPINNERS	SPOON HOOKS
SNELLED HOOKS	FLY SPOONS
CASTING SPOONS	LEADERS, ETC.

From 1918 letterhead

Carrie J. Frost, of Stevens Point, Wisconsin, began her fly-tying career in the late 1880s, at first as a hobby. In 1896, with a local female talent pool, the ex-schoolteacher built a nationwide clientele under the trade name of Miss C. J. Frost. Carrie was active until 1919, when the company became the Frost Fishing Tackle Company—absorbed in 1926 by Oscar L. Weber's Lifelike Fly Company. Miss Frost's last flies were sold under the "Water Witch" logo.

Any fly that can be traced to one of these five top women tyers would be a wonderful find, especially a lake fly with its peacock or ostrich herl "Victorian collar" tied around the head.

are also sought after. I used to think that retailer-packaged flies were worth less than maker-packaged ones, until I saw an incomplete six-pack of Kosmic-brand flies (illustrated in figure 14-2) bring $165. Of course the prestige of the Kosmic name may have accounted for the high price.

Collectible flies can be traced to personalities. Henry P. Wells, a New York tackle tinkerer and author, produced the classic Parmachene Belle after taking a big Maine brook trout on little Parmachenee Lake, just above Aziscohos. Louis Rhead and J. Harrington Keene, both transplanted Britons, advocated the "lifelike" fly. They were original and prolific tyers, and today their superb ties bring a premium.

After the turn of the century, a number of fly designers introduced the odd Quill flies, which were tied on extended duck quill bodies. The Darning Needle, tied by Joe Welch of Pasadena, California, is a prime example. Some of these creations had wire-framed wings shaped like propellers. Whether these were inspired by the Wright brothers is open to conjecture, but many of these strange spinning flies have found their way into collections.

FAMOUS WRITERS AND TYERS

"Famous" flies have by far the widest appeal among collectors. It's only natural to want to own a fly tied by the likes of Elsie Darbee or Colonel Joe Bates. Although the market is almost flooded with flies tied by the most productive of these celebrities, prices remain high and bidding is competitive. Given the number of tyers, the following list must be incomplete, but it does include a sprinkling of marine tyers, such as Mark Sosin, Chico Fernandez, and other saltwater pioneers, whose work I feel will soon become important to collectors.

- *Theodore Gordon* is, without doubt, the most venerated tyer of all time. The originator of the Quill Gordon and the Catskill style has legions of fans, but few of his flies remain with us. Most of Gordon's work disappeared under rocks or in bushes almost a century ago. Gordon tied a cast of thousands, however; he even traded a batch of flies to Ed Payne for a fly rod.

- *Rube Cross* tied Catskill-school flies for many famed anglers and his precise offerings occasionally surface.

- *Francis Sterns*, the first great female fly tyer for L. L. Bean, produced some of the most sought-after ties available today. Her work brings a premium.

- *J. Edson Leonard*, author of *Flies*, tied some exceptional wet flies that occasionally come to light. His flies are "correct" in all ways, very reminiscent of the classic era.

- *Elsie and Harry Darbee*, Catskill tyers extraordinaire, produced some of the most elegant flies of the classic eastern style.

- *Winnie and Walt Dette*, who began tying excellent flies back in 1929, were the founders of the modern Catskill school.

- *Art Flick*, of Streamside Guide fame, ties a precise Catskill fly of flawless quality.

- *Del Mazza*, of Utica, New York, ties flawless flies in the classic eastern style.

- *Ted Niemeyer*, a West Coast authority on antique and classic flies, spins realistic and superb nymphs that bring a premium when available. He also ties a wicked-good salmon fly.

FANCY SALMON PATTERNS, ANTIQUE AND MODERN

Up until around 1915 most salmon flies were imported from Ireland, England, and Scotland by American tackle houses, to be sold under the house name. Many tyers worked for Martin Kelly, John Forrest, Samuel Allcock, or the Hardy Brothers, but the most valuable antique ties are from the hands of George M. Kelson.

Scottish ties have always had a following. In 1876 Dr. A. G. Wilkinson claimed that Forrest & Sons of Kelso, produced such excellent salmon flies because they were "tied by the deft fingers of Scotch lassies, and that gives them an additional charm." No doubt, but please note that "Scotch" is a drink, not a nationality. The skill and time necessary to produce a fancy salmon fly, Scottish or otherwise, is reflected in the current high prices. The following tyers have sterling reputations in the art.

- *Alex Rogan* was the dean of the classic Atlantic salmon fly, as both an originator and an artist. His work is tough to acquire but worth the wait.

- *Poul Jorgensen,* author of several books and a talented tyer of some of the best traditional salmon "fancies," has an avid following.

- *Charles Defeo* was perhaps the most popular salmon fly tyer of his generation. Charlie's hands seemed almost magic. His flies, like Jorgensen's, bring a premium.

- *Bill Hunter,* from New Boston, New Hampshire, has taught many of the current generation's salmon-fly artists. His own work is crisp and authentic.

- *Paul Schmookler* has brought fly tying to the level of high art, using the most exotic of materials and paying the strictest attention to form and composition.

Two dry flies tied by John Alden Knight. Personality items like these always have great collector appeal.

CLASSIC STREAMERS

The streamer as we know it was born in Maine shortly after World War I, when one of the earliest known advocates, Bill Burgess of Minot, sold his small baitfish imitations at Art Read's Portland tackle store. Previously, American flies had only imitated aquatic and terrestrial insects. Between 1920 and 1930 hosts of patterns were introduced on the new theme.

Chief Needahbeh and Mrs. Carrie Stevens developed standard designs for the long hook that are still used today. You'll discover that a fly fashioned by one of these greats will come tough. Other streamer pioneers included Herb Welch—originator of, among other patterns, the Welch Rarebit—and Jane Craig. The Supervisor, appropriately enough, came from Joseph S. Stickney, a supervisor with the Maine State Warden Service.

By the mid-1920s the Maine streamer had become the darling of New England fishermen. Bill Edson's Light and Dark Edson Tigers were the hit bucktails of 1923. Bert Quimby of Windham and Gardner Percy of Portland become Maine's most prolific tyer-sellers. Quimby's name alone evokes a nostalgia for the old-timey Maine woods just short of that of Scott Attwater.

Gardner Percy had the longest career. With his wife and Mrs. Nellie Newton, Percy started his streamer business in the attic of his Woodfords home in 1926. Mrs. Newton, an experienced fly tyer from England, represented the Percy Tackle Company at the 1927 New England Sportsman's Show, giving streamer-tying demonstrations. By 1930 the company had moved to Congress Street, Portland, and had eight full-time tyers trying to keep up with orders from tackle wholesalers across America and Great Britain. In the 1940s Leon Thomas bought hundreds of Gardner's flies and gave them away at a small premium. Percy made streamers until the 1960s. I used a slew of 'em, throwing the cards and cello packs away. Duh! Today, his tyers are among more prestigious personalities in the business, a group that includes the following:

- *Chief Needahbeh,* a Penobscot guide, fished the Moosehead Lake region and developed the flatwing streamer.

This fly ad, from the April 1923 issue of National Sportsman, *touts Bill Edson's early Maine bucktail patterns, the Light and Dark Edson Tigers.*

- *Carrie G. Stevens*, of Upper Dam, Maine, is considered the founding mother of the modern casting and trolling streamer. Her original patterns included the Pink Lady, Governor (after Percival Baxter), Mickey Finn (after Carrie Finn's short husband), Silver Doctor, Colonel Bates, and Shang's Special (after Connecticut decoymaker Charles "Shang" Wheeler). Her most famous tie was the Gray Ghost, which remains an extremely effective pattern.

- *Mike Martinek*. Acknowledged as the finest contemporary tyer of American streamers, Mike is one of only a few who can duplicate the flies of Carrie Stevens. He has even surpassed her work, and throws away more wing material than he uses.

- *A. W. Ballou*, the originator of the marabou streamer, was a Massachusetts fly artist of great imagination and talent.

- *Herb Welch*, a tackle merchant in Rangeley, Maine, designed the Black Ghost and several other noted patterns. He was a friend of Carrie Stevens and sold her flies for years.

- *Omar Needham*, a well-known Rangeley cane rodmaker, also tied excellent streamers and classic eastern trout patterns.

- *Charles A. Hunt*, of Dresden, Maine, is a fastidious tyer of classic streamers. Even the head of a Hunt fly is a thing of beauty.

DEER-HAIR BUGS AND POPPERS

Few flies can compare to the originality and just plain impressiveness of those developed by bass and saltwater fly fishers. Only a few tyers are listed here, but all have produced great furry collectibles.

- *Ernest H. Peckinpaugh*, the father of the popper, was from Chattanooga, Tennessee. Along with the first panfish poppers and the Peck's Feather Minnow series, around 1915 he introduced the Night Bug and the Ozark Ripley—true bass bugs built with cork bodies. Ernest Peckinpaugh's work is tough to find and identify today, since illustrations of his complete line of poppers and sliders were never published.

- *Cal McCarthy*'s warped imagination produced the famous Cal-Mac—a miniature broom with hair wings swept back over the hook. Neither slider nor popper, this odd bug was sufficiently effective to remain on the market for years. Introduced shortly after 1910, Cal-Macs were tied in several colors and were pretty crude, but McCarthy's bugs were the forerunners of the commercial popper.

- *W. J. Jamison* was a Chicago bait manufacturer of considerable ingenuity. Sometime around 1910 he developed the Coaxer. This odd fly-rod bait had a cork body, felt wings, and feathers tied onto the hook shank. The Coaxer came in at least a dozen versions, and one found in good condition now could be the start of a great bug collection. By the 1930s "Smiling Bill" Jamison offered a wide variety of hair-bodied bugs and a superb Mouse. Jamison's Mouse Bait, heavily built up from deer body hair, sported cute little black eyes, whiskers, ears, and a stubby tail—a knockout. As Jamison said, "This lure is remarkable in its resemblance to a live mouse." Today this fly rod rodent is a premier collectible.

- *Benjamin F. Wilder* began developing the Wilder-Dilg bug in 1908, while bass fishing in

National Sportsman, *April 1923*

Florida and Maine. In the late teens he teamed up with William H. Dilg, a Chicago sports magazine editor who wrote countless articles expounding the fly-rod popper's effectivness, and who had an in with several major luremakers. Around 1920 James Heddon's Sons began to manufacture the Wilder-Dilg, which I believe was the original slider; it had a pointed face instead of the blunt face of a popper. Behind the slider head came laid-back palmered hackle and a grouping of matched feathers. The Wilder-Dilg was an immediate success, and was offered in both bass- and trout-sized versions. At least one style was built with "Spook" colors (a paint job that looked like a skeleton) and called the Wilder-Dilg Spook. The most collectible of these hardheads are the early ones, made prior to 1920 by Wilder himself.

- *Orley C. Tuttle*, of Old Forge, New York, produced the first commercial deer-hair fly in 1919. Tuttle's Devil Bug seemed to have been built from the better part of a deer hide. This fanciful winged creature flew in the nightmares of many a smallmouth angler and was so effective that it remains popular today. The new versions, tied by the Dardevle folks, lack the evil qualities of the originals. Orley also

National Sportsman, *April 1923*

tied what may have been the earliest fly-rod mouse. Tuttle's Mouse, introduced prior to 1926, had eyes, ears, and a bushy tail. Tied differently than Jamison's later version, the Mouse's fur was swept back to the curve of the hook's shank, where it was secured before forming the tail. The fantastic Devil Bug and Mouse established Tuttle as the earliest and most talented of the deer-hair artisans.

- *James Heddon's Sons.* After the introduction of the popular Wilder-Dilg, Heddon added a number of fly-rod lures to its line, including the No. 75 Fuzzi-Bug and a smaller No. 74 style — both soft-bodied hair bugs. Next came the weedless Pop-Eye Frog, tied with a cork body, glass eyes, and hair legs — definitely a keeper. Around 1930, Heddon introduced improved versions of the Cal-Mac — the No. 50 Series Bass Bugs; these were made in two sizes, the smaller being the 1½-inch Baby Bass Bug. Perhaps the toughest Heddon fly-rod lure to find from this era was the Westchester Bug, a giant hair-bodied jobbie a full 2¾ inches long. The Popper Spook had a long body and came in five colors, but it was a pig to cast. Heddon also produced the Float-Hi Bug, with "whiskers" fore and aft of the cork body. Supposedly, this one was easier to toss.

THE SALTWATER FLY

The marine fly was born in the 1930s with the Bonbright, but its real growth came after World War II. Some classic saltwater patterns are as large as feather dusters and require pretty substantial shadow boxes, but others are comparatively diminutive. Jimmy Albright and Doc Robinson pioneered the two distinct styles — small and outsized, respectively. With the current rapid growth in the popularity of saltwater fly fishing, flies by Kukonen, Sosin, and other saltwater gurus will appreciate in value in the near future. You read it here first!

Catherwood's Mackerel, an original pattern tied without the use of a bobbin by saltwater-fly-tying pioneer Bill Catherwood.

- *Joe Brooks*, a true saltwater pioneer, caught the first permit on a fly. Brooks tied many patterns for that species as well as for bonefish; the most notable was the Pink Shrimp, adapted from Phillips's Western Shrimp.

- *Bill Catherwood*, the "hair" apparent of the huge furry fly dynasty, is known for his Giant Killers. Bill, who ties without a bobbin using strong sewing thread, developed his intricate patterns—some 30 in all—to imitate specific saltwater baitfish. Catherwood influenced several modern tyers, including Dan Blanton and Bob Edgley, the developers of the Sea-Arrow Squid. Bill also tied the original Needlefish pattern, its long snout since copied by many others.

- *Paul Kukonen*, from Worchester, Massachusetts, originated a number of saltwater flies and was one of the first exponents of shrimp patterns. A first-rate tyer.

- *Dan Blanton* and *Bob Edgley*, two tyers from San Jose, California, have carried the Catherwood style to great heights. The Sea-Arrow Squid is one of their excellent flies.

- *Chico Fernandez*, of Miami, Florida, developed the Sea Fly series. His innovative minnow-style streamers are well worth collecting.

- *Mark Sosin*, another Florida-based saltwater pioneer and ESPN-TV personality, originated several effective ties, such as the Blockbuster, that have become marine classics.

- *Lefty Kreh*, the well-known author-lecturer, is the father of the Deceiver series of flies as well as the Ballyhoo.

Collectors have such a wide array of collectible flies and bugs from which to choose. For best value,

Artificial Helgramite.
(Conroy, Bissett & Malleson.)

Book of the Black Bass, *J. A. Henshall, 1881*

though, purchase only documented items, avoiding those "supposedly tied" by whomever. And remember that moths also like flies. The offspring of a couple randy little moths can destroy the value of any fly collection. Strategically placed mothballs help curb their appetites.

Often neglected by collectors are the old artificial insects made during the last decades of the 19th century. Big tackle houses such as Shipley's and Conroy, Bissett & Malleson sold these almost-flies for years. Variations included hellgrammites, bees, and crayfish. Collect 'em all.

FLY WALLETS AND BOOKS

Old fly wallets complete with their original contents are coveted items indeed. The paged leather wallet evolved during the Civil War era. In 1869 Genio Scott mentioned a

> Fly book with leaves of Bristol-board, or other stiff material, to which are attached short ends of elastic, with a hook to attach a loop, and a ring at the other end of the leaf for the hook. This plan of carrying flies without bending the gut was invented by Mr. Hutchinson, of Utica, New York, and the cards may either be attached to the book or laid in as leaves, so that the angler may merely take a single leaf of selected flies, and place it in his pocket-book for a day's fishing.

In 1881 Doctor James Henshall mentioned the Hyde, Southside (A&I), and Holberton (CB&M) fly wallets—all available in a variety of leathers, from calfskin to Moroccan, with parchment leaves and some type of spring arrangement. Over time there was little improvement on Hutchinson's 1860s version. Variations on the styles listed in the first edition of Henshall's *Book of the Black Bass* were made for a number of years. Fly books by Levison, Holberton, Bray, and Common Sense were manufactured in quantity, and examples survive today.

The Holberton book, developed by "Major" Wakeman Holberton, a New York tackle dealer, angling writer, traveler, and trout nut, had metal hook clips, but looped gut leaders were simply tucked under a partition and could tangle. The

In 1923, $8 would buy you a decent fly rod—or the Improved Common Sense Wet & Dry Fly Book, the Cadillac of American fly containers.

early 1920s the Common Sense folks introduced the Improved Wet & Dry Fly Book, which, in addition to leaves, had a built-in, compartmented dry-fly box covered by a transparent lid. It even came with a pair of tweezers to help you ease the dry flies from their compartments. Oh, to find one of these with tweezers intact!

TINS AND TACKLE BOXES

Old fly tins and tackle boxes attract many collectors. The original tins, dating back to the early 1800s, were made by talented smiths on tackle-house order. Most were simple hinged affairs containing a number of soldered bars that held strips of cork. After the Civil War, fly tins were coated with Japanned lacquer and ranged in size up to 8 inches.

Bray fly wallet, patented in 1880, had celluloid leaves with metal clips and springs. Between each leaf was a felt "page" to protect the flies and absorb moisture. Manufactured until at least the 1920s by Dame, Stoddard, the Bray book was among the best of its kind and a favorite of author J. Harrington Keene.

Levison's book, patented on March 11, 1884, had four or five leaves, hook clips, and, for loops of gut, a vertical spring not unlike a minuscule version of a screen-door spring. Unfortunately, only young anglers had the dexterity and acute eyesight needed to attach or remove the fly. Anglers like me were up the creek without a magnifying glass. William Mills & Son touted the Levison fly book, but I'd guess it died of natural causes.

The light and practical Common Sense fly book was introduced sometime after 1900. The original style had heavy cotton leaves that sandwiched flies in position. The Bray and Common Sense books were designed for wet flies and, later, streamers. By the

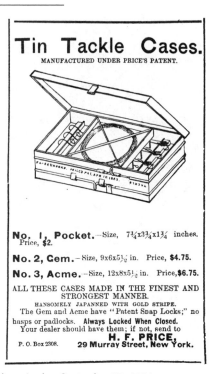

The American Angler, *September 27, 1884*

One of the earliest manufactured tins came from New York's H. F. Price, who received a patent for his Tin Tackle Cases on April 10, 1883. The Price boxes had "patent snap locks" and up to three tiers of compartments. These handsome tins, Japanned black with a gold encircling strip, are very collectible.

With the introduction of production stamping methods and aluminum, the 1880s brought us into the semimodern era of fly boxes. The most popular were made by Wheatley of England. Wheatley boxes had such advanced features as compartments with individual see-through lids. Today they bring a premium, but they're so practical that many anglers still use them streamside. American versions followed, of course, and some of the best had a hinged aluminum leaf to make two compartments for large and small wet flies. Many of these were sold by L. L. Bean and Weber of Stevens Point, Wisconsin.

Wooden tackle boxes have been with us since the 18th century, popular with fly fishers and bait fishers

An 1841 study of Theophilus South snelling hooks for the next day's outing. The etching also shows a ringed net and the elusive and extremely valuable potbellied creel, made from ancient times until the 1860s or so.

alike. Purism is a comparatively recent phenomenon. Many years ago, Rip Cunningham gave me a tackle box owned by a William Rogers of Boston. Its contents included many flies from Dame, Stoddard & Kendall, a Bray fly book, several phantom minnows, and two bobbers. Mr. Rogers was the typical all-around fisherman of his day; the bobbers reveal that he used worms on more than one occasion. This beautiful box is covered in leather and has a locking hasp with an 1871 patent date plus a "star" trademark.

Unless well marked, wooden tackle boxes are difficult to pinpoint as to maker. Illustrated here is a well-built joiner's special, made in Auburn, Maine, that was sold during the 1920s by the Stevens Tank and Tower Company, builder of wooden water tanks. John. W. Gilson of Stuart, Florida, built perhaps the most intricate wooden tackle boxes. Constructed from glued-up mahogany and finished to the nines, they were sold by Abercrombie & Fitch from 1947 until 1950, when Gilson sold the business to William and Charles Smith. The Smith brothers continued to make these beautiful boxes and sell them through A&F until 1956. Well-made wooden tackle boxes bring high prices today, but for unknown reasons, few appear at auction.

WILLOW AND SPLIT-ASH CREELS

Like wooden tackle boxes and fly containers, old creels elicit a lot of interest from collectors. The very oldest creels have a potbelly shape and were woven from willow, as in the example shown in figure 14-12. Some old British creels were actually made from copper, perhaps to serve as streamside boilers for impromptu lunches. The earliest "trout baskets," as

Creel, or Fish-Basket.
(A. B. Shipley & Son.)

John Krider called them, have center-hole lids. Most post-1900 versions have side-hole lids. Many more recent creels were actually made from wicker. Because this material isn't as strong as willow or ash and tends to discolor and rot, wicker creels were often fortified by leather strips.

The classic American creel came from the basketmaking trade, a cottage industry once dominated by various Native American tribes: in the East, the Penobscots of Maine; in the Midwest, the Hurons; in the Southwest, the Piutes of Arizona. A rare birchbark creel appears now and then to be eagerly gobbled up by collectors, but most Native American products were of split willow or ash.

Creels were not standardized items. A tackle dealer of the 1850s, such as Martin Bradford of Boston, would order creels by the gross from an outfit such as Adams & Lunt of Charleston, Massachusetts. Adams & Lunt contacted its man in Bangor, perhaps Charles Ramsdell, who passed along the order to the Penobscots. The finished creels were eventually shipped via wagon over the old Post Road to Boston.

Because they were made as piecework, each creel basket was unique, subject to the styles and whims of the Native makers. "Green"

white ash was heated and then pounded until it separated into layers at the growth rings. These were split off, cut to size with a crooked knife, and woven while still flexible—the same basic construction method used for the Adirondack pack basket. Later creels were also produced by Yankee housewives in much the same manner, with woven willow saplings, whole or round.

Giant fish baskets are sometimes found. Introduced in the 1870s, the camp or canoe creel became popular on eastern lakes and could hold the catch of two or more fishermen. Camp creels often had four wooden legs. In proper use, the bottom of any angling basket was lined with wet ferns.

The traditional fishing basket is a prime example of folk art and a real link to the classic era. When looking over a creel, check the basket carefully for recent repairs to the weave. Open the basket and push gently on its bottom while scanning its edges for breaks. If the clasp is missing, the creel's value drops precipitously.

TROUT NETS AND SALMON GAFFS

Today we use nets to carefully cradle trout while we gently remove the hook and release them. It wasn't always so. Nets were originally designed to shorten the path between stream and table. And a gaff . . . well, the less said, the better. The absolute worst of the lot was Marble's Automatic Gaff, a medieval torture device with doubled claws that snapped shut on contact. Despite their slaughterhouse overtones, antique gaffs can be among the handsomest of tackle accessories.

I once owned a beautiful gaff with a polished bamboo handle, a brass hook retainer, and a fancy brass finial that was removable to allow the hollow handle to accept an extra salmon-rod tip. This fancy tool came from England, which is famous for gaffs that can do everything except tap dance. Some were marked by their maker, usually Hardy Brothers. Variations among gaffs were legion.

There were just as many variations in nets as in gaffs, and nets can fold, collapse, and take down with

A direct ancestor of the wooden-framed trout net, this circa-1880 alewife dip net is being repaired by an old-time Nobleboro, Maine, fisherman.

the best of them. Old landing nets ranged from models with wooden handles and cheesy rims to those with bamboo shafts and nickel-silver hoops. Obviously, not all nets are equal in collectibility. One style—the teardrop net—didn't fold, telescope, or do anything much—except look great. Most of these are thought to be of fairly recent vintage. Actually, their lineage goes back to precolonial America. Nets like the one pictured here survived little changed from the Maritime Archaic Period (4000 B.C.) to the early 1900s, when they were still used by New England alewife harvesters. Today these nets are still built by steaming wood into a bow and attaching it to a wooden handle, much as they were in the stone age.

Fly-fishing versions often had handles made from exotic wood rather than a plain old piece of ash, and the hoop may have been glued up from bamboo. A few decades ago, Cummings made a number of

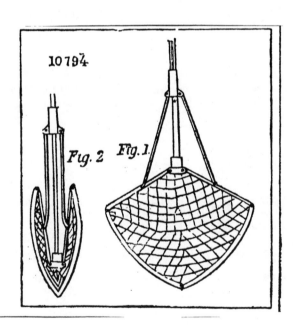

handsome bamboo bent-bow nets. Other wooden landing nets were made in Pakistan. My own net was steamed and fashioned by Maine woodsman Jeffrey Barter. I use this net every season, although these days the fish are getting smaller and fall through the mesh.

The collapsible net is comparatively recent. The first patent for an "improved landing net" was issued to New York tackle tinkerer Charles De Saxe on April 18, 1854, the same day he received the first surviving fishing-rod patent—No. 10,794 for the net, and No. 10,795 for the rod. A year later De Saxe patented the original weedless hook. Busy guy. The De Saxe net folded like a parasol for easy storage. The rights were assigned to Thomas H. Bate. One in original condition would today be a rare addition to any collection.

In 1875 Maine's Charles Nason invented the "Nason Patent Net Ring," made from a length of flexible spring steel. This became the forerunner of many collapsing spring nets. Around the same time Conroy, Bissett & Malleson pushed an odd affair that seems unrelated to any other style. In the 1881 edition of the *Book of the Black Bass*, Doctor Henshall praised the style as his favorite, calling the nets "the most portable and convenient, and are made with two or more hinge joints." The net had a brass hoop that folded into four sections, with a remarkably long two-piece handle.

Later net versions are a little easier to find. The closest relative to the Maine net is the Dorsal Fin, as pictured by A. R. Harding in his *Science of Fishing*. By the 1890s the spring-hoop style evolved into a jointed, folding model that was widely available. At the low-price end were the trout nets of the Richardson Rod Company; these were often given away to

No. 10,794.—CHARLES DE SAXE, assignor to THOMAS H. BATE.—*Improved Landing-Net for Anglers.*—Patented April 18, 1854.

This invention is for making the hoop or mouth of the net with joints or hinges, that the hoop may be folded up in a small compass, and thus be rendered more convenient, while, at the same time, its strength is not diminished. The arrangement will be readily understood on reference to the figures. Fig. 1 represents the net open and ready for use; fig. 2 represents it closed.

Long-Handled Landing-Net.
(Conroy, Bissett & Malleson.)

Another net style, the circa-1880 CB&M, made with a folding four-piece brass ring. Henshall liked this style, mentioning that the nets were "most portable and convenient."

schoolboys who sold 50 subscriptions to the *National Sportsman*. A Richardson net in good condition is today a nice piece, especially if it retains the thick black paint over the handle and hoop. The Harrimac or A. F. Meisselbach landing nets, with bamboo handles and German silver ferrules, were superb.

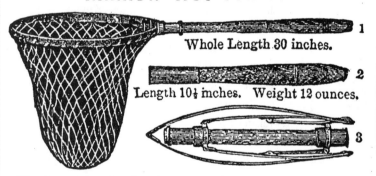

DODGE'S PATENT
Collapsing, Folding, Landing and Minnow Net Frame.

1

Whole Length 30 inches.

2

Length 10¼ inches.   Weight 12 ounces.

3

No. 1 represents the frame unfolded and expanded, ready for use. Can be unfolded and expanded with the net on the ring (in 30 seconds) by the same motion that is used to open an umbrella. The ring is 12in. in diameter, made of steel, broad and strong where it joins the staff, and narrow and light at the outer side. The other working parts are made of brass, nickel plated. Nos. 2 and 3 represent the frame collapsed and folded, staff unjointed ready to pack. Can be carried in tackle box, trout basket, grip sack or pocket.

PRICE COMPLETE, $4.00.

J. N. DODGE, 278 Division St., Detroit, Mich

Shades of Charles De Saxe. Dodge's patent net folds like a parasol.

All of these post–De Saxe–Nason nets are usually available at tackle shows and auctions. Prices vary according to condition and the quality of materials. Nets with original cotton or linen bags are more valuable than those with replacements. The authentic method of attaching a replacement bag was to run cotton or linen lacing through holes in the net bow, picking up each mesh along the bag's top. Field replacements were made by wrapping using a length of fly line around the hoop and through the bag meshes. One way to tell if a net bag is a recent "doctor job" is by examining the material. Nylon has been around only since 1940 or so. Characterized by its soft sheen, nylon would not be original on a net made before World War II.

Edson Manufacturing of Sturbridge, Massachusetts, made a nice net from aluminum and brass. A classic West Coast net, the Barnes, still occasionally appears streamside and is becoming quite an item with collectors, especially the model with a split-bamboo frame. The Barnes net, introduced around 1915 by Carlos Young, a small-time manufacturer at 230 Market Street in San Francisco, California, was carried by every major tackle house in the country and sold for around

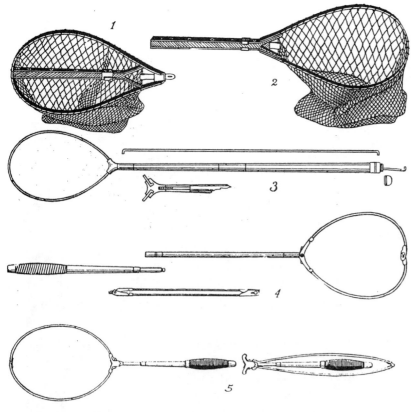

FOLDING PATTERN LANDING NETS.

1—Barnes, Folded; 2—Barnes, extended; 3—Dorsal Fin (head Springs out Straight and Slips into Handle); 4—Collapsing, Jointed Handle; 5—St. Lawrence.

The Landing Net Lineup of 1912, from A. R. Harding's Science of Fishing. *Number 3 shows a variation of Nason's 1875 Patent Net Ring, still stored in the handle some 37 years later.*

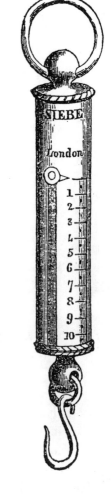

An 1841 fish-weighing scale. Age causes these old spring scales to exaggerate the weights of fish—kind of like anglers.

$3. Perhaps the toughest (to find) of the "newer" models is the wrapped-grip St. Lawrence net, which is probably of Canadian origin. All these antique nets folded for easy storage, a plus for collectors with space problems. But I think nets look best draped on the den wall. Or at least I used to. I had a nice old nickel-silver-appointed Harrimac net hanging in my tackle room until a lady guest remarked, "Oh, you catch butterflies!"

FLY-FISHING KNIVES AND TOOLS

Little fly-fishing gadgets—from line driers to knives—are big these days. Illustrated here are big examples of small-scale collectibles: a weighing scale from Theo South (1841), a fly tyer's vise and tweezers from Uncle Thad Norris (1865), and a nifty angler's knife from bassman J. A. Henshall (1881). Collectors tend to squirrel away just about anything shiny, and

many of these items can be stored in a desk drawer in a small apartment.

Some of the early line driers were actually quite large. Often called "knitty-knoddies," most were made of wood or, much more rarely, ivory. In the 1890s newer folding versions began to appear, usually made of plain or nickel-plated brass. Some were signed, such as by "L. T. Weiss, Brooklyn, N.Y., Pat'd July 22, 1919." Others were marked by the seller. Examples have been found bearing the stamps of Abbey & Imbrie, M. A. Shipley, and Edward vom Hofe.

Fisherman's knives can be used for almost anything, even denuding apples, thus their wide appeal (sorry). Most have tried to redefine "multipurpose." Abercrombie & Fitch sold a German-made knife that combined a blade, a scaler, a weighing scale, and

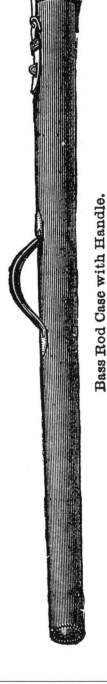

Bass Rod Case with Handle.

a priest. One of these sold at a 1991 auction for $450; for this, you peel very soft apples. Puma manufactured a combination knife-priest with a built-in balance scale, now valued at around $175. Marble sold an ugly folding fishing knife that recently brought $80 at the block. You get the picture. These things are skyrocketing in price. Pegley-Davies made an Angler's Knife complete with a ruler, hook gauge, screwdriver, scissors, file, can opener, cutter, corkscrew, and knife blade. Hardy Brothers made so many knife variations that its tools could make up an entire collection. Another Pocket Tool Outfit looked like a dealer's display for Exacto. Maybe it was.

Fly-tying tools from 1865. The vise looks more appropriate for extracting confessions.

The miniature baseball bats known as priests, such as Hardy's Fly-Fisher's Curate and similar

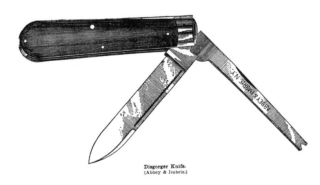

Disgorger Knife.
(Abbey & Imbrie.)

An 1881 Abbey & Imbrie angler's knife. What, no bottle opener?

devices used to club fish to death, actually have a following among collectors. Personally, I prefer more innocent collectibles, such as fishing award buttons and anglers' key chains and watch fobs. To each his own.

ROD AND REEL CASES

Years ago, Captain Terry Lewis walked into my old tackle shop with a smashing 5-inch-diameter leather rod case whose original name plate was marked "A. J. Campbell, Sangorville, Me." Good leather tackle items are not all that common, but one with your name on it is a billion-to-one collectible. This fine piece now holds three classic Maine-built rods. Old leather rod cases can be spectacular if in good or better condition. A leatherworker can repair an elderly rod case, but the repairs are difficult to hide unless the repairman is a true expert.

All the top tackle merchants of the decades flanking the turn of the century offered split-leather cases, usually marked by the seller on the bottom—"Wm. Mills & Son," for instance; or "Edw. vom Hofe." A scan of auction offerings over the past decade reveals that very few of these items survived. Surprisingly, most sell for far less than their true worth. Leather reel cases are also getting rarer—not that they ever really flooded the market. In the old days, few anglers were willing to fork over extra money for an expensive reel case. Most just bought their fly reels and went fishing.

Cost-conscious collectors into leather should keep the sizes of their favorite rods and reels tucked into their wallets when attending tackle shows, just in case an empty leather case turns up. New cases are available from superb leather craftsmen, but be prepared to write a large check—much larger than one for a comparable old piece. The wide gulf between prices for rare and current items places antique leather goods into the bargain category—at the moment.

You also might want to keep an eye out for "previously owned" aluminum rod tubes, which always come in handy; many fine rods on the market lack cases. Sure, you can case a rod in a brand-new tube, but it's nicer if tube and rod are contemporaneous. Some tubes—Milward's come to mind—have very nice brass caps and collars, providing ambience in a way hard to match in a new tube. And good used tubes are bargains, often selling for only around $20.

THE ART OF FLY FISHING

The wonderful old fishing paintings and prints are visual crowns to any classic-tackle collection, although current prices have become pretty stiff. A

The talented work of A. B. Frost is always in demand. This painting, titled Upstream, *is just one of several Frost originals rendered into hand-colored prints.*

little searching, however, can often turn up some charming prints for peanuts. This is especially true of calendar prints. On the other hand, original oil paintings by the likes of Kilbourne and Frost command serious cash. The fish carvings of Maine, Scottish, and English craftsmen are yet another interesting category of collectible art. These items vary from

crude folk art to intricately accurate renditions of once-living fish. Luckily, there's enough fly-fishing art around to suit the tastes—and pocketbooks—of just about everyone.

ORIGINAL OILS AND WATERCOLORS

Surprisingly, hunting scenes far outnumber angling among original paintings. This is just one reason for the high prices of piscatorial art; art, in general, is a pricey affair. Still, there seems to be no great shortage. Most major tackle auctions feature a substantial amount of original angling art. Some of the best artists include S. A. Kilbourne, A. F. Tait, A. B. Frost, Winslow Homer, T. A. James, Lynn Bogue Hunt, Vermonter Ogden Pleissner, and my neighbor, William Goadby Lawrence.

Two of the earliest American angling artists were Arthur F. Tait and S. A. Kilbourne, who were active from the 1850s to the 1880s. Tait's work typically portrayed Gentlemen casting into impoundments and streams using mid-19th-century techniques. This is handy for the student of antique tackle, but Tait paintings are quite rare. The same is true of Kilbourne's. Many of his original oils depict reposing fish, often with period tackle; they're reminiscent of Audubon's "sleeping" birds. Bainbridge Bishop painted similar oils—expired fish lying next to ancient fly rods and reels.

The late 1880s began to see the style of paintings, whether watercolors or oils, move toward more realistic form. A. B. Frost was perhaps the first of this newer breed of angling artist. His paintings expressed movement. His fishermen were doing things, casting or wetting lines; his fish were often airborne. Louis Rhead, the well-known fly tyer, also painted a few fishing scenes, although most of his work was in the form of etchings. Rhead's paintings were wonderfully colorful and animated; the water truly moved.

Modern artists include Ogden Pleissner, who worked exclusively in oils prior to 1934 but has since depicted many salmon-fishing scenes in watercolor. Also working in watercolors is Maine's Arthur Taylor, whose paintings vary from lively angling scenes to still lifes of classic tackle. For watercolors with a western flavor, we have Chet Reneson. The colorful works of these artists are in great demand today, and when one of their originals arrives at auction it's virtually gobbled up.

ANGLING PRINTS AND SKETCHES

Most of the more recent artists mentioned above have published numbers of fine watercolor prints, especially Pleissner and Taylor. Older prints by Bill Lawrence and Lynn Bogue Hunt surface from time to time. Generally, prints are far less expensive than original watercolors or oils.

Long before the Civil War, etched plates began to embellish angling books; soon, they were slipped into magazines. Etchings, then ink sketches, and finally watercolor prints documented the sportfishing scene and remained popular until photography assumed this role after World War II. This same time frame saw a proliferation of colored posters and calendars that advertised tackle; these are usually affordable piscatoria. Magazine covers, the most inexpensive form of angling art, can be purchased, matted, and framed for a song. Some of the best-known cover illustrators include J. A. Herrmann, Hy S. Watson (one-time editor of *Popular Mechanics*), P. B. Parsons, and Lynn Bogue Hunt.

Affordable reproductions of 19th-century angling prints are often available. Several Nathaniel Currier lithographs are knockouts, including Arthur F. Tait's 1854 *Catching a Trout*. This showed Daniel Webster, an unknown angling companion, and a black "boat boy" using over-the-hand rods. A second famous Tait print, *Brook Trout Fishing*, published by Currier in

ALLERTON LODGE.

A view of fishing author R. G. Allerton's lodge at Bugle Cove, Mooselookmeguntic Lake, Maine, with Bald Mountain in the background. Etched in 1877 by Thomas Moran.

MALE LAND LOCKED SALMON or QUANANICHE.
(Salmo Salar Sebago Girard.)

1862, showed an ardent angler using a below-the-hand reel mount (see figure C-8 in the color section). These historic A. F. Tait prints have been reproduced, and at least one nice set was colored in Italy.

When it comes to great fish prints, none can compare to the superlative chromolithographs of Sherman Foote Denton. Around 1880 Denton, then working for the United States Fish Commission, developed new methods of retaining lifelike colors in taxidermy. His work was soon acknowledged as equal if not superior to that of Peter Malloch, Great Britain's beloved fish painter. Early Denton watercolors were also published in various U.S. Fish Commission publications, and in 1895 a new series of fish prints was included in the first New York State Fisheries, Game & Forest Commission report.

These S. F. Denton prints were far superior to the work of previous taxidermists. Until 1909 subsequent reports included new renditions. In 1901–02 the New York commission published three different portfolios of Denton's work, which were eagerly sought by the leading taxidermists of the times. In all, 99 Denton chromolithographic fish prints are

known. Today, the most sought-after species include the brook trout (four different prints), salmon and landlocked strain (three prints), and brown trout (two prints), plus the bonito, bluefish, striped bass, and black bass (three prints). Other prints included a plethora of saltwater and freshwater panfish, as well as several other gamefish.

Sherman Denton's lifelike art has finally attained the status that it deserves, and the popular species prints now sell for three-figure prices. Just five years ago, Denton prints were common and very inexpensive. I have the two superfish of Maine—Denton's male brook trout, perhaps the most beautiful fish that swims, and his male landlocked salmon, rendered from a 13-pound ouananiche caught by fish culturist–author A. N. Cheney.

You'll run across other fish prints, including a few lively ones by Kilbourne. A smattering of other old-school painters, such as S. E. Eames and A. D. Turner, made prints and etchings. Newer prints of the fisherman-with-landing-net-and-jumping-fish style are seen in early-20th-century tackle advertisements. Winchester and Bristol always had a new calendar to fire the interests of anglers during the long, dreary winter, and calendar art can be attractively framed.

ANTIQUE AND CONTEMPORARY FISH CARVINGS

In the arena of three-dimensional collectibles, I'll mention taxidermy only briefly: Natural skin mounts dry out and crumble, fins fall off or break. That's about all there is to say. The exception is a specimen that was preserved under a glass bubble or inside a deep frame, rendering it immune to the perils of dusting. The excellent mounts of the Hardy Brothers and P. D. Malloch were preserved in this fashion and still look good a century later. Modern fiberglass mounts are hearty but they often suffer from the Colors From Mars syndrome. Pink tarpon and hot-green bass were not exactly Denton's ideal. The only realistic glass mounts I've seen were painted by New England Taxidermy in the 1970s. Contemporary paintwork, as seen in Pflueger's and Reese's fish, is just too damned odd for my taste.

Now on to the fish carvings. The earliest renditions came from mostly anonymous folk artists. Their carvings and colors may have been off, a trout may have looked like a bass, and the colors may have faded or been over-varnished to a muted brown, but these often crude renditions had an inherent charm that elevated them into true art.

Some old-time fish carvers actually signed their "objects de-ahht," as they say in Maine. One of the Pine Tree State's early carvers, Horace G. Hines, active in the 1920s and '30s, produced very fine head and fin details. His trout were affixed to fancy plaques or ensconced within dioramas. A contemporary of Hines was Robert Chapman, of Raymond, who carved folksy brook trout with metal fins. Arthur Hanslan (1880–1944) of Oxford also carved some folk-art-style trout. Phillip Sirois of Arrowsic, a later carver, specialized in brook trout that looked more realistic. Lloyd Thomas of Camden was one of the few Dirigo artists to carve black bass. Maine's last carver of note, and perhaps its finest, was Lawrence

Irving of Winthrop, who left behind realistic examples of brook trout and salmon.

Contemporary American carvers include George Warren from Cape Cod, Bruce Buckley of Costa Mesa, California, and Michele Oliver, Roy Ploughman, and Kevin Creaser. Stan Gibbs, of Sagamore, Cape Cod, is one of the top up-and-coming fish painter-carvers in the East. He's best known for his excellent striper plugs. My old collector-friend Steve Smith of Jamestown, New York, is one of the best fish artists of our time. His skill at intricate detail work is coupled with a good eye for color, and he has a steady hand. I envy him.

15 Collecting Vintage Fly Tackle

Collecting and using classic tackle is surprisingly popular—and addictive. It usually starts out innocently enough, but often ends in a mad compulsion to expand closet space. Fly-reel collectors are lucky; they need only a few shelves or a chest of drawers. Rods are another story. Even when broken down and cased, rods take up a lot of room. Some are attractively displayed under glass, either on a wall or in, as we say in Maine, a wicked-shaahp coffee table. More often than not, however, rods are stacked upright in dark little closets, and the one you want is always in the back.

A century ago, Doctor James A. Henshall amassed one of the earliest tackle collections. Modern collecting dates from the early 1970s, when Thomas & Thomas, Mary Kefover Kelly, and Martin J. Keane began dealing in "classic" fly rods, meaning those built with Chinese bamboo, modern guides, and no intermediate wraps. Later, many of us discovered the "old stuff": 19th-century wooden rods, antique fly reels, old flies, and artwork.

Many collectors gravitate toward a specific era of tackle construction. Tackle can be categorized as coming from the *smith age* (1800–65), the *expansion age* (1865–1900), the *classic age* (1900–55), and the *semimodern era* (1955–75). Musically inclined collectors might prefer to think of these last two as the Kate Smith Age and the Aerosmith Age.

Today's fly fishermen have rediscovered tackle from the classic age. Often selling for about the same price as a new graphite rod, a Tonkin-bamboo rod has a wonderful "feel" of great appeal to modern anglers. Since most classic-age tackle will be fished, I'll provide some pointers here on choosing good working instruments. Expansion-age tackle is also usable, especially the reels. Expansion- and smith-age beauties rank just behind the classic stuff in popularity, and some of them bring pretty heady prices, far exceeding those of classic tackle. Yet quality antique tackle remains a bargain today, awaiting astute collectors with a long-term view.

Within each age there are different criteria for defining a worthy investment. In the following pages, I'll try to provide a few guideliness to help you chart your course, and some notes on items that should increase in value.

 ## Reels

EARLY BRASS AND GERMAN SILVER WINCHES

The beautiful fly reels built around the time of the Civil War or earlier are much sought after, usually bringing higher prices than contemporary rods. Rarest are maker-signed examples by Ustonson, Haywood, Conroy, Billinghurst, Fritz Vom Hofe, and Crook. They're all available—for a price. Next in desirability and rarity are those reels marked by the selling house, such as those of T. H. Bate, John Warrin, Andrew Clerk, or Martin Bradford. Lesser reels make up the bulk of the offerings from this era and are priced accordingly. Examples include a variety of unmarked but great little models such as English stop latches and super little pump clampers, most of them made of brass and lacking a click. Even

From a Japanned salmon-fly tin to antique and classic reels, there's a wealth of fly tackle available to suit any collector's taste. When it comes to split bamboo, you can search for the old and rare, such as this Bangor Leonard, or for much newer models by Thomas or Granger—and take them fishing.

Two early American fly reels, both marked by J. C. Conroy & Company. At left, a raised-housing, German silver click reel. At right, a tiny all-brass No. 6 multiplier.

unsigned examples from the smith age bring healthy prices.

German silver is the collector's material of choice; early reels made from this metal are worth more than similar brass models. All other things being equal, however, the primary determinant of value is condition. For example, a pair of identical Conroy click-housing reels—one in good condition and the other in excellent—may bring $650 and $1,200, respectively. It's common to find a reel with feet that were filed down a long time ago to fit a particular seat. While this old practice does remove some of the original integrity, it doesn't appreciably mar the value of the reel, especially if done professionally by a tackle- or gunsmith. Bent feet or crank handles do reduce value, but these problems can often be corrected by a careful application of pressure, preferably by a professional. Buffing will remove minor scratches, but I advise against this. Buffing removes a century of patina; most collectors would rather see a few scratches than a mirror surface.

Regardless of age, any antique fly reel with broken or replacement parts, such as a handle or foot, is worth far less than one that is completely original. If the reel has a modern replacement part made by a professional craftsman, this should be mentioned as a provision of sale. In the same vein, if any tackle item was purportedly owned by a famous individual yet is unmarked to that effect, a letter of authenticity must accompany the piece, either written by the former owner (if still alive) or by a reliable and honest relative or dealer who can produce information to substantiate such a claim.

Antique fly reels are popular collectibles. They're small, take up little room, and don't deteriorate like wood, bamboo, and varnish. And they're charming little buggers, admired even by noncollectors. "Hey, nice reel! It's even smaller than my Zebco. Oops! I dropped it!"

CLASSIC REELS

The very best of classic-era reels are the high-quality German silver and rubber examples by Philbrook & Payne, H. L. Leonard, and Edward vom Hofe or his brother Julius; the rare perforated reels by Orvis; any fly reel stamped "Kosmic"; and the various models marked by William Mills & Son. To these we can add the later works of Zwarg and Walker, contemporary models such as Bogdans and Seamasters, and the original wedding-cake Fin-Nors. New or old, all these quality classics bring good prices and retain their value.

Any reel found in its original box is a good find, but when it's a Billinghurst? Well, let's just say the "spittle factor" increases—and so does the price.

The next group of classic reels includes the many Hardy models that are so popular on both sides of the Atlantic, both for collecting and for fishing. Other British reels by Allcock, Young, Farlow, and their contemporaries are midpriced collectibles that will always have a keen following. American reels in the same price range ($150–300) include the better products of U.S. Net & Twine, Montague, and Cozzone, as well as some Pflueger models, such as the Hawkeye and the Golden West.

Whether a pricey vom Hofe or a modest Pflueger, a German silver or hard-rubber reel must be inspected closely for buggered screws and dented rims. You'll need an even closer eyeballing to find hairline cracks in the rubber side plates. These cracks usually run from screw to screw in a low arch and may be hard to spot. Hard rubber also tends to "brown" as it dries with age.

One of the great windup toys of early-20th-century America, the Meisselbach auto reel is just one of several moderately priced side-mounters now collectible.

Getting the brown out isn't easy, and many collectors just wipe the plates with a light oil and call it done. Any antique hard-rubber fly reel will show browning. That's how you know it's an antique.

Julius vom Hofe, Montague, and Pflueger made hard-rubber classics with lesser nickel-plated accents. These are good, usable reels and excellent streamside image builders. The early Pflueger Medalist, with its agate line guide, falls into the collectible/fishable category, as does the nickel-silver-accented Gem. Antique Montague reels, such as the little Julius vom Hofe look-alikes, started off as low-priced collectibles but have recently gained in value. The wonderful Newark reels of Gus and William Meisselbach are top bargains in today's market. Modestly priced from the start, they still don't command high prices, yet these production reels are fascinating examples of their era with a growing collector following. I've fished a number of these warriors, content in the knowledge that a streamside disaster would not result in a financial one.

Many fine British reels remain inexpensive if in honest fished condition. Several models from Hardy, including the St. John, St. George, Uniqua Mark II, and Lightweight, are usually priced reasonably, and most J. W. Young and post-1900 Allcock reels are relatively inexpensive.

NOUVEAU REELS

Brand-new classic reels emerged in the early 1970s in tandem with a renewed interest in classic cane. The best examples of the breed are still made by Stan Bogdan & Son of Nashua, New Hampshire. Most of the new classics were very high-quality items, although a few had tiny flaws in design, good taste, or both. Some were limited editions or commemoratives. Although they're expensive enough, the reels don't bring the heavy prices their producers had hoped.

Commemoratives usually arrived as simple trout models. Leonard put out a raised-pillar reissue.

Thomas & Thomas made the Kosmic Commemorative, thinking it was reproducing a design by Julius vom Hofe. Unfortunately, original Kosmic reels were mass-produced by U.S. Net & Twine, so Thomas & Thomas spent two years in research and engineering to make a copy of a production reel. Carpenter & Casey tossed their classic into the ring, and it was perhaps the best of the bunch. A little later, so did Bill Ballen and North Fork; this reel actually had an adjustable drag.

The KOSMIC COMMEMORATIVE REEL
Thomas & Thomas - makers

Many of these are very nice reels and look like vom Hofes at a distance. They are all of decent vintage, but Chateau d' Rothchild 1948 they ain't.

A more recent introduction is the reel entombed in a satin-lined walnut presentation case. These containers look like miniature caskets, within which the reel lies in state until shown to an admirer wearing white cotton gloves. I almost expect frantic organ music from *The Phantom of the Opera* to begin playing as the lid slowly opens. These casket reels are never used or touched by human hands; merely loading a fly line destroys their "mintyness," as collector Bruce Boyden once termed it. To me, a new reel that must remain forever new reminds me of a new car with its rich Corinthian leather covered with vinyl. Take 10 percent off the price of these reels for sterility and another 10 percent for morbidity.

There are upbeat new classics available that are meant to be used and often purchased in that condition. The original Marryat and Bob Corsetti's Peerless models stand up to angling conditions and still look great. They cost less than the memorial-park reels because someone may have actually taken them fishing.

THE COLOR OF GOLD

The 1970s also saw the arrival of classic reels splendidly attired in gleaming gold. Although perhaps not in the best of taste, the reels were designed for endurance, strength, and longevity. Fin-Nor introduced the original model, the legendary Wedding Cake. Then Captain Bob McCristian started producing the gold Seamaster. Both models have an avid following. My friend John McBride is a typical collector of the new classic saltwater reels. John uses them on everything from 12-pound rainbow runners to sailfish; he often fishes with an older, gold-anodized Seamaster. Classic marine fly reels are an auction draw, and in excellent condition can be good investments. The original gold reel is an anomaly: it has *good* bad taste, and is certainly much tastier than the new Seamasters, which come anodized in a shocking blue.

$UPER CLA$$IC REEL$

These days, when it's becoming increasingly difficult to find a good hamburger, it's nice to know at least one thing is still made the way it used to be. A century ago a chap could plunk down a month's pay on a top-quality reel by Philbrook & Payne. Fifty years ago, the last of the old-time quality reels were still available, from Edward vom Hofe's sons. Now these have become $uper Cla$$ic$. Today, brand-new Super Reels can be had—for a price. Old or new, the following $ix $uper Cla$$ic$ are true examples of investment tackle. They represent the highest-quality reels of each angling generation—functional, yet sporting a timeless beauty.

- *Marbleized Philbrook & Payne,* the original mouthwatering gem responsible for the classic reel "look." These are always tough to find and expensive, because Frank and Ed never made enough.

- *Ben Hur Trout Reel,* Talbot's single and beautiful contribution to fly-tackle collectors. It's so rare that only a few are in collections.

- *Hardy Bougle,* the odd raised-pillar version of the Perfect. The small 3-inch pre–Mark II model commands the highest price.

- *Edward vom Hofe Perfection,* not originally Edward's most expensive model but now the most coveted vom Hofe of them all. The classic of classics with a price tag to prove it.

- *Zwarg Model 400,* most sought after in the little 1/0 size. This multiplier looks like a single-action reel; it's so smooth that I used one for an entire day before discovering why I could gain line so fast.

- *The Bogdan Reel* was conceived around 1970 and surpasses all other modern classics in design and looks. Size doesn't seem to matter; salmon models have a waiting list as long as the trout versions.

Contemporary high-quality classic reels are rugged and dependable, a joy for fresh- or saltwater use. They're currently made by Penn, Fin-Nor, and (shown here) Peerless.

There are other Super Reels, such as the various Leonard bi-metal models. The criterion is simple: Each model has an appeal that far outstrips its availability. The vom Hofe Perfection, for example, is not all that rare, but the demand for this classic assures a four-digit price. Super Reels are major investments, bought by collectors as others might buy commodities on the futures market. Of course, classic reels are a lot more fun than pork bellies—even the casketed ones. But give me an old Pflueger any day. I can take it fishing.

Grading a Classic's Condition

Most dealers follow a standard tackle-grading system based on the condition of the rod, reel, or fly. The following guidelines can help determine whether a piece is worth top dollar or a figure substantially less.

Unused   Original factory condition, sometimes called mint. Absolutely unfished, and can the seller prove it? Reels have their original box, case, or papers. Rods come with tube and bag, hanging tag, or plastic grip wrap. A fly is still in its original unopened package or accompanied by the tyer's card or autograph. I suppose a creel would not smell of fish. These are top-dollar items.

Excellent   Reels have their original patina; they have no scratches, dings, corrosion, pitting, or chips; they're mechanically perfect. Rods may have very minor age crazing (minute hairline fissures) but perfect wraps, no missing guides, and all sections present and to length; they're in their original bag and tube. No restoration. Flies are unfished but may not be in their original pack. Creels have every weave of willow intact, plus their original latch and strap. Classics in excellent condition bring 75 to 80 percent of mint-condition tackle.

Very Good   Reels may have a small scratch or ding, light wear on nickel plating, or minor foot alterations, yet are mechanically sound. Polishing is allowed but frowned upon. Rods may have minor flaws on the varnish or wear at the edge of a few wraps; they have no missing parts and are good enough to take fishing. On classics, one short tip is allowed, but all other sections must be to their original length. Antique rods may have a few replacement wraps. A professional restoration is acceptable. Smith-age rods with more than two tips must have one tip to the original length. Flies may be slightly out of shape. Items are worth 50 to 60 percent of the top price.

Good   Reels may have some buggered screw heads; most of their nickel plating is present; a foot may be altered; small chips or dings, pitting, or corrosion may be present; they're mechanically good enough to use. Rods may have flaws in their varnish, on soft or alligatorized varnish, or varnish overcoats; they may be missing a few wraps or guides; one tip or the mid may be short; the bamboo shafts must be sound and fishable after refinishing. Smith-age rods may have more than one section slightly short. A fly will have minor pressing damage, and a creel may have a broken weave or two. The item is worth 30 to 45 percent of its highest value.

Poor   Reels may have cracked side plates, heavy corrosion, or broken or missing parts; they're not mechanically sound. Rods may have delaminated shafts, replacement parts not correct to the original integrity, or several short sections. These are "beaters," yet are still good for original parts. Flies may be used, have rusty hooks, or be damaged by moths. They're users. A. J's Handy Hint—never stuff a trout into a creel in poor condition. Values are marginal—5 to 20 percent of the same item in unused condition.

These are general guidelines, and more or less follow those presented by the National Fishing Lure Collectors Club. In respect to rods, the NFLCC has a simplistic Grading System and definition of Restoration. According to NFLCC guidelines, a rod that has more than one replacement section is a "reproduction," exceeding 25 percent of its originality. The same would be true of a complete revarnish job. Many older rods have replacement sections and refinished shafts, yet aren't "reproductions."

The price of any tackle item is also determined by its degree of rarity and its appeal to collectors. Any newer (post-1900) classic in less than excellent condition is not an investment piece, although it may be perfectly usable. In the same vein, antiques (1865–1900) in less than very good condition should be avoided. Smith-age tackle, built from the 18th century until the end of the Civil War, retains value even if its condition is as low as good. And last—don't spend good money on bad tackle.

In truth, classic reels fit into everyone's price range. Most models are neither scarce nor inordinately expensive. We can use, display, or hoard them. And when the time comes for a trade or resale, a well-chosen classic reel can bring a good return.

 # Rods

ANTIQUE WOODEN RODS

I find those wooden rods of yore intriguing in a way that no classic rod can touch. And, as seasoned collectors say, "They are tough." It's almost impossible to find a complete rod built before 1850. Few examples were ever made, and most appear to have succumbed to neglect or abuse. Even British rods, made for centuries prior to 1850, seem lost to time. The American tackle business, working at its bench-work pace, produced so few pieces before 1850 that the chances of snatching one up are slim indeed.

Antique rods built between 1850 and 1865 are more available, although still scarce. In many cases rods of this era go unrecognized, even by experienced collectors, whose eyes may not be trained to spot the signs of age. Reels of this period are much easier to spot, and because the materials are inherently more durable, they've withstood hard knocks and streamside crashings to come to us in reasonably good condition. This is worth repeating: In tackle collecting, top condition means everything.

Wooden rods fall into two categories: antiques and historic. Antiques have a complete butt and mid-section, at least one tip, and are usually unsigned by the maker. These rods make great wall hangers, and the slightly daft may enjoy using them in the field—as long as the trout are small. Some antiques are found in better condition, perhaps housed in a wooden form or with extra tips stored in a bamboo tip case. These exceptional pieces are worth far more than their current prices would indicate.

Historic rods are usually engraved in script with the name of the maker, the tackle house, or the original owner; they were often special presentations. The maker's name, if present, usually appears on the butt cap, while the owner's name is usually found on

the fixed reel band. These pieces are ultra-rare and invaluable, although currently they often bring lower prices than classics built 100 years later. This will change as collectors come more and more to appreciate these elder fly-fishing implements.

Any scripted antique rod, or dated one, with most or all its tip sections and at least one tip to original length, is a great find. So far very few such rods are in private collections, although a number are found in the inventory of the American Museum of Fly Fishing in Manchester, Vermont, and at the Anglers' Club of New York. That there are such rods in museums should motivate the private collector to continue the search.

At present, historic rods can run from $400 to $1,200 in very good shape. Signed examples in good condition command lesser prices. Considering the rarity of any historic rod, its present value is a bargain, much like boarding a time machine and returning to

This signed and dated Conroy butt cap would be an ultra-rare and irreplaceable piece of Americana—even if it weren't fitted to the only known two-handed Yankee trout rod.

old Boston to shop at Richardson & Dexter's "Sportsman's Warehouse."

ANTIQUE SPLIT BAMBOO

Antique split-bamboo rods—those built between 1865 and 1900—are far rarer than anyone thinks. The supply of surviving antique rods from Murphy, Leonard, Wheeler, Orvis, the Pritchard Brothers, Landman, Chubb, Nichols, Divine, the Conroy partnerships, and Thomas, Edwards & Payne combined is very small. For every hundred classic rods sold on the market, only one good antique will surface; even fewer come to us with their original trav-

Details of the God Bout River Rod, engraved "1852." Unfortunately, the rod went through a lodge fire and was rebuilt with later-issue ferrules and snake guides. The original butt cap, not shown, indicates an English origin.

eling gear and tips to length. Remember, the youngest of these are a century old, and most show the inevitable abuse of time. Still, the good ones, even the really great ones, are presently selling for half their actual worth. Dealer Bob Corsetti, who handles a great many rods every year, puts it this way: "These antique rods are greatly undervalued. They represent the best investment in tackle today. It's simply a matter of maturity in the tackle-collecting field. Plus they are quite scarce!"

Even pivotal historic bamboo rods from the likes of Murphy, Norris, Krider, Leonard, and Wheeler remain on the 50 percent off rack, as do those models at the other end of the antique bamboo era, the incredible Kosmic, Thomas & Edwards, Ed Payne, Brewer-built Thomas, and Landman rods. All these rods are priceless. And today, at least, they're bargains.

CLASSIC PRODUCTION CANE

A production rod is defined as a model built on machinery and finished by more than one individual. Born around 1900, the classic production rod has several key features, including a Tonkin shaft instead of a Calcutta, and snake guides in place of the older loose-ring guides. In the 1920s classics began to lose the intermediate wraps spaced along the shaft. A tad later, old-style full-metal sliding-band seats were canned in favor of newer threaded versions.

Production-rod quality can vary in the extreme. Listed in ascending order of general quality (and disregarding variations within each company's product line), I'd start with Montague then move on to Horrocks-Ibbotson, South Bend, Heddon, Phillipson, Orvis, Cross, Granger, Edwards, Leonard, Young, Thomas, and, finally, Payne. That's my list, anyway. Other collectors may disagree about the exact order of the midrange, starting with Heddon. But just about everyone agrees that the Payne rods rank highest in attention to detail.

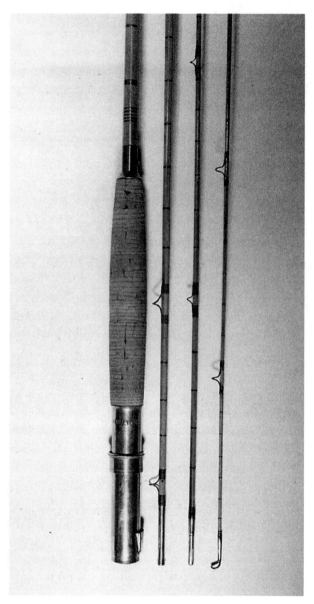

By 1900, the basic features of the classic fly rod were defined. This delicate 7½-foot Isaak Walton illustrates the progression from antique to classic production rod. Built by Fred Thomas, it has the new look, but the shafts were made from Calcutta cane. Bought for less than $200, it tosses a 4-weight line and is still fishing.

I love to watch the purchase of a classic bamboo rod at a tackle show. The prospective buyer usually picks it up and rotates it two or three times, as if trying to bore a hole in the ceiling. Then he starts to twitch it, looking for evidence of casting ability. In truth, the only way to know how a classic rod will cast is to load it with line and cast it—preferably into a wind blowing at 10 knots or better. Anything less is guesswork. Unfortunately, guesswork is the method by which most rods are purchased. It needn't be so.

Here are a few tips that will help eliminate trial and error.

First, check the rod for original integrity. Are all the sections there? Most older classics were built 3/2—meaning three pieces, two tips. From the 1930s to the present, many rods were built 2/2, and a few, such as the Orvis 99 and Madison, were built 2/1. Next, check the lengths of the sections by standing them all upright on a flat surface. Unlike many antiques, which often have longer or shorter tips, production rods have all sections to length. The exceptions are a few oddballs, such as two-piece staggered-ferrule rods with tips much longer than the butt; Pezon-Michel rods come to mind.

Next, check for a possible replacement section. Is the color or grain of one of the sections obviously different? A rod with a replacement section is worth substantially less than one with all sections intact. This is mitigated in part if the replacement section is from a recognized maker or, better, was later fashioned by the original maker. If a rod isn't "right," a dealer will usually say so.

A great deal of a classic rod's value is contained within the original maker's varnish. Original varnish jobs, 20 or more years old, will show early signs of crazing under magnification. These minuscule cracks are at the surface of the coating, do not detract from the rod's value, and are impossible to duplicate by a refinisher. If a rod is refinished, it helps if the work was done by a key craftsman from the original rod gang—a Paul Young rod refinished by Bob Summers, for instance, or a Wes Jordan rod refurbished by Orvis.

A rod with missing or short tips or a short or replacement mid is worth a mere fraction of the same rod in excellent original condition. To a lesser extent, chipped varnish, missing wraps, small fractures, hook digs, and missing guides all lessen the value of a rod, as do loose ferrules. If you feel or hear "ticking" when you sweep the rod from side to side, the rod has one or more loose ferrules. Such a rod should never be used without tightening the ferrules, as the bamboo shaft may break where it meets the metal. A good rod with any of the above defects can recover a portion of its value if refurbished by a talented restorer.

Many rods can be classified as "beaters" or handyman specials, and these can be fixed up for use

Montague Manitou Fly $35.00

The supreme achievement of the oldest and largest makers of split bamboo rods in the world. This rod consists of three pieces and an extra tip hand made of six strips of the most carefully selected Tonkin cane, aged and heat treated to secure an action found in no other rod and resulting in a rich deep brown color. The ferrules are of 18% nickel silver hand welted, perfectly fitted, serrated and wound over and made waterproof. The reel seat is of rich black hard rubber with spiral locking nickel silver band and butt cap. The solid cork grip, shaped to fit the hand, is of fine satin finish cork. The first guide and the tops are of genuine agate, smooth and well mounted. The intermediate guides are file hard steel and a hook ring is provided just above the grip. At guides, ferrules and grip the rod is wound with pongee and purple silk and then heavily varnished by hand with a special waterproof varnish. Packed in a heavy canvas bag and the whole in an aluminum rod case with watertight screw cap.

Made in the following styles

8½ ft. light trout	9 ft. medium trout	9 ft. bass
9 ft. light trout	9½ ft. medium trout	9½ ft. bass

Montague Red Wing Fly $25.00

A rod of highest quality consisting of three pieces and an extra tip made of thoroughly seasoned, selected brown, heavy Tonkin cane in six strip construction. The ferrules are serrated and welted and chased and, with the hand made reel seat, are of 18% nickel silver. The grip is of solid cork, well shaped. The first guide and tops are of genuine agate and the intermediate guides of file hard steel and there is a convenient hook ring at the grip. The rod is wound with red and yellow silk and nicely varnished. Packed in a partition canvas bag and then in an aluminum rod case with watertight screw cap.

Made in the following styles

7½ ft. light trout		9 ft. bass
8½ and 9 ft. medium trout	9½ ft. medium trout	9½ ft. bass

Montague Trail Fly $20.00

A high grade rod consisting of three pieces and extra tip made of selected brown Tonkin cane. Ferrules are hand welted and serrated and, with the patent waterproof reel seat. are of 18% nickel silver. The solid cork grip is nicely shaped. The first guide and tops are of genuine agate and the intermediate guides are of file hard steel. A hook ring is mounted just above the grip. Wound with yellow and dark red silk. Packed in a partition canvas bag and aluminum rod case with a bayonet cap.

Made in the following styles

8½ ft. trout		9 ft. bass
9 ft. trout	9½ ft. trout	9½ ft. bass

Page Three

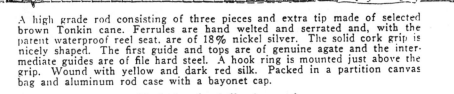

From a 1940-era catalog, here are the top three Montague rods—good, fishable classics, especially the 7½-foot Red Wing.

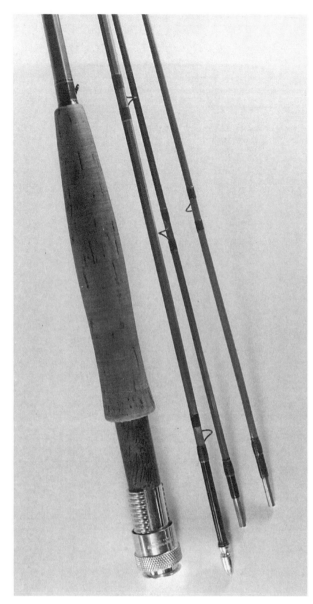

It may be a production rod, but a Payne in pristine shape is really too good to take fishing. This is an 8-foot, two-piece Model 202 in excellent condition, a late one made for Abercrombie & Fitch.

on the stream at modest cost. Certainly a well-worn Heddon No. 10 or a Thomas with one remaining short tip is still a very usable and enjoyable rod. And if a production rod of little collector value breaks through honest use, there should be no tears lost in Mudville. Well, maybe a few.

Through the years, I've fished a number of fly rods with tips 1 or 2 inches short. They caught some fine fish and were a joy to use—especially after experiencing the staccato whiplash of second-generation graphite. I waved 9-footers until my casts got sloppy, then dropped down in length to rest my wrist. This,

of course, is why classic rods 8 feet or shorter command the highest prices—this and the fact that they're much rarer than their longer brothers.

Production rods needing major surgery or replacement sections can be purchased for parts. The going price is $5 per good section. If you aren't the handyman type, it's best to let a dead dog lie. Some rods are just worthless. I kept a pile of steel rods for 10 years before I finally sold them for scrap at 50 cents a pound.

HANDCRAFTED $PLIT-BAMBOO CLA$$IC$

Without a doubt, the most sought-after classic rod is a Garrison, followed closely by a number of other high-priced zeniths of the rodmaker's art. In a scene right out of a Humphrey Bogart film I once sold one of these "high-classics," took the money, and flew to Mexico. That pretty much says it all. Martin J. Keane is the foremost authority on these lofty sticks, and his *Classic Rods and Rodmakers* should be on every rod collector's bookshelf. It is The Source for a wealth of material on the classic rods from our best craftsmen. Collecting these fine bamboo rods is frighteningly expensive, but it must be gratifying because it has an amazing following. Here's just a partial list of some of these great rodmakers:

Sam Carlson	Walt Carpenter
Lyle Dickerson	Everett Garrison
Pinky Gillum	George Halstead
Gary Howells	Letcher Lambuth
Morris Kushner	Edwin Powell
Walton Powell	Lew Stoner
Thomas & Thomas	Frank Wire

Add to this list such contemporary makers as Bob Summers, Mark Aroner, Jon Parker, Darryl Whitehead, Mike Clark, George Maurer, Ron McKinley, the late Tony Maslan, Bob Taylor, Steve Jenkins, Dave Klausmeyer, Homer Jennings, and David LeClair, and you'll see that there's a vast array of new split bamboo available, although there's a bit of a wait involved. Handcraftsmanship takes time.

The collective output of this large group of rodmakers provides a good selection at any major auction, in mail-order tackle listings, and directly from the makers. Some of these rods are almost too good to use. Entire volumes could be written about them, although the author would quickly run out of glow-

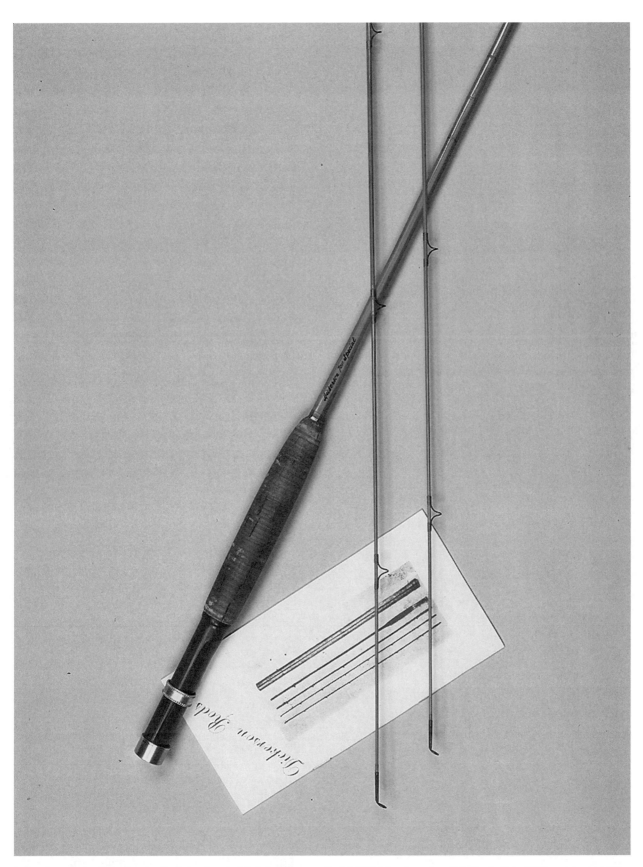

A superb example of the handcrafted classic rod, this Lyle Dickerson Model #7011 is a two-piece 7-footer built in 1952 and marked "Special" on the shaft. It's accompanied by a 1960 Dickerson catalog.

ing adjectives. Perhaps such a book is best left for the future, after the present speculation about which new rodmaker will gain superiority, die first, or find a more financially rewarding occupation sorts itself out.

SYNTHETIC SHAFTS: 50 YEARS LATER

Hard to believe, isn't it? Yet the fiberglass fly rod really was introduced a full half century ago, right after World War II. I have no idea where those 50 years went, although I was around the entire time, but the notion that fiberglass would ever get "old" never entered anyone's mind. Certainly not mine. This begs the question: Are glass rods antiques? Well, what do we call a 1945 Evinrude? We call it an antique outboard.

Antiques or not, glass rods are definitely collectible, especially some of the earlier limited runs and those rods that were aesthetically pleasing and excellent in casting. The first ones, built by the original Arthur Howald process, were the Shakespeare Wonderods. Unfortunately, they weren't handsome affairs; their only redeeming feature was a unique spring-closing reel seat. Next we have the Conolon rods from the early 1950s. These were better looking, but were made with insufficient resin; the blanks would literally saw themselves in half until they broke. Any Conolon rod (these were eventually sold by Garcia) still in its bag and fiber tube and showing little signs of use is worth far more than its original price—and that was a lot of money back in the 1950s.

During the ensuing years, the makers finally figured out how to build a rod that wouldn't self-destruct. Some of the best-casting fly rods of the 1950s–'60s era came from Montague, but even these often snapped after lengthy use. Montague, which made an entire line of reasonably pleasant fly rods of up to 9½ feet in length, used the original-style Montague decal just above the grip. It also made a better grade of private-label glass rods for outfits like L. L. Bean.

The age of glass came to an end in the mid-1970s with the ascendancy of graphite. After that, fiberglass fly rods featured fewer guides and inferior fittings, and were relegated to the beginner/panfish market. This leaves us with just one great Glass Decade. From 1965 to '75, many makers tried to produce

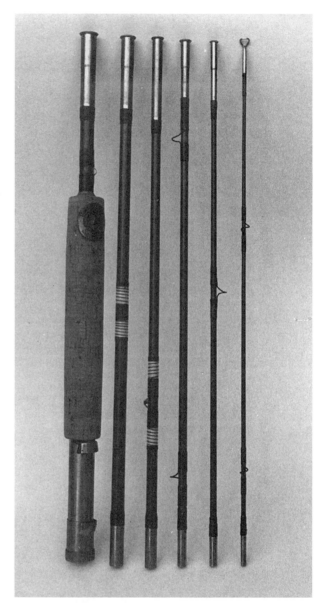

A six-piece Wright & McGill Trailmaster pack rod, originally cased in a fitted bag and short aluminum tube. Note the Goodwin Granger reel seat, still used 40 years after it was patented.

excellent casting tools; some versions even attempted to rival the elegance of production bamboo. In many cases, "elegance" didn't weather the transition from Tonkin to glass particularly well—especially if the attempt at elegance included spiral wraps over gold or silver tape.

A list of excellent casting rods of the Good Glass Era would include the Browning Silaflex and the Cortland 2000. These were at least nonoffensive looking; not so the products of other makers. Heddon's top-of-the-line series, the Mark IV, was rather tacky, including a cork grip, tenite, and walnut

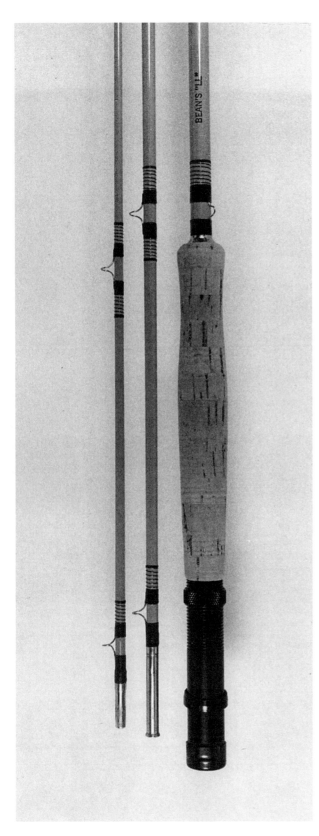

A very nice glass Phillipson Expoxite, an 8-foot, 5-weight three-piecer marked with the premier Double LL label of L. L. Bean. The pristine condition of this rod adds a great deal to its value. (The James R. Babb Collection)

inlays—all on the same rod. Shakespeare, long troubled by a tacky white blank, introduced a Woodgrain fly series in the early 1970s—a bad idea gone worse, complete with gold foil underlayments and a "teak wood finish" blank.

Wright & McGill made a number of quality glass rods, including the two-piece Aristocrats. Its Granger series featured a 6½-footer for a 6-weight line. Built on a straw-colored blank with dark brown wraps, the Granger rods had a charm that eluded other manufacturers. Even these, however, suffered from spiral wrapping, an insufficient number of guides, and an inherent tip-heaviness in the larger models. And, of course, the little 6½-footer wouldn't shoot a light line.

Many Wright & McGill glass rods bore the patented Goodwin Granger nickel-silver reel seat. Perhaps the most unusual was the multipiece Trailmaster. The No. 4 TMF, at 7½ feet, was a four-piece pack rod that came in a 23-inch aluminum tube and "inner liner" (rod bag). Tossing a 6-weight line, it was a great eastern rod for places like the White Mountains. The western version of the Trailmaster was the No. 6 TMF, which was packed in a 15-inch case and came in six sections that turned into a 6-foot, 9-inch rod for a 7-weight line.

Garcia's rods were built on advanced Conolon blanks and were designed by some of the best fly casters of the day. The Garcia Lee Wulff Special, the Five Star Special, and the parabolic Garcia-Ritz models were fine casting instruments, and a great deal of research went into their design. Unfortunately, all featured ugly blue blanks and accents, including the famous one-piece 6-foot salmon version. These otherwise fine fishing rods are proof positive that bad taste can spoil collectibility.

TOP GLASS

Call me conservative, but fly rods should be brown, dammit. Fortunately, a number of makers from this era managed to produce rods that were both fishable and aesthetically pleasing, among them Phillipson, Montague, Conolon, Berkley, Browning, Scientific Anglers, and Fenwick. Most of their blanks looked like chocolate or caramel cane—from a distance.

Bill Phillipson was one of the first rodmakers to use epoxy adhesives for laminating blanks. His rods came in three grades. The top of the line were the Epoxites, available in nine models. The 6-footer (No. EF60) for a 4- to 5-weight line was the classic midge rod, and perhaps the first little glass rod to handle tiny lines successfully. Lower-priced Phillipsons included the Royal Wands, a midpriced series of excellent rods that seemed to toss nothing but 6-weight lines, and the Master series. Although his least expensive, these rods still had German silver ferrules. The best of this series was the No. MF76L, a 7½-footer that took a 5-weight line.

During this same period, Scientific Anglers,™ Fenwick, and Berkley also produced classics that were both fishable and appealing. The early System rods by 3M Scientific Anglers,™ with their refined tapers and chocolate-toned blanks, were some of the best glass production rods ever built—and they carried the price tags to prove it. The amazing System Four rod—7 feet, 2 inches long yet weighing only 3 ounces—was a classic, as were other models in the original series.

The early Fenwick Ferralites are also future collectibles. Today, many fly fishermen who like a slow rod with good delivery still use them. Unfortunately, a white diamond wrap that really belonged on a salt-water trolling rod marred the Ferralite's appearance. From the same era came two rods by Berkley that were among the finest of working glass collectibles. Berkley's Parametric rods were built on a blank with exceptional damping qualities, and, although a bit garish, they could do things that no other glass rod had ever done. To the value of the Parametrics as fine casting tools is added a plus for collectors: If my aging brain remembers correctly, they were made for only two years. Parametrics gave way to Berkley's Specialist series, which were toned down to become great classic glass rods, perhaps the most elegant of the era. The 6½-foot Specialist took a 4-weight line—very light for a production rod; the 7-footer pushed a 5-weight with authority and weighed only 2½ ounces. The early System rods, Fenwicks, and Berkleys were the Cadillacs of their day, selling for what seemed like astronomical prices in 1965. Today they're worth far more.

Conscientious builders of quality glass tried to preserve the feel of bamboo, and a few were successful. Custom versions by Phillipson, Russ Peak, Vince

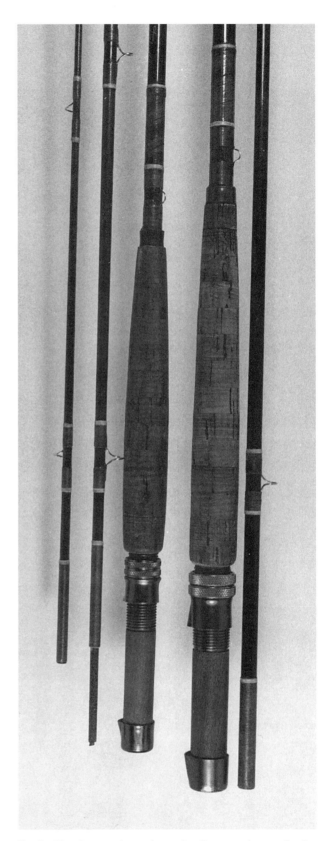

The Berkley Parametric, perhaps the finest-casting production glass rod ever designed, and the forerunner of the Specialist series. Shown are a three-piece 6½-footer (left), and an 8½-foot two-piecer.

Cummings, Art Neumann, and (ahem) yours truly made their mark, especially the light-line designs and smaller rods. Some, such as Peak's Golden Zenith series and the German silver–ferruled Cummings models, are now cherished classics. These small-stream rods will never go out of style or favor, and with their high quality can only appreciate in value.

Major split-bamboo rod manufacturers made their own versions of the perfect glass rod. In many cases, the only differences between the Berkley-Fenwick-Phillippson offerings and the hoity-toity versions were refinements in the grip and reel seat, and copious quantities of German silver. Thomas & Thomas hit us with the Kaneglass; other models were built by Hardy, Orvis, and Leonard. The list could be endless, but the original supply was not. To retain their value, good glass rods must meet the same criteria as classic cane. Unless you're just looking for something to fish with, avoid any fiberglass rods in less than very good condition.

A PHITE BETWEEN TWO GRAPHS

James Watt discovered carbon graphite several centuries ago, but it was the aerospace industry's search for stronger, lighter, stiffer materials that produced the material from which modern fly rods are made. As adhesives and fibers improved, first-generation graphite gave way to stiffer, lighter, second-generation graphite, which in turn gave way to even stiffer and lighter third-generation graphite. Many anglers—especially those who drink a lot of coffee—cast best with graphite. But for those whose temperament and style are a tad slower, the hyper-fiber seems dazzlingly fast, especially compared with the more relaxing rhythms of bamboo or even fiberglass.

Graphite rods were introduced in 1973, by Jim Green of Fenwick. Over the next decade, other manufacturers joined the fray and the material became ubiquitous. Good classics from the era included the Fenwick HMG, the Shakespeare Graflite, and early Cortland models—the latter two a tad tasteless. High-end classics included the exquisite Orvis Limestone Special, the Leonard Graftex, and the elegant rods of Thomas & Thomas. In the future I expect early Sage rods will join this list, as will the rods of Joe Fisher.

THE GREAT BORON RUSH

As graphite progressed through its R and D stage, designers tried adding another synthetic fiber to blanks in an effort to slow things down. Boron was called the most amazing fiber since cotton. Engineer Don Phillips's first boron blanks were termed "delicate" by the angling press. The first production models reached the angling world around 1974, as a calmer alternative to graphite. Our own R and D rods, made in our old shop at Church Street, East Boothbay, were marked "Clearstream." The company president was Bruce Odell, and I served as chief designer and sweeper.

I was directly involved in The Great Boron Rush and remember it well. Boron was available only in longitudinal strips, and needed another material to provide support. That other material was graphite. Boron rods were at least 75 percent graphite, something few makers bothered to mention. The boron-graphite composite rods were a tad heavier than straight graphite models, but they loaded in a slower rhythm that was pleasing throughout the cast. And this pleased the dickens out of fly fishermen who liked a "bamboo feel" but wanted the added sensitivity and durability of a synthetic shaft. Boron was a wonder material—at first.

I began using the blanks designed by ex–Leonard craftsman Ted Simroe and sold under the "Rodon" label when they first came out. The blanks were great, but the "bor-core" ferrules weren't: They never lasted beyond 1,000 casts. The blank's maker, Dick Kantner, agreed to produce blanks for us using another style ferrule without the boron spigot, similar in some respects to the Fenwick ferrule. We built about 100 R and D rods with the new ferrules, and with reel seats by Allen Putnam and Research Engineering. We also built a one-piece series that eliminated the ferrule problem entirely.

Major companies began climbing aboard the boron bandwagon. Shakespeare was the price leader, and sold a rod with a very ugly grip for a mere $60. For that price, the company must have sprinkled a bit of boron over the blank while repeating mystic incantations. Other makers produced rods with both more boron and more class. Ted Simroe's Rodon rods were well appointed, as were those by Orvis. In all, perhaps a half-dozen rod companies tried the material and half of those went into production. But boron never sold in the quantities needed to turn the nickel, and the last makers ceased before the end of the boron decade, around 1985. That leaves us now with far fewer boron classics than early graphite examples. With the inevitable carnage produced by car-door closings and cow-pie slippage, very few boron rods are likely to survive. In the near future, boron rods will start to get "tough."

Esoterica

Certain fly-fishing collectibles are so scarce or just so odd that they can't be pigeonholed. Boron rods probably fall into this category. Sometimes rather new fly-fishing equipment becomes collectible merely because a famous person owned it or made it. Other items may become fashionable and escalate in price after years of languishing in the wings. Following are just a few examples of some items that may catch on in the future.

MORTISED RODS

Made from 1870 until the turn of the century by both Americans and Brits, these beautiful antiques were some of the most labor-intensive split-bamboo productions ever built. Strips of cedar and mahogany merged intricately with the six strips of bamboo to produce a grip and swelled butt that was just a killer. Makers included Leonard; Varney; E. P. Bartlett; Hardy; Chubb; Montague; Divine; Conroy, Bissett & Malleson; and Conroy & Bissett. As collectors come more and more to appreciate antique rods, the mortised rod will stand high in the ranks.

COMPLEMENTARY FIREARMS

What in the world do firearms have to do with tackle collecting? Suppose you have a "complete" Leonard collection. What better cornerpiece could you add than one of Hiram Leonard's multishot rifles, shotguns, or pistols? The list of gunmakers who made super tackle contains some impressive names: Samuel Phillippe, Solon Phillippe, Morgan James, William Billinghurst, John Krider, Hiram Leonard,

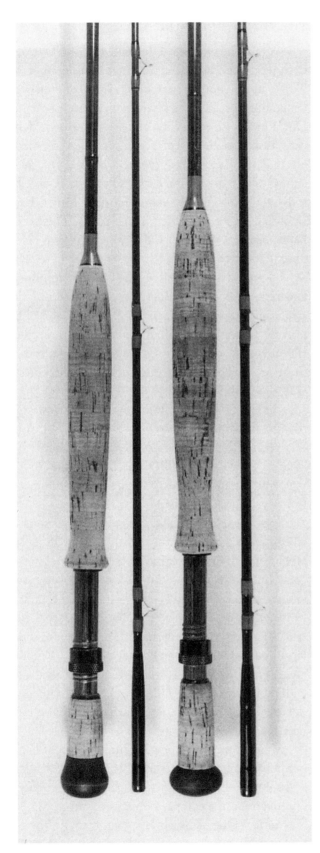

Two Clearstream Oceanic series boron rods, a 10-weight model (left) and a large 13-weight stick for heavy fish. Few companies made boron fly rods, and they'll be future collectibles.

Charles E. Wheeler, Gilbert Bailey, and Charles Nason. Something to think about.

PERSONALITY PIECES

I doubt that Clark Gable ever considered that his matching Orvis Rocky Mountain Special rods would eventually sell for 10 times their original price. And I assume Grover Cleveland attached less value to a Divine rod he received as a present than did the collector who later bought it. With "personality" tackle, the quality of a piece is less important than who owned it. This phenomenon isn't unique to tackle collectors, of course. But it is growing rapidly in popularity, and collectors are busily building collections of the fly tackle of the rich and famous.

19TH-CENTURY ETCHINGS

Original etchings are far less appreciated than they should be. They are rare and, if by the right artist, they are good. My nod goes to T. Sedgwick Steele, active in the 1870s, as the top dog in the field. George E. Johnson was also one of the top illustrators of this period, and Thomas Moran etched a number of great eastern plates. These little works of art accurately depict fishing in prephotographic times, and the renditions of period forests and streams are true classics.

D. J. BERNERS, MAKER

A lot of famous angling writers made their own tackle, providing more possibilities for starstruck collectors. We can track the practice back to the Prioress of Sopwell Abbey, who made her own rods. Ike Walton whipped out some wicked-good snelled hooks, and Theo South fashioned whalebone rod tips. Good luck finding these! Later authors made fishing stuff, too. Between penning *Fly-Rods and Fly-Tackle* (1885) and *The American Salmon Fisherman* (1886), Henry P. Wells found time to make his own rods and reels. As the designer of the fly-rod grip that bears his name, Wells had a good handle on how fishing tackle should function. A Wells-thought-out fly reel appears in figure 15-18, which shows its classic contracted spool and raised pillars. According to Wells, he made reels from German silver and brass, plus at least four examples from aluminum. Are there any left? Another author, Edward Ringwood Hewitt, made

Bing Crosby owned this Fin-Nor wedding-cake reel. I expect its drag has a particularly mellow tone. Personality pieces have an avid following.

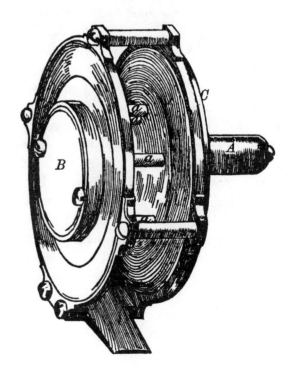

A look at the Henry Wells click reel. Somewhere, perhaps in a New York brownstone's closet, the Wells model should still exist.

reels, too, and several of his have survived. Perhaps a few of Henry's creations did likewise.

⚞🐟 The Paper Trail: Aids in Tackle Identification

Collectors have access to a wealth of new books to help them identify tackle production years and model names. Older volumes, such as the two Henry P. Wells books just mentioned, offer information on fly-tackle developments and trends. All that we know of production years, rarity, and subsequent value comes from paper—books, magazines, advertising, and letterheads—whether hot off the press or 400 years old.

ANGLING BOOKS

Angling books reach farther back into antiquity than the oldest known rods or reels. You can pick up a copy of John Dennys's *The Secrets of Angling* (1613) for as little as $15—a reprint, of course. Other volumes, more informative for fly-tackle buffs, cover the years from 1800 to today. Recently I discovered a new (1970) edition of Sir Humphrey Davy's *Salmonia or Days of Fly Fishing*, first published in 1828. (Aside from being a fly-fishing purist, Sir Humphrey was a brilliant chemist, linked to our modern angling scene as the discoverer of the element boron.) Early editions of Davy, or any other author, are more expensive, but many classic angling works can be purchased for prices in the $75–150 range. For mid-1800s tackle grit, I've found valuable material in the works of Thad Norris and Genio C. Scott. Some 19th-century volumes, such as Henshall's *Book of the Black Bass* and Murray's *Adventures in The Wilderness*, not only mention tackle trends but contain period advertising by makers and sellers such as Conroy, Bissett & Malleson, and William Read & Sons. Twentieth-century volumes give us information on classic-era tackle. To gain knowledge on fly tackle in its many forms and functions, I encourage you to read.

J. C. CONROY & CO.,

IMPORTERS AND MANUFACTURERS OF

Fish - Hooks, Fishing Tackle, &c.,

No. 65 FULTON STREET,

NEW YORK.

J. C. CONROY & CO. (late J. & J. C. CONROY), 35 years manufacturers and importers of Fish-Hooks and Fishing Tackle, in all its branches, inform their friends, amateurs, and the trade generally, that they have usually on hand, and are continually making, to supply deficiency, a very complete assortment of

THE BEST ANGLING MATERIALS,

as well as a desirable stock for the Wholesale Trade.

———

In competition, J. C. C. & Co. have been

AWARDED THE ONLY GOLD MEDAL

EVER GIVEN IN THE UNITED STATES,

AND NINE SILVER MEDALS

FOR

THE BEST ARTICLES OF AMERICAN MANUFACTURED
FISHING TACKLE.

In an 1869 copy of Murray's Adventures in The Wilderness, *I discovered this rare J. C. Conroy & Company advertisement, proclaiming that its products received the first gold medal ever given to American tackle in an American competition. This probably occurred during the Franklin Institute Fair of the Mechanic Arts, held in Philadelphia.*

PERIODICALS AND MAGAZINES

Perhaps the most interesting sources of tackle information are the many weekly and monthly periodicals geared to the sportsmen of bygone days. The oldest info comes from the rare and brittle newsprint pages of New York's *Spirit of the Times*. The *Spirit* was first published by William Trotter Porter and then carried into the late 1860s by George Wilkes. This broadside-style paper was followed by *Forest and Stream*, published from 1874 until 1900 in its original weekly format and from the early 1900s to 1920 as a monthly magazine. (Throughout this book I've reproduced tackle ads gleaned from 100 pounds of bound *Forest and Streams*.) Other antique-era source periodicals include *Chicago Field, The American Field,* and *The American Angler.*

New publications of the classic tackle period included the *National Sportsman, Hunting & Fishing, Outdoor Life,* and *Field & Stream.* These magazines, especially issues between 1920 and 1950, contained ads that pinpointed the introduction of various classic rods and reels built by Heddon, South Bend, Pflueger, Bristol, Horrocks-Ibbotson, Phillipson, and just about everybody else in the trade. This is affordable research material: Older periodicals sell for $12–20, and newer mags run $4–8. As a bonus, you'll find great articles by John Alden Knight, Lee Wulff, and Homer Circle. In bonus round two, magazine cover art in good condition, typified by the trout and bass action paintings of P. B. Parsons, can be matted and framed. And when you're through with old weeklies, you can use them to start the morning woodstove fire. That's what happened to the great majority of newsprint periodicals, making them tough to find today.

CATALOGS AND LETTERHEADS

Catalogs are more expensive than periodicals, but they contain "compressed grit." You get hundreds of descriptions and illustrations of rods and reels and a good selection of accessories—all pinpointed to one specific year. For bulk tackle info, they shine better than George Bush's thousand points of light. Tackle-house catalogs, such as those from Abercrombie & Fitch, featured various makers that changed with each decade—perhaps Thomas rods in 1925, followed by E. W. Edwards models in the 1930s. This helps a collector answer the age-old questions: Who, what, and when. Top classic catalogs by Edward vom Hofe, F. E. Thomas, E. F. Payne, Heddon, and even Pflueger now sell for more than $100 a pop. Pre-1900 catalogs are even more expensive. For collectors with a serious need to know, catalogs are the best single source of hard fact.

Maker and dealer letterheads, such as a masthead from Abbey & Imbrie or a letter penned by Charles Orvis, have an avid following. For tackle buffs, these scrids of paper serve a dual function. Their mastheads carry product lines, company officers' names, and even the dates of various partnerships. These informational tidbits can be real eye-openers, such as John P. Moore's Sons' byline, "Established 1823," placing this tackle shop as one of New York's oldest. The bonus is the rare letter signed by a known personality in the tackle industry. Checking through the 30 or so letterheads in my collection,

Old periodicals come in very handy for collectors' research. In putting together this volume, I used 100 pounds of bound Forest and Streams, *issues dating from 1875 to 1900. Other sources included* The American Field, Chicago Field, *and the old* Spirit of the Times.

I discovered two gems: one signed by consummate tackle salesman Herb Frost, and another handwritten and signed by Wisconsin fly tyer Carrie Frost—no relation to Herb, but definitely Frosting on the cake. Herb's letter, a plea to the Pequea Works for BB split-shot sinkers, is reproduced in chapter 13. Right now, tackle letterheads are cheap, running $4–20. Signed letters by pioneers such as Orvis or Leonard bring much higher prices.

COLLECTOR'S BOOKS AND MAGAZINES

In the past two decades, the collective knowledge about classic and antique fly tackle has been greatly advanced by books from ardent collectors such as Marty Keane, Mary Kelly, Jim Brown, Steve Vernon, Frank Stewart, Dick Spurr, Mike Sinclair, and Al Munger in the United States, plus Jamie Maxtone Graham and Graham Turner of Great Britain. Worth their weight in gold, these references belong in every collector's library and are listed in this volume's bibliography. A few of these works are out of print, such as Jim Brown's little volume on United States reel patents. Usually published by specialty houses, most tackle reference works are still available but sometimes difficult to find. You will find addresses of the two most prolific houses, Centennial Publications and the Thomas B. Reel Company, in appendix I.

The "little magazines" are a great help; these run feature articles by some of the country's most knowledgeable tackle historians (ahem). The first tackle collector's periodical was the now-defunct *Antique Angler*, published by Paul J. Webber in the 1970s. Today it's been replaced by Brian McGrath's *Fishing Collectibles Magazine*, now published in Maine, the original old-tackle honey hole. For esoterica and historical research we have the American Museum of Fly Fishing's appropriately titled quarterly, *The American Fly Fisher*, and the *Old Reel Collectors Association (ORCA) Newsletter*. These publishers and organizations are also listed in the bibliography. Overall, the wealth of knowledge gained from paper, from the oldest angling book to the most recent magazine issue, is not only accurate but often entertaining.

 # Finding Vintage Fly Tackle

After surviving the speculative bubble of the 1980s, the market for collectible fly tackle has become more or less stable and a lot more democratic. There are collectors for everything from giant ocean reels to Hardy Toothpicks who are willing to spend anywhere from two-digit figures with a decimal to the right to four and even five figures. Much of the best stuff now moves through auctions and dealers, but a few good items are found in flea markets and collectible shops, hiding in the shadows of broken toasters.

THE JUNK SHOP GAMBLE

Antiques stores and even lawn sales can be productive. One of my best finds—six Abbey & Imbrie brass reels, a revolving-plate Hendryx grilse reel, and a tiny "Trowbridge"-marked trout reel—I got wholesale from an antiques picker who stored his wares in five outbuildings. Check out this type of semi-junk-store dealer, but don't bother with the hoity-toity establishments; dealers who display 18th-century paintings and quaff mead from real Paul Revere tankards don't bother with that lowly fishing stuff. The retailer who handles funky wooden items and Americana is more apt to handle old tackle. Recently I stopped at such a place and discovered a salmon gaff, priced at $50 and bought for $40. Not just any old gaff, mind you; this one was hand-forged steel with an original spike or "spear" on one end of its ash handle and a tapered gaff hook on the other. It's over 150 years old, and it's now *mine*.

Buying tackle from a generic collectibles dealer can be a gamble. You have to develop an eye for quality, and don't go by the price on the tag. Many of today's antiques peddlers are convinced that all "old tackle stuff" is worth big bucks; after all, in the late 1980s, the nationwide wire services carried articles about the infamous $22,000 Haskell Minnow. Now, flea markets are more likely to yield low-end Montague rods with broken tips for $75 (actual value—$0), or rusted old Union Hardware skeleton reels for $25 (actual value—$0). Don't trade $100 for nothing.

TACKLE SHOWS AND MEETS

Collectors—especially new ones—are much better off attending regional shows. Here you'll find tables set up by 50 to 100 sellers, from hobbyist-traders to big-bucks dealers. At tackle shows most sellers are knowledgeable, their wares are priced competitively, and the selection is massive. This is where you meet the best dealers—amiable Bob Corsetti selling fly rods, sly Tony Flynn pulling a $1,000 reel out of his pocket, bibliophiles Susan and Steve Starrantino selling angling books, and the inimitable Brian McGrath offering subscriptions to *Fishing Collectibles Magazine*, which incidentally publishes a schedule of upcoming swap meets in every issue. I always try to attend Ben Clark's Northeast Antique Tackle Show in the spring and Tony Flynn's Northeast Sporting Collectibles function in the fall. Both are held in Boxborough, Massachusetts. The Florida Antique Tackle Collectors hold a show every February; the 1996 extravaganza included an auction by Henry Broggi. Other such affairs occur from Pennsylvania to the West Coast.

Shows and auctions, listed in appendix I, are conducted by individuals and also under the auspices of the National Fishing Lure Collectors Club (NFLCC). Despite the Bassmaster connotation of this organization's name, its membership includes many classic fly tackle collectors. Recently I attended the NFLCC's Region Two Antique Tackle Show in Allentown, Pennsylvania. The highlights were two new acquisitions—a venerable and rare Haywood winch, and a brass curved-crank "wedding-cake" reel by John Bernard of London. Many shows and auctions are held in conjunction, such as the Northeast Antique Tackle Show and Lang's Spring Auction, held in the same hotel and making for a two-day affair. Each season, spring and fall, thousands of collectors converge at shows, many conducting hotel-room swaparamas the evening prior to the function. Excellent fly-tackle buys are made in this fashion.

CLASSICS BY MAIL

Between shows, many dealers sell via mail order. Scads of classic fly rods and reels are sold every year by Martin J. Keane, Bob Corsetti, Dick Spurr, Len Codella (originally with Thomas & Thomas), and Carmine Lisella. Through the years, I have bought many a classic rod through the mail, sight unseen. Seldom have I been disappointed, but dealers always offer an inspection period, allowing purchasers to return the item for a refund. Some dealer catalogs tend toward, er, uh, rather *glowing* descriptions of their tackle. This type of sales hype goes back to Thomas Conroy and Herb Frost—so read between the lines. You will find a handy list of dealers' addresses in appendix I.

Trade meets are excellent places to find bargains and swap information. At Ben Clark's Northeast Antique Tackle Show, a collector checks out what appears to be a Denton chromolithograph, a red snapper—one of our more obscure fly-rod gamesters. Snappers take weighed nymphs, drifted slowly.

GREAT TACKLE AUCTIONS

I love attending tackle auctions. Here's where the entire fraternity, from new collectors to old farts, meets to outsmart or outbid their friends. Auctions are the highlight of each passing season, and I've picked up some wonderful pieces at these functions, including a superb Edwards Mount Carmel salmon rod ($150), a sumac-handled antique Orvis ($275), a tiny J. C. Conroy multiplier ($400), and a once-in-a-lifetime example of C. F. Murphy's craftsmanship ($1,000)—all purchased for less than their actual value. It pays to attend these spirited events.

The first large fishing-tackle auction was held in Kennebunk, Maine, in the summer of 1985. "Colonel" Richard W. Oliver, the original auctioneer, continued the functions until 1992. Oliver's famous winter events, the heady High Roller's Auctions, began at Marlborough, Massachusetts, in 1988. Henry Broggi, an excellent tackle auctioneer, started at about the same time as Oliver and still runs a number of sporting auctions each year. Henry's Southbay Auctions are usually held in Huntington, New York. Richard Oliver's semiannual events were succeeded by the present auctions of tackle sage Bob Lang. Hard-core collectors as well as casual classic-tackle fans soon mastered the subtle nuances of purchasing by bid. Today these auctions offer quality tackle, usually described accurately enough in the catalogs for buyers to bid by phone. A number of British collectors bid on-line, and so do the Japanese. Absolutely great tackle outlets, major auctions are listed in appendix I.

If I could offer just one tidbit of advice, I'd urge new collector-users to attend the sporting shows and tackle auctions and introduce themselves to others by trading business cards. You are a stranger but once.

Get to know other collectors and their tastes. Ask questions, even stupid ones. Just don't ask the same stupid question twice. Get to know all kinds of collectors, even the bass-lure types. These folks run into all kinds of tackle, such as Heddon fly rods or old brass fly reels, and often they want to get rid of them, through either a trade or a sale. Remember that the combined tackle collectorship of the NFLCC has swollen to almost 5,000 individuals. If each member trades or sells 10 tackle items per year, that's an annual turnover of a half-million collectibles.

Fishing tackle is probably the most collected item in the United States. When a show is over and you're back home, stay in touch with other collectors by phone or e-mail. Everybody in this racket runs into items that don't interest him—items that might be just what you're looking for.

16 Maintaining, Repairing, and Using Classic Fly Tackle

Many anglers have discovered the pleasures of everyday fishing with classic fly tackle. Even antique rods and reels see occasional (and very careful) service, and a few brave souls rise to the age-old challenge of taking a large fish on a single hair from a horse's tail, using colonial-style fly outfits—reproductions, to be sure, but with real horsehair lines and leaders, and 2-pound tippets at best.

Given proper care, classic tackle of any age should last an angler's lifetime. Rods in marginal fishing condition—especially classic split-bamboo production rods—can be brought back to life with a little work. In this chapter I hope to cover the many little tasks required to maintain classic rods and reels, as well as minor and slightly less minor repairs, and handy tips on fishing veteran tackle in both fresh and salt water.

 Reels

Reels are a lot like old car engines: They like oil, and many a classic reel hasn't seen a drop of it in years. After oiling the working parts of a fly reel, small squeaking noises magically cease and the handle usually turns freely. Light oil applied to the exterior of a metal reel will help eliminate green and brown oxidation while retaining the patina. The same light oil will replenish the naturals oils found in side plates made of hard rubber, which dries over time and through exposure to the sun.

LUBRICATION

You need an oil that won't build up or "gum" as it evaporates—and all oils will evaporate given time. I prefer 3-in-One or Nyoil. They are light, won't cake, and are very traditional—the 3-in-One brand was introduced in 1883. Nyoil goes back even farther. First made in New Bedford, Massachusetts, by William Nye, it originally contained pilot whale oil plus three other lubricants. Developed for Arctic whalers, Nyoil won't gel even in frigid temperatures—good stuff for early-season anglers from Montana or Maine. Although now whale-free, Nyoil remains the best of all oils.

Use a little light oil to lubricate the click pawl and post, and the handle post; put some in the main-bearing oil port; and give the reel a general wipe-down with it after use in fresh water. Wipe a little Nyoil on hard-rubber side plates whenever they appear dull or dried out, whether or not the reel has been fished. Many collectors use grease on a fly reel's spool shaft, since grease doesn't evaporate and lasts an entire season. But use grease sparingly, and each year clean it from the shaft and spool cavity before regreasing.

It isn't a good idea to oil the handle post on antique reels that have wooden knobs, as excess oil will discolor the wood and possibly swell it, causing it to bind or crack. If a wooden reel handle turns freely, leave well enough alone and just give it a polishing with a carnauba-based wax, such as Butcher's.

Old angling photos are collectible. This one depicts famed Maine guide C. Ross McKenney fly fishing above a beaver dam near Shin Pond around 1920. McKenney, author of Language of the Forest, *headed up the Dartmouth College Outing Club and took Teddy Roosevelt into the Maine woods. A gourmet camp cook, the old guide died on October 12, 1971. Ross McKenney was fishing with "classic fly tackle." He just didn't know it then.*

THERE'S NO SCREW LIKE AN OLD SCREW

Reels that haven't been used for decades may prove challenging to disassemble and lubricate because the side plate screws won't budge. Rather than bugger the screw heads, detracting from the reel's value, try spraying the screws with WD-40. Wait a day and try again to start them. If one or two screws are still stubborn, soak the entire reel in kerosene or Hoppe's No. 9 gun-cleaning fluid. Allow plenty of soaking time—days, if need be. Soaking will eventu-

ally break free the screw threads, and the reel can then be disassembled, cleaned, and properly oiled for use or storage. Don't forget to remove the wooden crank handle of any antique before immersing the reel in this bath.

Be sure your screwdriver is seated all the way into the slot before you attempt to back out any screw; some screw heads have slots smaller than the average screwdriver. While it's true that dinky little jeweler's drivers will fit into these smaller slots, they're just too small to provide the torque needed to start a stubborn screw. Instead, buy a good-quality screwdriver and grind the blade until it's thin enough to fit the small slots; then grind the sides so that the driver is no wider than the screw head. This eliminates the chance of scratching the reel's side plates and rolling the edges of the screw slot—flaws that so easily devalue a fine old classic.

FRESH AND SALT WATER

Through the years, I've used a number of antique and classic reels in fresh water—from bone-chilling fishing for ice-out landlocked salmon to warm, breezeless evening angling during mayfly hatches. I've paired a Conroy multiplier or an old wooden Nottingham with a matching antique rod. With classic rods, I've used a plain old Pflueger Medalist spooled with 6-weight line, a Meisselbach Featherlight for 4-weight work, and a tiny Hardy Flyweight Silent Check loaded with a 3-weight double tapered. Any of these reels or their cousins will provide the same excellent service today as they did decades ago—in fresh water. Salt water is a different story.

The marine environment is tough on reels, even brand-new models built for the job, such as those by Penn, Fin-Nor, or Seamaster. A classic workhorse that's held up well for me over the years is the Rogue fly reel. Although built some 25 years ago as a steelhead reel, the Rogue is mostly anodized and excellent for stripers, bluefish, dorado, and bonito. The Rogue is relatively inexpensive on today's market, and it's a good bet for general saltwater work shy of tarpon and billfish. The Rogue reel holds a 9-weight fly line and a copious amount of backing.

Whether a reel fished in salt water is a modest Rogue or a pretentious gold model, though, it needs attention immediately after every use, not two or three days after. Hose down the reel with a fine, almost wimpy spray of fresh water, turning the reel so that no area is missed. Wait a minute for the dried salt to go into suspension, then rinse again thoroughly with a second light spray. A light spray is far better than a rushing torrent, which can actually drive salt into the internal working parts of the reel, usually through the space between the spool and the side plate. It you're in an exotic tropical location where a freshwater hose may be unavailable, take your classic saltwater reels (and rods) right into the shower stall. Here, you can do your classic reel a favor while singing your favorite classic aria.

Even if well cared for, expensive older salmon reels may not survive much punishment in salt water. I've used an Edward vom Hofe 4/0 and a Zwarg 3/0 in the salt, the latter for a decade. Both reels had parts made from nonanodized aluminum, such as spool ends and their balanced crank handles. These areas pitted no matter how much TLC they received. So beware. Any high-classic salmon reel used in salt water should get a freshwater shower at the first opportunity.

After a freshwater rinse, wipe off the reel with a terry towel. When almost dry, spray it with a water-displacing lubricant such as WD-40, CRC, or the newly developed Tacklekeeper from Kennebec River Fly & Tackle. Or you can let the reel air-dry completely then wipe it down with Nyoil, as mentioned earlier. For long-term storage after use, remove the fly line and backing, then reclean and oil the spool. A reel used in salt water that's stored unrinsed and unoiled will be discovered much later under a pile of white stuff, the aluminum areas pocked with large pits that totally destroy the piece's value. The best place to ruin such a classic—reel or rod—is a nice humid spot, like an attic or a basement. That'll do the job some wicked-fast.

Some classics just can't take the punishment meted out by the large fish found in salt water. I dis-

The Meisselbach Featherlight was a popular reel, and many are still found in good used condition. Today a Featherlight is a sound choice matched with a 4-weight split-bamboo rod.

covered this when I taxed a venerable Pflueger Medalist by repeated landings of 10-pound fish on an 8-weight line. The pressure against the line and backing warped the Medalist's spool so that the reel practically froze. At first I thought the problem was just a defective old spool, and I scrounged up a replacement, but the second spool did the same thing. Moral: Classic Medalists don't make good midweight saltwater reels. I suspect the list of reels that don't belong in salt water far outstrips the list of those that do.

ATTEMPTING REEL REPAIRS?

I'm a former gunsmith and have repaired simple reels from the smith age, mostly by replacing parts—such as a click pawl or a handle—with parts "borrowed" from a similar reel by the same maker. This works out pretty well, but these are repairs to be entered into slowly, not brashly. And when the piece is finished any potential buyer should be told about the replacement crank, or whatever. If you'd like to try your hand at minor reel repairs you'll need a few simple tools, including small bastard files, riffling tools, various pin punches, needle-nose pliers, leather buffers, a substantial vise, and the trusty old ball-peen hammer.

Antique brass reels often have a bent reel foot, as do newer brass classics, such as Meisselbach Amateurs and Experts. If the reel's foot is slightly bent, leave it alone. Chances are the reel's value isn't too diminished by a slightly bent foot. If the foot is bent to an extreme, however, you may want to take the chance of trying to straighten it. It is a chance: The foot could break, leaving you with a rather worthless reel. If you're a gambler, remove the foot from the reel and place it on a round steel bar or pipe that has the same circumference as the underside of the foot. Fasten the ends of the foot to the bar with tight wraps of plastic electrical tape, then tap the bent portion to shape with a 13-ounce hammer. Pretty delicate business, eh? It works, though—most of the time.

Brass and German silver antiques are sometimes found with missing crank knobs or perhaps with just the ghost of an ivory, bone, or walnut handle that split when the reel was dropped years ago. If you have a few extra handle knobs kicking around (don't we all?), and one seems to fit, historically and physically, you can remove the crank and attempt a repair. File the knob post on the underside of the handle until it can be tapped out with a blunt pin punch. Check the replacement knob for bore and length, leaving room for easy turning when finished. Slide it onto the post, and tap the post back onto the crank. Using a ball-peen hammer and anvil (or the top of a vise), peen the underside end of the post until it's tight. This particular repair sounds easier than it actually is. In truth, it's what I call a "delicate hammer job."

Finding replacement parts for newer classic reels, whether they be Medalists or Meisselbachs, is a lot easier. Just look for junker reels of the same make and model that have usable parts. This reel-repair stuff isn't for everybody, and, frankly, I hate doing it myself. Professional reel repairmen can do a much better job than me—and they probably don't rely as heavily on hammers.

 # Rods

For natural reasons—meaning natural wood, grass (bamboo), and oil (varnish)—rods are more difficult to maintain than reels. A rod's components will dry out, melt, warp, or even rot. With demand for classic rods continuing to grow, it's essential that new old-rod aficionados learn how to keep these valuable pieces in top condition.

HEADING OFF STORAGE PROBLEMS

Whether you choose to display your tackle or store it away is up to you. I hang nameless antique rods on the wall and display a number of reels on shelves. The rods do pick up dust and dry out, and, in time, the ferrules loosen on the shafts, but these aren't big problems with unmarked rods of little value. They still look good. An alternative is to display rods in a shadow box, but this can get out of hand. An old rod may run 12 or more feet long. Unless you're blessed with unlimited space, your rods will have to be displayed broken down.

A shadow box cures the dust problem, but the rods still dry out over time, causing the varnish to craze and eventually crack. One way to slow this drying process is with an occasional application of wax. I use Butcher's Bowling Alley wax. Apply it sparingly with a soft cloth, wait 15 minutes, then wipe down the rod. But go easy: Avoid too much rubbing of elderly rods that have tender wrappings and delicate tip sections.

Rods not on display can be stored in their canvas bags and tip cases or in aluminum tubes. Nothing protects a rod from gradual deterioration so well as a tightly capped aluminum tube. Damp or dry air is excluded, and the varnish retains its essential oils. This isn't true of the canvas sacks that came with so many rods built before 1920, however. With these canvas-bagged classics it's a good idea to treat the varnish with a little wax. Tips stored in a closed bamboo tip case will retain the varnish's natural oils better than will the butt and mid, which are stored in the bag's outer compartments.

A good aluminum tube will solve this problem if the rod is small enough; this is unlikely with an 1890–1920-era piece equipped with a bamboo tip case, however. Most classic-era rods, whether two- or three-piece, will slide into a modern tube with no trouble, although in the case of a short three-section classic you may have trouble finding a case that's short enough. Fortunately, you can always cut one to length. The bottom cap of most tubes is simply pressed on. All you have to do is stick a 4-foot piece of pipe in the tube, whack it a couple of times, and the cap should pop off. Then just cut the tube to length with a hacksaw, smooth the cut end with a fine file, and tap the end cap back on. Your "custom" case is ready.

Most collectors store rods upright in their cases or bags. But the American Museum of Fly Fishing has a collection of more than 1,400 rods, and they rack them horizontally, like wine. I suspect that once a month someone stumps through rotating each rod case 180 degrees, like a monk in a champagne cellar.

Rotating a horizontally stored rod and case will cut the risk of "bag marks," those minute checkerboard patterns that mar the finish of so many old rods. Bag marks are usually caused by heat, such as is found in an attic. Excessive heat will melt a classic's varnish very quickly. It's a terrible feeling when you try to slide a rod from its bag and it doesn't move. Rods are best stored in the living area of a house, preferably on the first floor.

Cool, damp areas are also murder on varnish, but the damage doesn't stop there. Dampness or mildew rots silk windings and delaminates the strips, eventually allowing the entrance of dry rot right into the bamboo. In warm, damp coastal climates, collectors must remain vigilant, especially if they're storing rods in areas with little air circulation such as closets. If mildew forms on rod bags or varnish, consider buying a dehumidifier; or you can place a bucket of moisture-sucking crystals directly in the storage closet.

A BRAND NEW BAG

While many rods have come down through the years relatively unscathed, the same can't be said of their bags. The bag may be missing altogether, or it may be so damaged and frayed that it no longer holds the rod sections. Sometimes classic-tackle dealers carry new cloth bags of the proper size; Bob Corsetti, for instance, tries to have them on hand. But bags

When you're not using it, keep a classic rod in its case. Here we see an 1840 angler doing just that.

aren't exactly a high-profit item and they're difficult to keep in stock. I solved the problem by making my own. Most of the materials used originally are still available, including the canvas used for 19th-century salmon and trout rods. To work with heavy canvas you'll need some time and a good sewing machine. I use a rugged Singer, and I keep spare heavy-duty needles handy.

Standard three- and four-compartment rod bags are the easiest to make. You'll need just over a yard of cloth, chosen to match the original sack if it still exists. Colors varied from maker to maker; the most popular were khaki, forest green, gray, and dark blue. It may be difficult to find pure cotton dyed to the color you need, and you may have to settle for a nonauthentic cotton-polyester blend. The really fussy (if they don't mind a messy and thankless job) can buy white muslin cotton and dye it with RIT. After World War II, many bags were made from rayon. Good luck finding this material. You'll also need heavy-duty cotton-polyester thread in a match-

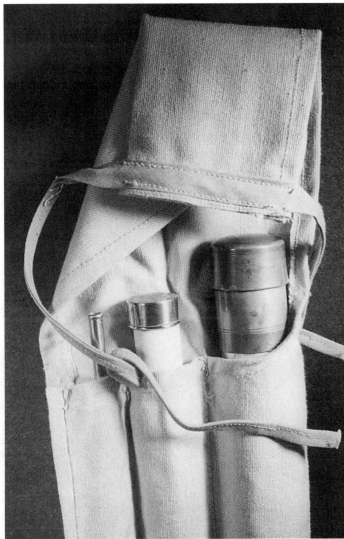

A new three-compartment canvas rod bag designed for the old style of "put up." Note how the seaming tape has been folded and sewn one last time to form the finished tie.

pencil. Cut the material with a good pair of shears or a fabric cutter.

Now it gets harder. Fold the top of the fabric to form a ⅜-inch hem, and sew it as straight as possible. To anchor the thread, start and end each line of stitching by switching the machine into reverse for a half inch, then continuing. After you sew the top hem, fold the material lengthwise, seam side out, and sew the entire length of the side and bottom as straight as possible, using a few strategically placed pins to keep the seams from creeping. You should have all the seams showing on the outside of the bag—which is actually the inside. Now, use a length of bamboo or a broom handle to turn the bag inside out, pushing from the bottom. If you want to add a traditional hanging loop, sew it in while the bag is still inside out. Mark with a pencil the individual parallel section compartments, then stitch them closed with the sewing machine.

CANVAS BAGS

You can follow the same basic procedure to make a reproduction canvas bag appropriate for a rod from the 1850–1920 era, complete with a folding top flap—to protect the butt, mid, and tip case—and upper and lower bag ties. I prefer to make bags to the Bangor pattern, with a folded top and ties designed to encircle the contents before tying. With the British pattern, the upper ties emanate from the same direction, a rather odd way of tying anything.

The traditional bag colors of this era were light khaki and undyed, off-white canvas flecked with darker cotton. I'm lucky enough to live in a coastal village, where the local fabric outlet carries traditional cotton canvas for the boatsy types. Inland collectors may be able to find canvas through ads in boating magazines. You'll need at least 6 inches of material for the flap. The flap gets a ¼-inch hem, as described above for the simple rod bag. Flip over the material before sewing the bottom and side seams, and, when the bag is reversed with your handy length of bamboo, the flap's hem will be hidden correctly on its underside. The actual number and width of compartments depends on the sections of the partic-

ing color. You're on your own trying to figure out how to load and feed the thread through the half-dozen little arms that sprout from various sewing machines.

Determining the exact size of the finished bag isn't as easy as you'd think. If you have the original, you can open it up with a thread-removing tool and measure its width. If you don't have the original bag, guesstimate around 9 inches wide, unfolded, for three compartments suitable for a two-piece two-tip rod; or 11¼ inches, unfolded, for four compartments to fit a three-piece, two-tip model. Better too loose than too tight. To determine the length of material you'll need, measure the longest rod section and add ¾ inch. Lay out the material on a table, measuring with a plastic or metal yardstick, and marking with a

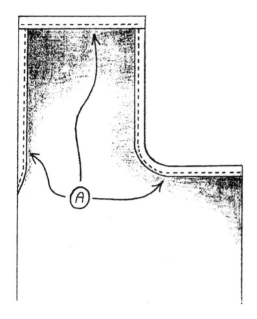

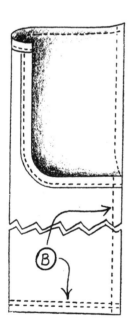

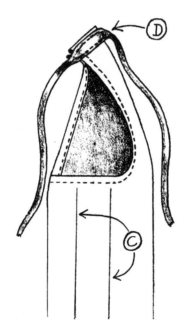

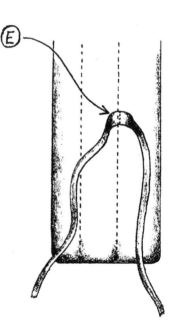

A New Bag for an Old Classic
Use the following dimensions to make a three-compartment replacement canvas sack for an early classic rod that has a tip case. The example shown here was designed for a 9-footer in the style of F. E. Thomas.
Bag Dimensions Before Seaming

Overall Length	37 inches
Overall Width	11 inches
Flap Length	7 inches
Flap Width	6½ inches
Seam Width	⅜ inch

(A) Sew the seam along the flap edge, as shown.
(B) Fold the material, leaving the seam outside, and sew a seaming stitch along its entire length. Then make a double seam along the bottom. The bag can now be turned inside out.
(C) Use a straightedge and pencil to scribe individual compartments, and run a stitch along each scribed area.
(D) Fold the top flap together, like a hood, and sew the top shut. Place the top tie over it and double-sew into place.
(E) Flip the bag over and sew the lower tie 7 inches up from the bottom on a compartment line.

ular rod and whether or not a tip case still exists. After you sew the compartments, fold the flap double, with the entrance to the compartments on the inside. Finish the sack by sewing the flap top in this doubled position and adding the ties.

Reasonable bag ties can be fashioned from ⁷⁄₁₆- or ⅜-inch triple-folded "seam material" or "seaming strips." Cut a couple 2-foot lengths and sew one at its midpoint along the full width of the end of the flap, in two separate sewn lines. Attach the lower tie on the same side of the bag, 8 or 9 inches from the bottom, by sewing vertically along one of the compartment lines. As an added touch, fold the long tie ends one more time and sew them to a shoelace width.

A reproduction canvas bag, made in the original style from the original materials, not only will protect your antique rod from further deterioration or breakage but will also add a great deal to its overall desirability.

EASY FLY-ROD REPAIRS

High-quality rods needing repair—Leonards or Thomases, for example—are best left to experienced craftsmen with the skills to do them justice. But there are many production rods by manufacturers such as Montague, Divine, Chubb, Bristol/Edwards, South Bend, Granger/Wright & McGill, Phillipson, and so on, that may not be particularly valuable but that do make great casting tools. These can be refurbished by the home hobbyist with a bit of patience and skill. Of course, not all production rods within a given manufacturer's line are created equal: Some may be too cheesy to bother repairing; others may be rare, high-quality models best reconditioned by professionals.

Take Montague, for example. Its extensive line of models ranges from trash to treasure. Such low-end rods as the Sunbeam or Mt. Tom are good ones to stay away from. The cane is worthless, and the cheap, plated fittings are thin and tend to bend or split. As has often been said, these rods weren't any good even when they were new.

Next up the Montague ladder are models like the Flash, a good candidate for home repair into fishable condition. Higher-quality models such as the Rapidan and the Fishkill are better left for the skilled home craftsman with a Flash or two under his belt. Montagues of Red Wing and Manitou caliber are

worthy of professional-level workmanship. You can acquire the skills needed to do justice to a Flash in a short period of time, and the model is a good choice for repair practice. The Fishkill demands a finer touch, attainable with further practice and knowledge. When you enter the Red Wing stage, you're approaching semiprofessional status.

Bamboo Fly Rod
No. F-12—$16.50

● A custom-built rod of selected Tonkin cane. Six strip construction. All strips are cut and prepared from a single stalk of heat-tempered bamboo. The result? A perfectly balanced rod of uniform toughness and resiliency.

Other Bamboo Fly Rods available at prices of $10.00 to $68.00.

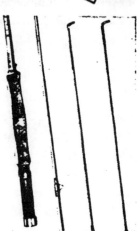

FREE CATALOG! You ought to know all about the popular Rods, Reels and Lines offered by Bristol. Send for illustrated catalog.

THE HORTON MANUFACTURING CO.
634 B HORTON ST., BRISTOL, CONN.

For an all-around fishable fly rod, the Bristol F-12, built by the Edwards family, is hard to beat. Even 8½-foot models aren't tiring and can be fished all day.

As your interest grows, you'll find it worthwhile to collect various small parts such as guides and ferrules from junker rods too far gone for restoration. Difficult-to-find guides include English snakes, German silver Divine types, plain standing guides, trumpet guides, Richardson guides, and, most elusive of all, the old loose-ring style from the 1800s. Odds and ends like pre-Perfection tiptops and agate salmon-rod tops will also save the day. Another handy item is the old "spare tip jar," sprouting sundry orphaned fly-rod tips.

Over the years of fixing up "handyman specials" for fishing I've often replaced original guides with modern snakes. Old rusted guides can cut up an expensive fly line's coating in just one outing. Obviously, I wouldn't do this to a fine, rare rod, but remember that most production models were built for everyday fishing by average anglers. These modest rods can be modernized for today's fishing without worrying about destroying the value of a premium collectible.

Bringing a classic workhorse back to snuff with a little rubbing, a replaced wrap or two, or perhaps some new varnish, then going out and catching a fish with it is extremely satisfying. Rods of this ilk are common on tackle lists, usually cataloged as being in good or lesser condition. Such rods may have missing or broken wraps, or, worse, may need to be completely refinished. A lot of work, to be sure, but then rods in this condition sell for far less than they would if in top shape. The repair steps detailed below will bring rods in from poor to good condition back to a usable state.

Rods in very good condition cost more, but they'll have all their wraps intact. Often, all that's required is a careful check for frayed wraps, which are easily corrected by a new coat of color preserver and a localized coat of varnish. Many rods in this class have bag-marked or slightly alligatored varnish and need a careful rubdown with polishing compound. I'll try to give some handy hints on these various repairs, from strategic small touch-ups to an all-out refinish or restoration. Here, first, are some of the materials and tools you'll need for these general repairs.

FINISHING MATERIALS

Before varnishing, rod wraps receive a few coats of color preserver. This is still available from rod-making suppliers; the most commonly used brand comes from threadmaker Gudebrod, Inc. Color preserver tends to evaporate quickly, but it can be thinned with lacquer thinner. The old-timers used clear shellac, and so do I. It's cheaper than color preserver, and I can buy it at local hardware stores. At the moment I'm using Zinsser Bulls Eye shellac. To penetrate the threads, it's important that clear shellac be thin and watery; it's thinned with denatured alcohol.

Now let's look at varnish. In 1892, J. Harrington Keene wrote that "the best coach-varnish is unapproachable for rods. It should of course be applied by means of a camel's-hair brush, in a room where no dust is flying about." Sage advice. Coach varnish was the professional rodmaker's choice throughout the 19th century and well into this one. But coach varnish went out of favor along with the coach. Today, marine spar varnish is its closest living relative. Many classic rodmakers used Valentine Super Val Spar until it was discontinued. A similar finish is still available from Pratt & Lambert, called No. 61 Spar Varnish. Pettit makes a good varnish known as Z-spar. Birchwood Casey's Tru-Oil is an old standby, as is Interlux Captain's Varnish.

You can mix your own "coach varnish" by adding a very small amount of Formby's Tung Oil Finish to any of the abovementioned oil-based products. The use of tung oil in varnishes goes back to the 19th century and lasted throughout the classic varnished-rod era. The whiff of tung oil is unmistakable when you open any cased Leonard, Payne, or Thomas rod—or almost any other traditional, high-quality rod. Tung oil varnish is softer than straight-linseed-oil-based versions, but it remains flexible for years. These varnishes are best cut or cleaned from brushes with the very best spirits of turpentine. Universal thinners like Thin-X are not completely compatible with tung oil varnishes.

For complete refinishing or for varnish buildup, use the best camel's-hair brush you can find, and pick carefully through the brush dealer's display to find bristles that are firm and don't stick out to the sides. I like Grumbacher brushes; they don't seem to shed

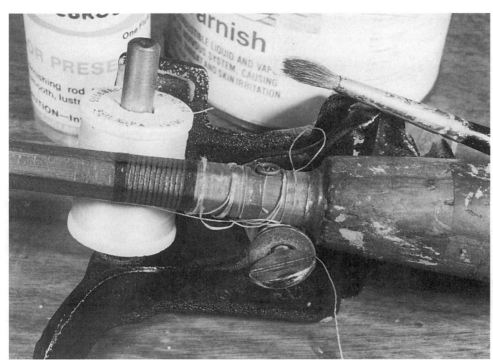

This Bristol F-12 needs a new wrap at the hook-keeper ring—not at all difficult to do. Items essential to the process include a Thompson thread winder, thread, a knife, color preserver and varnish, a small brush, and a bit of time.

and needs to be water-proofed before the fly rod is used. Varnish applied directly to the frayed area will darken the thread, destroying the original maker's true colors. Thus you should always use color preserver or clear shellac for the initial coating of any exposed threads, not varnish.

Inspect a newly purchased rod for any fuzzy areas at the edges of each winding, where it meets the varnished shaft. Any fraying should be singed with an alcohol lamp, rotating the rod to one side of the lamp. If the wraps remain intact after this "defuzzing" process, coat just the exposed area—not the entire winding—with clear shellac. It's important to retain the original color of the thread, so apply at least three extremely light coats, wiping off any excess with your finger. Allow the area to dry for half a day before you apply the next coat. After the last coat of shellac, wait a day, then apply a thin brushing of varnish only to the area coated by the color preserver. Wait another day and brush the entire wrap with a final coat of varnish. Once dried, the repair is completed. To tone down the shiny refurbished area to match the rest of the rod's finish, the wrap can be strategically rubbed with fine polishing compound.

hairs. This is important, since loose hairs can ruin an otherwise perfect finish. Each hair must be picked out, and if even a minute passes between the varnish application and their discovery, the loose bristles will have done their dirty work. For applying thin varnish overcoats to existing rod finishes, for varnish touch-ups, and for shellac, use a stiffer sable brush. Sable is more expensive, but it will spread a thinner coat of any finish—important in those kinds of jobs.

Brushes in sizes 2 through 10 (⁵⁄₁₆ inch wide to ⅜ inch wide) will cover the needs of rod sections from bamboo tips to large wooden butts. Use the narrow brushes for touch-ups and small wraps and the larger brushes if you're covering a wide area. For most bamboo-shaft work, a ¼-inch-wide brush is ideal. If a brush is too small you'll spend more time loading it than you will getting the varnish on the rod.

PRESERVING FRAYED WRAPS

A classic rod with slightly frayed wraps is showing the first signs of deterioration. As the rod is used, water will seep through the frayed area and collect beneath adjacent wraps, further weakening the silk, even in spots that appear to be protected by varnish. The area of a frayed wrap is a prime candidate for rot

REJUVENATING A TIRED FINISH

Finishes that are lightly bag marked, crazed, chipped, alligatored, or scratched often require little more than a few careful hours of polishing to bring the rod back to a fishable state. (Before you attempt any rubdown, however, check the rod's wraps for areas unprotected by varnish. If you find exposed threads, take one step backward and coat them as described above under "Preserving Frayed Wraps.") The trick to restoring a finish is to use a light hand

and a lot of polishing compound. You need only some 4/0 steel wool, some old soft rags, and two grades of gunstock rubbing compound: For the initial rubdown, use No. 3F Brownell gunstock compound; for final polishing, use the finer-grit Brownell No. 5F polishing compound (or their equivalents). I prefer the Brownell compounds because the vehicle that carries the grit actually softens many linseed oil spar varnishes, making the work go quickly. With the older, soft tung oil finishes, you'll find the going a little more difficult; you'll have to go easy, taking care not to rub off the original finish from the high hex areas.

Starting with the rubbing compound, thoroughly moisten a piece of 4/0 steel wool and rub the rod shaft briskly along the flats. The steel wool should be literally dripping with compound. Note the really poor areas of the old varnish and work these well, but stop rubbing after a few minutes and clean the compound from the rod with a soft rag. If areas need more work, add more compound to the steel wool and continue. When the rod starts to look really smooth, pick up a clean tab of steel wool and switch to the finer-grain polishing compound. Rub it in liberally, as before; let it dry before wiping it off with a fresh rag. Now look over the rod carefully. You may have to work a bit more on tenacious bad spots before going back to the fine grit.

Remember that it's very easy to cut right through several coats of varnish in no time if you're heavy handed. This phenomenon will be most noticeable where the windings meet the shaft. With split bamboo, rub just the flats and avoid the sharp edges at the glue lines. Always work slowly, checking on the final sheen frequently. Rubbing is an art that can only be acquired with practice, but if done right it results in a dazzling finish that looks as smooth as silk. Now give the polished shaft a light coating of wax. Wait 20 minutes, rub down the wax, and the rod should be ready to take fly casting.

REPLACING WRAPS

The most common problem afflicting older fly rods is broken or partially missing wraps at the guides or ferrules. Fortunately, such rods can be easily repaired, provided that the rest of the wraps are sound and the varnish is in good condition.

You need to choose a thread of a size and color that closely approximate the original. This is fairly easy with newer production-built fly rods, but gets a little more difficult with older antiques. Production rods were generally wound with size oo or A silk thread in basic red, brown, light tan (pongee), tan, yellow, and salt-and-pepper colors, with or without tipping in a contrasting color. Many older rods—say, from the 1930s back to the post–Civil War era—were wrapped with "old red," which is still available in dark red silk and is reasonably duplicated in nylon by Gudebrod's Candy Apple Red. Sun often bleached old red wraps to a rose shade. Most rods built before 1865 were wrapped in black, sometimes with coarse cotton thread, roughly corresponding to size C or D.

I see no problem with using nylon thread (Bill Phillipson used it on his rods). It resists rot, has less fuzz than other threads, and, when prefinished with a color preserver, looks wicked-sharp on production rods built after the 1930s. For older or more valuable rods and antiques, it's best to use silk.

The tools needed for rod wrapping are minimal and cheap: a sharp knife, some masking tape, a Thompson adjustable-tension thread winder, and a set of rollers made from a pair of nylon furniture casters. Mount the rollers back-to-back on a short board; they'll come in handy as a third hand while you're wrapping the opposite end of a bamboo butt section.

To get started, you'll have to unravel the offending wrap very carefully. First, apply masking tape to the varnish on both sides of the wrap (this keeps the varnish from lifting). A delicate knife cut across the guide foot will remove most of the original wrap. Clean excess scrids of varnish from the "hump" of the guide with a fingernail.

Some people prefer to remove wraps with varnish remover. To protect the varnish, carefully tape the shaft, right where the windings meet the varnish. Then carefully coat the wraps with varnish remover. This will soften the wraps, and you can pluck them away with your fingernails. Once you've cleaned the area to the natural bamboo, you can reattach the guide.

Mount the Thompson thread holder to a table edge and adjust the tension so that the thread leaves the spool at a smooth, constant rate. Attach one leg of the guide in its original position with a small piece

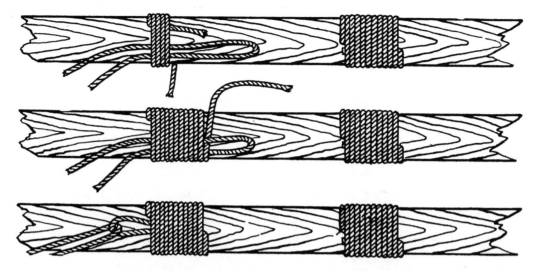

Regulation Method of wrapping a Rod.

To wrap a rod I use the same winding method shown in this 1912 illustration, but I deploy the pull-through loop when there are just six or seven rotations to go before I reach the end of the wrap.

of masking tape, but don't let the tape wrap around the rod and stick to itself. Start the new wrap right where the shaft's varnish meets the bare area by tucking the end of the thread toward the inside of the wrap while rotating the rod. The following wraps will wind over this thread end. Continue wrapping for six neat turns and closely trim the end with a knife. Continue the wrap up over the guide foot. Stop about five turns short of the spot where the original wraps ended or met the inside varnished area.

While holding the rod section in place with tension on the thread, slide a 4-inch loop of precut thread under the wrapping thread and against the shaft. The loop should point toward the eventual end of the wrap. Continue rotating the rod to load the last five turns over the loop. Press a finger against the wrapping at the loop, so that it won't unravel; cut the thread, leaving about 3 inches of tag end, and pass that tag through the loop. Grasp the loop firmly and pull it carefully back under the five wraps. It will draw the tag end of the wrap with it. Trim this tag of thread as close as possible to where it emerges from the wraps, remembering that a little twinge from the alcohol lamp or even a cigarette lighter will remove any errant fuzz.

This wrap should be neat, with no gaps between the individual winds. If you spot a gap, work it up to the solid area using your thumbnail before you continue. Most wrapping difficulties occur while you're trying to make the initial tuck. An excellent wrap is the hallmark of quality workmanship.

A new winding should always meet the old rod varnish and never wrap up over it. This will make it easy to brush on the several finishing coats of color preserver and varnish. Careless strokes brushed up over the old varnish will look like hell. One easy way to stay neat is to leave the retaining tapes in place until the revarnishing is complete. Any small imperfections where new and old varnish meet can be blended with rubbing compound.

TIPS ON WRAPPING GUIDES AND KEEPER RINGS

On a stripped rod's shaft, guides leave behind "ghosts"—little outlines of their feet. Replace the original guide feet exactly on their ghosts, using two thin bands of painter's tape to hold each foot in position. Start the wrap at the outside edge of the ghost, where the original wrap met the lighter-colored blank. With snake guides you can wrap up over the foot, remove the tape strip, and finish off at the point where the old threads stopped—visible on the blank as the other end of the ghost.

Antique loose-ring guides, which evolved into the classic hook-keeper ring, are a different story. Before the late 1890s wrappers stopped at the end of the first foot, removed the tape on the other foot, and continued wrapping. They brought the thread under the other foot and ring, then tucked it against the preceding wrap with their finger. After wrapping the shaft under the ring hump, they wound the thread up over the remaining foot, bringing the wrap down the foot and finishing it about ⅛ inch beyond the foot end.

To speed production, though, wrappers switched to a new two-part winding style (shown below, left), first on loose-ring guides and later on snakes. This was the same wrap we now use on modern snakes and strippers. Thomas and Edwards often wrapped their keepers in this new style, but they also used the old one-piece wrap. The idea of using an obsolete loose-ring guide as a hook keeper is attributed to William E. Edwards, sometime around 1898.

Shellacking and varnishing a keeper ring or flip ring has its problems. The loose ring often drops down into the fresh varnish and sticks to the finish as it hardens. To solve this problem, make a jig from some light stainless wire, as shown on page 329. Bend one end of the wire back parallel to itself so that you can tape it against the grip (for a hook keeper) or the shaft (for a loose-ring guide). Bend the other end of the wire into a little hook to lift the loose ring out of your way when you varnish. With the jig in position, the guide or keeper can be completely shellacked and varnished. When the finish is dry, the jig can be removed and tucked away for the next repair job.

REPLACING TIPPING AND INTERMEDIATE WRAPS

Many rods feature guide wraps that have a thin band of tipping in a contrasting color. Most tipping is only six threads wide and fits into the slim space between the end of the main wrap and the beginning of the varnished area on the rod's shaft. To allow for this space, install the pull loop in the main thread approximately a dozen turns before the edge of the varnish. When the main thread reaches the varnished area, unwind six turns, pull the end through, and trim. Start the narrow tipping at the "tuck," the same point where you pulled the main color's thread under its wrap. Begin the tipping by making two turns of thread. Then install the pull loop. Continue the tipping until the wraps meet the varnished area, then finish the wrap in the usual manner.

Intermediate wraps, found on antiques and pre-1930s classic rods, are thin windings—sometimes only ⅓₂ inch wide—placed along the shaft between guides and usually spaced in decreasing increments from butt to tip. Intermediates are so narrow that they are tough to wrap on a varnished shaft and impossible to wind on a slippery, unvarnished bamboo blank. Start wrapping an intermediate as you would tipping—butting it against the existing varnish with two turns of the thread, and installing the pull loop with the third wrap. Now make one more revolution, carefully trim the tag end with a sharp knife, and continue over the loop with two more turns (or whatever the original area will allow). Cut the thread and tuck it into the loop thread. While "pinning" the wrap with a finger, briskly pull the loop through and trim. Since it's only six threads wide, this is a difficult wrap to master, even more so during a complete rod-refinishing project. I suppose that replacing tipping and intermediates builds character.

TIPTOP REPAIRS

Installing a new tiptop is easy. Original German silver tops often break where the ring meets the barrel. Sometimes the top is just too worn for active casting with expensive fly line. Agate tops are impossible to replace unless you have one squirreled away in your parts bin. Many rodmakers—such as Edwards and Granger—used standard Perfection tops, which are still made and available from any good rodmaker's-supply house.

Remove an old broken top by briefly rotating it over a match or cigarette lighter. The top will slide off easily (wear gloves; it's hot). To secure the new top, coat the bamboo or wooden tip with a little match-heated Gudebrod ferrule cement. A modern substitute is a quick-melt glue stick. I usually carry one of these to exotic tropical locations, just in case a top loosens from the intense heat. After applying the glue to the rod tip, heat the replacement top while holding it lightly with a pair of needle-nose pliers. Immediately run the lighter over the previously applied glue on the bamboo tip, and, while the top is still hot, slide it over the tip. Make sure that the tiptop is on straight and lines up with the guides before the glue cools and hardens. Allow excess ferrule cement to cool for 10 minutes and remove it with a thumbnail—one of the handiest tools at a rod repairman's disposal.

REPAIRING HOOK DIGS

Occasionally you'll find a rod with a minor hook dig—a small chiplike fracture in the cane caused by a collision with a fly whizzing past at 70 miles per hour. Hook digs may be so tiny that they're almost undetectable with the naked eye, or they can be much larger and hard to miss. Big or little, hook digs are weak points on the shaft. A common method of repairing a hook dig is to wrap the entire area, including ¼ inch on either side, with size 00 white silk thread. When treated with varnish, this thread turns clear and almost disappears. Before you wrap over the hook dig, though, remove the varnish from the damaged area.

STRAIGHTENING A ROD SECTION

After years of angling, many an old fly rod develops a set—bending down and to one side or the other, depending on whether the angler was right handed or a lefty. Sets also develop on rods used for trolling. Sometimes just the tips have a set, but often the entire rod arches in a display of honest wear. Wooden fly models with a set cannot be cured, but with split-bamboo rods we're luckier, especially if the rod is an older classic built with early hide glue. Fly rods made with resorcinol glue cannot be straightened, but then they're less likely to take a set in the first place.

Any rod that really needs straightening must first be stripped of its guides, wraps, and varnish—quite a production, and I wouldn't hesitate to just go ahead and fish a wand with a mild set. To correct a short set, perhaps a foot long or less, run the bent shaft area, with the set's concave side up, over a rodmaker's alcohol lamp until it's quite warm to the touch. Then rotate 180 degrees, rest the section at an angle on a tabletop, and rub the heated area briskly and forcefully with the palm of your hand. Tips are delicate: They take less time to heat and can get overstraightened, actually pointing off toward the opposite direction. Go easy. You can *try* to straighten a set without stripping the rod, but I don't even bother with this any more: 9 out of 10 times the heat will blister the varnish, so I strip the section(s) prior to heating.

My favorite method for curing a longer set is to hold the shaft horizontally over the surface of a woodstove—the Mainer's Cure. Lacking a woodstove, you can accomplish the same thing by using a long cast-iron griddle designed to fit over two burners of a gas or electric range. Wear gloves for either method, and keep the shaft within 2 inches of the hot surface. You can heat almost an entire rod section this way, making the process relatively quick and easy. Work down the heated set with your palm until the shaft is no longer hot. The trick is to use just enough but not too much heat, so that the glue holding the bamboo strips just starts to "flux." In this semiplastic state, the strips will bend to the desired direction with a light rubbing force.

Any rod that has taken a set once will do so again, as will be pointed out by Fred Thomas on page 340. Perhaps the easiest way to put a set back in a rod is to use it for streamer fishing. Also, most anglers tend to fish from just one side of a river, and, to make matters worse, they strip a sunken line. This is tough on a classic rod. Change sides of the stream to work the rod in the opposite direction. When playing a fish, turn the rod upside down, guides facing up, during half a prolonged fight and when the fish is struggling before the net. Frankly, I've lived with sets for a long time, and only the extreme ones are worth removing.

TIGHTENING LOOSE FERRULES

"Ticking," the nasty sound that comes from the ferrule area when the rod is wagged back and forth, is a serious problem. A rod that ticks should not be fished until the problem is corrected. A ticking ferrule is loose either in its wood-to-metal fit or in its metal-to-metal fit. A little careful wiggling will identify the problem. It might be both.

Ferrules that are loose on the shaft should be removed, cleaned inside, and reseated and glued. If the ferrule was pinned, this little antagonist must be removed before the ferrule can be heated and removed from the shaft. Some makers, such as Montague, "dead-pinned" the ferrule to the blank; under magnification you'll see this as a tiny pin seated flush on the ferrule just before it reaches the shaft. A dead pin has to be tapped below the ferrule wall and farther into the shaft before the ferrule can be removed. Most available pin punches are too big to fit rodmaker's pins, and you'll probably have to grind one down on a wheel to make it small enough to tap the pin through without marring the ferrule. In a pinch, you can make a pin punch from a common steel nail.

Makers such as Thomas used a ferrule pin that extended through the ferrule and showed on the other side. These are easier to remove. Tap down the pin ¹⁄₆₄ inch, so that the same length protrudes from the opposite side of the ferrule. Grasp the pin carefully with a pair of fine needle-nose pliers and pull it straight out. Some makers, such as Granger, didn't use pins at all. Arrr, they be the smart ones!

To remove the ferrule and reseat it with fresh glue, use the same heating procedure recommended above under "Tiptop Repairs."

Wear a glove! I like the new hot-melt glues for this. Allow the heated ferrule to cool for 20 minutes before you try to remove excess glue from the shaft. You'll have to remove the thread wrapping at this location and rewrap it once the ferrule is tight to your satisfaction.

Although the problem is rarely seen on the best nickel-silver ferrules, cheap production ferrules often lose the fit between their male and female portions. This can be cured in a radical way. Leave the male ferrule on the shaft and chuck it into a quality electric drill chuck, such as a Jacobs. Don't plug in the drill; you're only borrowing its chuck. You need two reference points to make the repair accurately. Position the shaft so that one flat is up, using a snake guide as the reference point. Slide the shaft into the

The original Goodwin Granger rod, as well as the later Wright & McGill versions, are now very popular. Still a great value, the classic Granger is a joy to fish.

chuck so that one of the three chuck lands is also facing up.

Using the chuck key, tighten the chuck forcefully by hand. Now remove the male ferrule from the chuck and try it in the female ferrule to see if the slop has been removed. If the two ferrule sections are still loose, return the male to the chuck, making sure that the reference points—the bamboo flat with a snake guide, and one of the chuck lands—are in the same positions as before. Install the chuck key and tighten a little more. Sooner or later the male ferrule will come tight to the female. Of course, what you're actually doing is warping the male so that it's tight in three spots and flattened in three others. This cure will last for a season or two.

REFINISHING

Sometimes a rod's finish is just too far gone to restore and you have to start fresh. There are probably more ways to varnish or revarnish rods than I could possibly mention within these pages. Most classic makers employed the traditional method of brushing, using a fine camel's-hair brush. Every rod-maker had his secret method, which usually accompanied him to Valhalla. Following is a brief overview of the steps involved in revarnishing a rod using techniques that have worked for me.

Before removing old varnish, wrappings, and the guides, draw a sketch of the rod sections on a sheet of paper, noting little details such as the distances between guides, the number of intermediates between guides (if present), and the exact colors and locations of tipping, butt wraps, or signature wraps. This diagram will later come in handy, since memory can falter.

To remove old varnish from a split-bamboo rod, make several light applications of mild "no-wash" paint stripper, which is gentle on bamboo. After five minutes, carefully wipe off the stuff with 4/o steel wool or a rag and repeat the whole procedure. By now the varnish has flaked away from the wrappings and the guides will come off easily; store them in an egg-carton parts bin in the same sequence you took them off the rod, from the stripper guide to the first

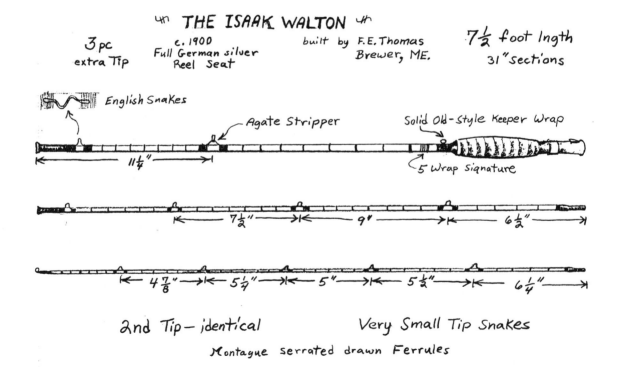

Before stripping any classic or antique rod, make an accurate sketch of it that includes all the details you'll need to put the wraps and guides back on again. This particular rod, an Isaak Walton, has 97 intermediate wraps and five signature wraps.

guide below the tiptop. By the third application of stripper, the rod should look absolutely bare and about seven shades lighter. Finish by wiping the shaft with lacquer thinner and burnishing with 4/0 steel wool. At this point the rod is ready for tacking and varnishing.

THE GHOST EFFECT

Once the rod is completely stripped, look closely along the shaft with a magnifying glass. This will reveal an amazing visual aid that will help you to reposition the guides and rewrap the fly rod: "ghosts" of almost every original wrap, even little narrow ones such as intermediates or signature wraps. Combined with the diagram of the wraps and their colors that you made before stripping, these ghosts will guide you precisely to the areas where you should position the new wraps. Ghosts are usually darker than the surrounding cane, and sometimes may be tinted to the color of the original thread. In some cases, you may find ghosts of the original maker's wraps, giving evidence that a previous whimsical refinisher paid no attention to the rod's original integrity. If you find ghosts in areas that contradict your rod diagram, note them on that diagram, and consider restoring the guides to the original maker's positions.

PREPPING AND VARNISHING

With the rod stripped and ghosts noted, you can take the first step in refinishing—applying a very light coating of varnish to all the shafts. Before varnishing, though, remove all dust and dirt by wiping down the bamboo sections with a dry tack cloth or a clean, soft cloth moistened with spirits of turpentine. Then pick a dry day with a relative humidity of less than 60 percent; if you work in humid conditions, the varnish job will be less than excellent and the coating may take forever to dry. Varnish should also be warm; set the can in a pan of warm water.

Varnish must have the proper viscosity—thick enough to flow easily from the brush. If it's too thick, thin it with a bit of mineral spirits. You'll get the best results by applying several thin coats. Lay on one coat of varnish before trying to rewrap the bamboo shafts, as the thread will slip on the slick, bare bamboo. Wooden rods should also receive a prewrapping coat of varnish.

New or replacement rod windings need to be treated with a few undercoats of color preserver, either the acetate-based Gudebrod brand or good clear shellac. Thinned with lacquer thinner (Gudebrod) or denatured alcohol (shellac), a preserver will retain the sheen of silk thread prior to an application of varnish. Not all makers used color preserver, so check the original wrap colors closely before using one. Some wrinkles on shellac are covered earlier in this chapter, under "Preserving Frayed Wraps."

For varnishing, choose a room with the least possible amount of dust. Pour some varnish from the can into a medicine dispensing cup, then re-cover the can tightly to keep the varnish solvents from evaporating. Now, with rod sections poised for finishing— well tacked and clean—sit down and wait 5 or 10 minutes, checking the rod shafts for ambient dust particles and avoiding sudden movements. You want to wait until the room's dust has settled.

Now, load the brush and flow on the varnish in 4- to 6-inch increments per brushful, always working back into the previously varnished area and never stopping. Any interruption will show in the final finish, so move along the shaft at a steady pace, using the ferrule areas as fingerholds and rotating the blank in the same direction throughout the application. After each increment, check for minute holidays (missed areas). Make sure that the bamboo is covered but not overloaded. An overload will cause a sag and ruin the entire job.

DRYING SECTIONS AND COMPLETE RODS

When the shaft is varnished, set it in a corner at a slight angle and wait five minutes or so. Then upend the section and let it stand in this reversed position; this helps check sags before they develop. Motor-powered drying jigs are available; these can be adjusted to hold sections of varying lengths by the ferrules and tips. An entire rod or just one or two sections can be slowly rotated on the jig, completely eliminating any worries about sags.

An entire rod can be assembled and mounted horizontally in a homemade jig, using plastic rollers at each ferrule and at the reel seat. Starting at the tip, varnish the rod and rotate it by hand until you reach the butt wrap. As the varnish dries, the rod should be rotated by a slow-moving rotisserie motor. It's possible for the rod to "walk" while rotating, moving the rollers beyond the metal ferrules and ruining the drying varnish. With care and practice, however, and by using the ferrule welts as retainers, this rotation method works great and leaves a very clean finish.

When you're done, clean the brush in spirits, wash the bristles in soapy water, rinse and shake thoroughly, and stand the brush upside down in a jar to dry. As the rod dries, leave the room carefully and slowly to keep down the dust. If you're varnishing in your home, wait until the end of the evening, when household activity is at a minimum. One small word of warning: Don't place the freshly varnished rod in

After three coats of shellac and a final coat of spar varnish, the hook-keeper wrap on this Bristol Edwards is finished. Complete refinish jobs are more difficult, but the rewards can be great. You'll end up with a rod that will bring hours of streamside pleasure.

an area frequented by cats. I won't go into the gory details of what cat hairs do to tacky varnish.

REWRAPPING AND SHELLACKING

After 10 or 12 hours, most tung-based varnishes will be dry to the touch and the rod ready to be handled. Wait at least three days, however, to start rewrapping the original or replacement guides. Use the wrap ghosts and your diagram for an accurate guide. When the windings are finished, give them a

coat of well-thinned color preserver, removing any excess by immediately rotating the wraps along a fingertip. If the shellac is too thick, the wraps may end up with dark areas that look like watermarks. A proper coating will lighten as it dries, allowing the sheen of the silk or nylon to show through. Shellac or color preserver dries rapidly, and two coats can be applied in a day. I apply a total of three coats then wait a day before going back to the spar varnish.

If you're brave and about to rewrap the 200-or-so intermediates on an antique or early classic rod, go slowly, allowing ample space to place your fingers as you turn the rod. I wrap half the intermediates between the guides, then stop and apply a thin coat of shellac to them. Working with an entire shaft loaded with delicate, unprotected intermediates carries the chance that you'll undo some of the little wraps that you've just done. After the preserver dries, you can proceed with the other half of the rewrapping job.

SUCCESSIVE VARNISH COATS AND RUBBING

After rewrapping and applying the shellac, you're ready for the next coat of finish—a thin coat of spar varnish applied to the entire rod, including the wrappings. Wait 48 hours, then apply another coat. Often, this is all you need do for a rod finished to production standards, especially if 00 thread was used for the wraps. If the individual threads are still showing, give them another coat of varnish, extending the coat no farther than the ends of the wraps. In all, you've applied two very thin coats of varnish over the shaft and five base and final coats to the thread areas. If the rod looks "done," it is.

Any further finishing usually involves final polishing with fine pumice. Varnish needs a week to dry completely, so don't try rubbing too early. Be very careful with polishing compounds and use a soft, wet rag; change rags and pumice often. After completely refinishing a rod, I like to wait a good 10 days before fishing it. And now we come to the fun part.

Fishing the Classic Rod

The final pleasure of collectible tackle is fishing it. Many anglers firmly believe that no rod casts more enjoyably than one of good classic cane, preferably paired with a good classic reel. Every angler has a favorite rod and maker. As you might have guessed, I like Thomas and Edwards. Others sing the praises of Granger or Heddon. In the end, though, it doesn't really matter who made the tackle—as long as you get a kick out of it.

A classic reel, like a fine old Pflueger Delite or an imposing Zwarg, will stand up to today's fish just as it did decades ago. Good reels tend to last a long time. Collectible rods with synthetic shafts—early fiberglass and boron—will take more abuse than their natural-fiber cousins, especially around salt water, which is not kind to bamboo at all. Wooden rods must be babied during even the shortest of casts, but then using wood is an eccentric pastime. Most of us will fish classic cane. For anglers raised on graphite, cane rods require a bit of understanding, so let's look at a few simple rules to fish by.

FIRST CAST

Fewer bamboo classics are broken while angling than while being assembled or broken down. The rod is always longer than we think it is. Tips poke into trees, get slammed in car doors, and find all the innumerable positions that put them in harm's way. A little extra control while putting the rod together can help keep it to length.

To assemble a standard three-piece classic, seat the reel firmly on the butt. Hold the midsection in your left hand. With your other hand, grasp the tip section close to the male ferrule, align the guides, and push the male and female together *without twisting*. Attempts to rotate partially seated sections can seriously damage the joints and cane. This is why the British twist-lock ferrule was such a bad idea. For the butt and midsections, follow the same procedure. When running the fly line and leader through the guides, rest the butt section on something that won't muck up the end of the seat or reel. A tuft of grass makes a good spot; gravel, mud, and sand don't. The outfit should now be ready for the fly, and that's "anglers' choice," as Sewell Dunton would say.

Fly fishers who normally fish graphite are often taken aback when first casting a split-bamboo rod. The slower rhythm of bamboo requires an adjustment in timing. But once you get the feel of this mellow stroke, the classic cane rod becomes a joy that can spoil you.

The better old fiberglass models have a similar feel, although they're a little less lively than cane and transmit the strike from fish to angler at a slower rate. Split cane's higher rate of transmission makes hook setting easier. Graphite is faster still, and boron heads the list. I have no idea where wood stands. Probably in a corner.

Cane rods can be found to suit any taste, with a wide range of actions, from the old super-slow, wet-fly rod to a crisper dry-fly taper. A rod built to a wet-fly action isn't really wimpy, though; just slow and

The early-1930s Montague logo—a fish impaled on a claw.

powerful enough to rip a sunken line off the water. Different types of bamboo show big differences in the rates of recovery. Old Calcutta sticks are extremely mellow. Classic Tonkin rods vary with the makers' tapers. Rods from Leonard, Payne, Thomas, and Young can fall at both ends of the speed spectrum, from extremely fast to very slow. These makers built the largest number of rods designed for 4-weight lines, and Paul Young was the master of the tiny midge rod. Granger rods are generally agreed to exemplify the classic medium action. Heddon rods are slightly faster, and Edwards sticks are often faster than many fiberglass models. The impregnated rods of Orvis, Sharpe, and Phillipson have yet a different feel, due to the Bakelite filler that protects them from deterioration and saltwater damage.

Try rods of varying actions to find the ones that suit your fishing style and temperament. And don't forget ergonomics. For instance, an angler with large hands may fall in love with the delicate feel of a vintage Orvis Superfine only to find that, over the course of the day, the slim grip gives him cramps. Those with small hands will find just as many rods with oversized grips. I once fished a 5-foot, 9-inch Divine seemingly built for Conan the Barbarian. Despite its short length, its grip was a full 8 inches long, and, although it was the classic and usually comfortable Wells-style grip, the rod just never felt right. No matter how famous the rod or how great the action, if it isn't comfortable to fish, it isn't the rod for you.

A JERK AT ONE END

The tantrums of a fly caster snagged on a rock or a sunken log are highly entertaining—as long as he isn't you. But if you're going to fish split bamboo, it's

Fig 43.

Fig 44.

The famous 1880s Henry P. Wells grip as he designed it, detachable and permanent. It wound up being a permanent fixture indeed, one still used on many modern fly rods.

time to master your temper. The wild, frenzied thrashings of a jerk at one end of the line can fracture or even snap a cane rod. The proper—albeit unhappy—solution is to point the rod at the snag, pull the fly line tight, and pop the leader.

When you're learning a bamboo rod's limits, start with conservative casts and work up from there, gauging the rod's total arch as you progress. At some point the rod will limit out, and casts will become difficult without becoming longer. This limit will be well within the boundaries needed to effectively cover any water, however. Both Hiram Hawes and his wife, Cora, could extend a fly line more than 100 feet using Leonard rods. A comfortable medium distance, say 50 to 70 feet, is easy to achieve. Leave the Lefty Kreh long-range stuff to graphite.

The extreme casts shown in the movie *A River Runs Through It* were staged by professional distance caster Jason Borger under ideal conditions. And Borger wasn't casting a vintage Granger or Montague, either, but a brand-new hexagonal graphite rod built by Walton Powell specifically for the movie, using original 1920s full-metal reel seats. I examined one of these movie rods and found it had a Thomas seat marked "Abercrombie & Fitch, Yellowstone Special."

If you really want to go for distance, consider the big two-handed rods. Casting long distances with these is relatively easy, using the ancient Scottish figure-8 line pickup at the end of the drift—the old speycast. Anglers after big anadromous critters will find learning the speycast worth their while, since currently, quality two-handed rods are priced in the bargain-basement range. Even light Payne or Leonard salmon rods sell for a quarter the price of shorter models.

Setting the hook on a small fish requires no more than a simple lift of a bamboo rod. With larger fish, lift the rod only in a slight sideways bow, and set the hook with a pull of your stripping hand. This efficiently transmits the striking power down the line to the fish and keeps the rod from arcing into a dangerous bend. The shock of a large fish, such as a salmon or a striper, may be more than cane can take, so when you're fighting a heavy fish from the reel keep the rod tip lower than you would with a graphite rod.

Years ago I conducted a dead-lift test to see just how much weight a fly rod could handle before it hit its limit. Cane didn't do well, and models with nickel-plated brass hardware bent at the ferrules. Graphite made a better showing, and glass seemed to head the list. No fly rod, built from any material, can lift much weight until it reaches the 10-weight line class. Be warned: Cane rods will not stand much pressure.

Matching line weight to the average classic rod is fairly easy. The vast majority of classics take 5- to 7-

If you use a weight-forward line, start with one size smaller than indicated; you may overtax a cane rod with a long rocket-tapered line. The weight-forward equivalents are HDF, 5-weight; HCF, 7-weight; GBF, 8-weight; GAF, 9-weight; and GAAF, 10-weight.

Once you become comfortable with a bamboo rod, nervousness disappears and casting becomes a joy. I've used split-cane rods off and on since 1952, when I was 10 years old. That first rod was a Montague Sunbeam, and I broke one tip in the station-wagon door. Without an extra tip, I learned how to be very careful. I never broke another rod until I conducted my lifting tests, some 30 years later.

THE TAKEDOWN AND SHORT-TERM STORAGE

At outing's end, it's time to take down the rod, and, again, bamboo calls for a different procedure. Snip the fly, wind the line and leader back onto the spool, and set the rod's butt down on a clump of grass, or whatever, just as you did when you put it together. Hold the second section just below the ferrule with your left hand and reach up with your right hand, grasping the tip with your arm almost fully extended. Now push with your upper arm, while keeping your lower arm static. The two sections should slip apart

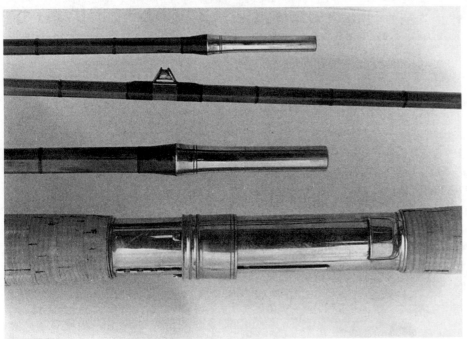

Some of the best buys in classics today, two-handed salmon rods were originally much more expensive than short trout models. Many anglers have rediscovered the speycast, which takes a lot of the work out of fishing these long wands. I wouldn't use this rod, though. It's a valuable Kosmic 15-footer.

weight lines; the greater portion toss 6-weights. Most rod dealers list the recommended line weight in a rod's description. Many models have the appropriate line written on the shaft or hanging tag, but these recommendations will follow the old system, in which lines were measured by diameter, not weight. Here are the modern equivalents of the old double-tapered lines: HFH, 4-weight; HEH, 5-weight; HDH, 6-weight; HCH, 7- or 8-weight; GBG, 9-weight.

under steady, controlled pressure. Follow the same procedure for the mid and butt joints.

To ensure easy setup and takedown, keep the ferrules clean inside and out. Avoid heavy oil or grease. Old-timers rolled the male ferrule on their noses before putting the sections together. A better solution is to wipe a small amount of WD-40 into the female ferrule with a cotton swab and remove it with a clean one. Once clean inside and out, ferrules work fine

without the application of odd oils, which may collect dirt and push a male up into a female.

With the rod taken down, remove the reel and wipe off the shaft with a clean, soft cloth. Slip the rod into its bag and lower the whole affair into the aluminum tube, making a circle around the opening with the fingers and thumb of one hand. This will cushion the rod from the tube's sharp edges.

After you arrive back home or at camp, remove the bagged rod from the tube and hang it to dry. Most rod bags have a little loop put there just for this purpose. Leave the tube open to dissipate any retained moisture. The next day, the rod can be returned to the tube for permanent storage—or it can be used again.

THE ANGLER'S DELIGHT

Some Suggestions on the Selection of a
Good Fishing Rod and its Care. Of Use to
the Novice, and Perhaps of Interest to the
Veteran Angler.

by Fred E. Thomas

Y ou ask me for some suggestions as
to the selection and care of a rod,
and here are a few points that
should always be remembered, if
a man wants to have a good rod to
begin with—and wants it to stay a good rod.
If, in the first place he purchases a first-class
article, there is no reason why, with ordinarily careful attention, his rod may not last him
for many years.

In purchasing a rod, the first thing is to
insist that you want a good one, a rod made
by some reputable maker, whose name on
the finished article is a guarantee that it will
be well made and fit to stand the strain to be
put on by the fisherman and his fish. Be sure
that you are buying a good rod, that it may
stand the test at the critical time.

In the first place, see that it is not nickel
trimmed, as these are sure to be brass with a
plating over them, and any brass trimmings
will surely crimp or bend. They are not stiff
enough for the needs of a rod with which to
catch the immense brook trout caught in the
waters of Maine. To be sure, nickel will
always stay bright without effort, and gives
the rod a stylish look, while German silver
will grow somewhat dull, but the latter can
always be readily polished with little effort.
No poor rod will be found with German silver mounts, because the makers of cheap
rods can't afford to use them.

See that the ferrules are of solid German
silver and thick and strong, as they will have
to stand great strains, and if not made right
and put on right, will weaken the rod and
cause it to break just at the end of the ferrule,
and always at a critical period in the landing
or playing of the fish.

Put the rod together and test it by bending
it against the ceiling, watching it carefully for
any signs of weakness—detected by any
unevenness in the curve of the rod from tip
to butt. Then try it by working in the hand by
whipping, from side to side and up and
down, to see if the ferrules are tight, as a loose
ferrule will work serious mischief. If the rod
stands all these tests, then you need not fear
to purchase.

But having made the rod yours, there is
still much to be remembered if you would
have it serve you as it should and will, if properly looked after. Care for your fishing-rod as
you would care for any other fine example of
workmanship. The most delicate piece of
bric-a-brac is not more deserving of care than
your faithful rod, which to be most useful
should receive the best of care, that its service and life may not be shortened.

After a day's fishing, don't leave the rod
out all night, standing against the piazza or
wall of the camp; for instance while you go
in for supper, and then forgotten till you are
ready for the next day's fishing. Indoors, it
can be laid upon hooks, not far apart and all
on an exact line and sufficient to support the
rod on a level its entire length. Here your rod
may lie and rest, in a dry but not warm place,
through the night. If you have no place to
support the rod from unequal strain, then
unjoint it, wipe it carefully and put it away in
a dry case—never put it into a wet case
unless an emergency, and then remove it for
drying by wiping at the earliest possible
moment. Never allow yourself to leave the
rod standing, wet. Be sure that it is always dry
when you set it aside.

A practice which is the result of thought-
lessness, not to say ignorance, is frequently
the undoing of rods that otherwise would
prove most serviceable for years—I refer to
leaving a rod, fully jointed and rigged, lying
in a boat or canoe, exposed to the merciless
rays of a hot sun. The mere weight of the delicate tip as it hangs over gunwale, thwart or
seat, will bend it. The heat of the sun will
help the weakening process, so that in a surprisingly short time the rod will have a tip

that, bend it back as you may, will never again be as true and as strong as it was before. If you must leave the rod in the boat or canoe, unjoint it, wipe it and put it away in its dry case where the sun can't get at it, and where each joint will be supported as it lies, by the case.

A fly rod, being made so much more delicate and yielding than the heavier bait-rod, ought never to be used in trolling. But there are times when one wishes to troll, and has no bait rod handy. At such times, always troll with the rod raised slightly above the water, just enough that the spring for striking the fish will be there, and yet not enough to give the tip a strain that can possibly be avoided. Then, the rod won't be dragging the weight of the line all day long, a process which is sure to give the rod a twist, especially on a sunny day. This advice applies equally well to the care of the bait rod, which will last longer and be in better condition when the fish takes the bait, than if it was subject to a constant strain.

Sometimes you will, when anchored and without time to up anchor and move the boat or canoe, have a big trout dart directly under your craft and literally wind your slender rod around the canoe's bottom. Such a strain is a splendid test of a rod's endurance, but will invariably bend the second joint badly if it does not break it. Until such time as you can get your bent joints replaced, which should be done if possible, this trouble may be partially corrected by removing the rings and replacing them exactly opposite to their former positions, when the strain on that side of the rod will, after a time, stretch the wood fibers on that side and straighten the rod so that the bend is not noticeable. But as these sudden strains must, by the laws of Nature, stretch the outer or longer circle of the bamboo composing the rod, so any future strain will cause it to bend in just the same way, no matter how evenly you may have straightened it after the mishap. As the second joint is the one that

really does the work of the rod, that joint is the one principally affected by such rough usage, and this past winter I have replaced two such that were spoiled last season in just the way I have mentioned. Both rods were made by me, and neither was broken, but both were badly twisted.

Finally, when you put the rod away for the season, be sure to wipe it thoroughly, mak-

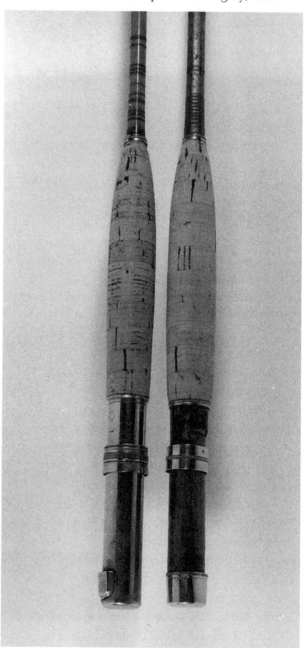

When you think of Thomas rods, you think of upcountry and streamer models—but Thomas made a lot of rods for lines down to 4-weight. These two Dirigo 9-footers work best with a 5-weight line.

ing sure that no moisture is on any of it or in the case, then put away in a cool, dry place. Too often a warm place is selected, because it is thought dryer than where it is cool, but to keep the rod in the best possible condition, it should never be allowed to become even moderately heated. If these directions were closely followed, few anglers would find their rods in unsatisfactory condition when they came to use them, after a fishing trip or after a season's inactivity. And except for breaks, they would rarely need repairs.

Maine Sportsman, May 1901

FISHING WITH DANIEL WEBSTER

Vermont rodmaker S. W. Goodridge had this to say about wooden rods: "In camping out unjoint your rod after a day's fishing, and place it in a dry place. A fly rod in the night dew will lose all its life and elasticity so as to be nearly useless for the next morning's fishing." Proper care will ensure that a fine old rod, bamboo or wood, will remain in good condition. Don't lean it in a corner, put it away wet, or treat it as harshly as you would a synthetic rod. Overlining a classic rod can also have dire consequences, putting undue strain on the blank and ferrules. A little care goes a long way, and once you've learned the rote, the rod will be better for it.

The ultimate streamside looniness is fishing with an ancient wooden rod, a way of seeing the sport of fishing through Daniel Webster's eyes. Some present-day anglers have even adopted pre-Websterian methods. John Betts, a pioneer in the use of synthetic fly materials, has taken a giant step backward, returning to jodhpurs and fishing in the style of Walton. He braids his own horsehair lines, ties The Dame's Twelve original fly patterns, and catches some nice fish. So, too, does Ken Reinard, who fishes in the 18th-century fashion, making his own period clothing, lines, and rods. Known as The

Colonial Angler of 1770, Ken has written an excellent book on the subject, *The Colonial Angler's Manual of Flyfishing & Flytying*. A few years back Ken became disoriented—perhaps he lost his ghillie—and was caught fishing 1770-style in the Pennsylvania State House pond. He was "arrested" and brought before the governor during the Governor's Ball. Some anglers will try anything to catch a difficult fish.

Perhaps the one thing that separates fly fishing from other forms of angling is its tendency to make the sport harder, more of a challenge. On a bet, Norm Crisp once tried to catch a big trout on a horsehair tippet, long considered the fly fisher's ultimate challenge. In 1836, our good man in Scotland,

LANDING A DOUBLE.

When lifting a larger fish or a "double," as pictured in this 1883 Doctor James A. Henshall illustration, don't let the rod form a dangerous curve. And when playing a fish, turn the rod upside down occasionally to avoid a permanent "set" in the shafts. This rod looks like an antique Orvis, one of Henshall's favorites.

Thomas Tod Stoddart, mentioned that the Reverend Mr. Paterson of Galashiels took a full-sized salmon on one. Crisp had rough going at first, discovering all the foibles associated with the elder sport—brittle hairs and scabby ones. Taking a cue from Vidal Sassoon, Norm finally washed his horsehair in shampoo and added a treatment of conditioner. And he caught a beautiful 16-incher on a single horsehair tippet. After the fact, he discovered that the Charles Cotton types actually used tippets made from two or three hairs, preferably from a well-washed and -kept stallion.

Of course, if you want to go fishing with Daniel it's a lot easier to equip your early-1800s rod with some modern 2-pound-test tippet material. It beats running around the pasture with a pair of snippers. I tried a "short" wooden antique, only 11-something feet long and one slow S.O.B., the kind of rod that you tend to baby. After all, it was 150 years old and could have disintegrated at any moment. I kept my casts short and managed to catch two little brook trout—about average for me. And it was a real kick, although my angling cohort, Barry Gibson, thought I had finally plunged off the deep end. If you decide to trek off to the Neversink with one of those old wooden jobbies, release a fish for the old Orator, but make it a small one.

I have tried to provide a few pointers on fishing the classics without sounding preachy. The ultimate test of rod, reel, and angler is in the enjoyment of fishing. Through the years I've had a wonderful relationship with the classic rod, experiencing some good rises and taking some memorable fish. Most were released, though a few were eaten in a ritual

that extends back beyond old Aelfric. It's been a journey, this fly-fishing game, growing from Cotton to Fitzgibbon, maturing through Norris and Scott, to become the sport of our own lifetime.

Appendix I

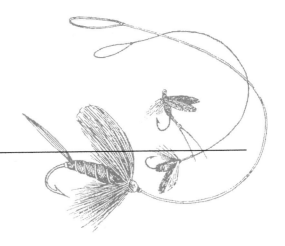

Classic-Tackle Contacts

Nationwide Dealers

Centennial Classic Sales
Dick Spurr
256 Nashua Court
Grand Junction, CO 81503
http://www.gorp.com/bamboo.htm

Classic Rods & Tackle, Inc.
Martin J. Keane
P.O. Box 288
Ashley Falls, MA 01222

Heritage Sporting Collectibles
Len Codella
587 East Gulf to Lake Highway
Lecanto, FL 34461

Jordan-Mills Rod Company
Carmine Lisella
11 Wesley Road
Congers, NY 10920

Rods & Reels
Bob Corsetti
17 Massasoit Road
Nashua, NH 03063

Vintage Tackle Service
A. J. Campbell
Box 66
Boothbay, ME 04537

Supplies

Angler's Workshop
1350 Atlantic Avenue
P.O. Box 1044
Woodland, WA 98674

Dale Clemens Custom Tackle, Inc.
444 Schantz Spring Road
Allentown, PA 18104

Shows and Auctions

Northeast Antique Angler's Show
Ben Clark
P.O. Box 81193
Wellesley Hills, MA 02181-0002

The NFLCC National
P.O. Box 13
Grove City, OH 43123

Richard A. Bourne Auction Gallery
P.O. Box 141
Hyannis Port, MA 02647

Southbay Auctions, Inc.
Henry Broggi
485 Montauk Highway
East Moriches, Long Island, NY 11940

Lang's Fishing Tackle Auctions
Bob Lang
31R Turtle Cove
Raymond, ME 04071

Publications and Organizations

Fishing Collectibles Magazine
Brian McGrath, Publisher and Editor
P.O. Box 2797
Kennebunkport, ME 04046

The American Fly Fisher
P.O. Box 42
Manchester, VT 05254

National Fishing Lure Collectors Club
P.O. Box 13
Grove City, OH 43123

Old Reel Collectors Association
c/o Arne Soland
849 NE 70th
Portland, OR 97213

Museums and Libraries

American Museum of Fly Fishing
P.O. Box 42
Manchester, VT 05254

Maine State Museum
State House
Augusta, ME 04333

Atlantic Salmon Museum
Doaktown, New Brunswick, Canada

The Milne Angling Library Room
University of New Hampshire Library
Durham, NH 03824-3592

Appendix II

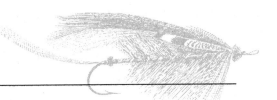

268 Fishing-Related U.S. Patents Prior to 1873

LIST OF PATENTS GRANTED BY THE UNITED STATES TO THE END OF 1872, FOR INVENTIONS CONNECTED WITH THE CAPTURE, UTILIZATION, OR CULTIVATION OF FISHES AND MARINE INVERTEBRATES.

1. HOOKS.

Date of patent.	Number.	Inventor.	Subject of invention.
July 28, 1846	....	Englebrecht & Skiff.....	Spring-hook. When fish pulls on bearded hook, a catch slips holding the spring-hook, which, being let loose, strikes the fish and holds him until taken off by the angler.
Aug. 21, 1847	....	Staunton Pendleton....	Spring-hook.
Aug. 21, 1847	....	Job Johnson	Spring-hook.
Aug. 15, 1848	....	Ellis & Gritty..........	Spring-hook. When fish pulls on the bearded hook, it slips the spring-hook, which strikes and holds the fish fast.
Aug. 15, 1848	....	W. P. Blake.............	Spring-hook. When fish pulls on bait, it springs open, holding the jaws of the fish apart.
Sept. 5, 1848	....	W. Jenks	Spring-hook.
Mar. 20, 1849	6,207	Job Johnson	Spring-hook; twenty-one different kinds of hook and method of attachment.
Oct. 8, 1850	7,709	Warner & Gaylord	Spring-hook.
April 6, 1852	8,853	Julio T. Buels..........	Trolling-hook.
April 11, 1854	10,771	.....do	Spring-hook.
April 11, 1854	10,761	Henry Siglers...........	Combination spring-hook.
Jan. 19, 1855	13,081	Richard F. Cook	Spring-hook.
Jan. 19, 1855	13,068	Charles De Saxe	Trolling-hook. Has a spring shield that covers the point of the hook when fishing among weeds.
Oct. 9, 1855	13,649	Job Johnson	Spring-hook.
April 22, 1856	14,706	Julio T. Buels..........	Fly or trolling hook.
July 14, 1857	17,803	Donald McLean........	Self-setting trap-hook.
Sept. 20, 1859	25,507	Riley Haskels	Trolling-hook.
Feb. 12, 1861	31,396	W. L. Morris...........	Spring-hook. When the line is pulled, a catch slips off the ends or levers of the hook, and a spring draws the bearded points together into the fish.
Sept. 20, 1864	44,368	N. A. Gardner	Spring-hook. Two bearded hooks, forming part of a coil-wire spring with eyes, and rod having a line-eye for setting and releasing the hooks.
Nov. 7, 1865	50,799	Germond Crandell......	Combination double-lever hook.
Dec. 19, 1865	51,651	Davis & Johnson	Spring-hook.
Nov. 20, 1866	59,844	Jacob King, jr	Spring-hook.
Nov. 20, 1866	59,893	C. O. Crosby...........	Fish-hook, (flattened in bend.)
Jan. 9, 1866	51,951	B. B. Livermore........	Plain hook, with wire loop, to prevent fish from stealing the bait.
April 24, 1866	54,251	Johnson & Howarth	Spring or spear hook.
May 15, 1866	54,684	W. D. Chapman........	Trolling-hook, with spring and fly, the latter easily removed.
Oct. 2, 1866	58,404	W. C. Goodwin.........	Plain hook; spiral spring around the hook to press the bait down to point.
Jan. 1, 1867	60,786	E. R. & J. W. Rhodes....	Spring-hook.
Feb. 12, 1867	62,042	B. Lee, jr..............	Hook; the shank made in form of spiral spring.
Aug. 27, 1867	68,027	F. Angilard............	Lever-hook, so arranged that when fish pulls at bait another hook strikes it and makes fast.
Nov. 12, 1867	70,868	A. J. Leinhart..........	Spring-hook.
Nov. 12, 1867	70,913	Elisha Sterling.........	Extension or trap hook, with two or more hooks; one on swivel to hold bait, the other to grapple the fish while pulling from the water.
Sept. 24, 1867	69,221	D. Kidders.............	Spring-hook. Two hooks, one on either end of piece of wire or shank; when baited they are pressed together, but will separate when the fish bites.
April 28, 1868	77,365	R. A. Fish	Hook.
June 30, 1868	79,446	J. B. Christian	Trolling-hook, with artificial bait.
July 21, 1868	80,151	A. A. Dennett..........	Spring-hook.
Jan. 26, 1869	86,154	Martin Heltz	Hook, with an eye to attach hook.
Sept. 14, 1869	94,893	Francis Kemlo..........	Lock-hook.
Sept. 14, 1869	94,894	......do	Lock-hook.
Sept. 14, 1869	94,895	......do	Grapple-hook, with guard to prevent fish from getting loose from barb.

Compiled by Spencer F. Baird, U.S. Fisheries Commissioner. From Condition of The Sea Fisheries, 1873.

REPORT OF COMMISSIONER OF FISH AND FISHERIES.

List of patents granted by the United States, &c.—Continued.

HOOKS—Continued.

Date of patent.	Number.	Inventor.	Subject of invention.
Oct. 12, 1869	95,755	F. T. Augers	Spring-hook. When set, the three hooks are close together; but when the line which is attached to the middle hook is pulled, it loosens the outer hooks, which expand in the mouth of the fish.
July 5, 1870	104,930	W. D. Chapman	Propeller or trolling hook.
Feb. 21, 1871	111,898	L. Arnold	Mode of attaching hook to line.
May 30, 1871	115,434	W. D. Chapman	Propeller or trolling hook.
Aug. 8, 1871	117,719	L. Arnold	Mode of attaching hooks to lines.
Nov. 21, 1871	121,182	J. H. Mann	Trolling spoon-hook.
Feb. 20, 1872	123,844	G. Sinclair	Trolling-hook.
July 16, 1872	129,053	E. Pitcher	Hook, with double spear, to thrust down into fish when caught by hook.

2. LINES, GRAPPLES, TRAPS, &c.

Date of patent.	Number.	Inventor.	Subject of invention.
Mar. 3, 1868	75,075	D. C. Talbots	Fishing-tackle for anglers.
April 7, 1868	76,489	T. B. McCaughans	Fish-trap.
May 12, 1868	77,893	Joseph Koehlers	Fishing-apparatus.
Dec. 22, 1868	85,199	E. B. Beach	Fish-trap.
Nov. 4, 1856	16,014	Elmore Horton	Spear and grapple.
Dec. 9, 1856	16,217	Levi Van Hossen	Fish-trap.
May 25, 1858	20,343	Jacob Gail	Grapple.
Jan. 18, 1859	22,644	Robert Gray	Fish-trap.
Mar. 8, 1859	23,154	Daniel Bowmans	Fish-trap.
Aug. 2, 1864	43,694	A. J. Leinharts	Fishing-tackle.
June 3, 1862	35,476	T. W. Roys	Method of raising whales.
Jan. 30, 1872	123,164	O. M. Faller	Fishing-apparatus.
Sept. 7, 1872	131,439	Harcourt & Cottingham	Fish and animal trap, made of wire, each wire of gate made movable.
Oct. 22, 1872	132,476	C. Lirandais	Trap-net, used in shallow water, umbrella fashion, to lie on bottom, and having springs to close by trigger.

3. REELS.

Date of patent.	Number.	Inventor.	Subject of invention.
May 26, 1838	....	Arrmah Tiffaney	Reel, for anglers' use.
July 26, 1838	....	......do	Reel or mackerel-latch, used in fishing from vessel.
Aug. 5, 1856	15,466	John A. Baileys	Reel.
Feb. 10, 1857	16,626	Edward Deacons	Reel.
Aug. 9, 1859	24,987	Edward Billinghurst	Reel.
Feb. 28, 1860	27,305	Mark S. Palmer	Reel, with guard to line to prevent it from clogging the reel.
Feb. 9, 1864	41,494	Andrew Dougherty	Reel, with brake.
July 5, 1864	43,460	W. H. Van Geison	Reel, with stop-attachment.
July 12, 1864	43,485	Darwin Ellis	Reel, with stop.
July 12, 1864	43,546	T. W. Cummings	Reel, (mode of attachment to pole,) by spring.
Aug. 29, 1865	49,663	W. M. Stewart	Reel, set inside of rod.
June 19, 1866	55,653	Anson Hatch	Reel, skeleton, similar to Billinghurst.
Aug. 7, 1866	56,937	A. B. Hartils	Reel.
Nov. 26, 1867	71,344	Julius Von Hofe	Reel.
Sept. 22, 1868	82,377	W. H. Bradley	Reel, with two concaved disks.
Nov. 3, 1868	83,740	J. Stetson	Clamp or reel, used for hand-line fishing.
Feb. 23, 1869	87,188	Francis Xavier	Reel, pivoted and made to screw into rod, bridge, or stick in ground, with bell attachment.
Mar. 23, 1869	88,026	C. S. H. Foster	Mackerel-latch.
Oct. 12, 1869	95,839	James J. Ross	Reel.
Nov. 9, 1869	96,652	P. A. Allmaires	Reel, set into rod.
May 31, 1870	103,668	G. C. Sheldons	Reel; more of a kite-string holder.
Mar. 7, 1871	112,326	E. L. Decker	Mackerel-latch.
Nov. 14, 1871	121,020	Silas B. Terry	Reel, with friction device.
June 18, 1872	128,137	A. H. Fowler	Reel, made of rubber skeleton.
June 14, 1873	134,917	George Mooney	Reel, manner of attaching to rod.
June 23, 1873	135,283	Charles L. Noe	Reel, similar to Billinghurst patent.

4. RODS.

Date of patent.	Number.	Inventor.	Subject of invention.
Apr. 10, 1854	10,795	C. Desaxe	Rod, made hollow, to contain float, lines, &c.; peculiar float.
May 20, 1862	35,339	Julius Von Hofe	Rod; tip has a sheave or pulley on the end.
Oct. 16, 1866	58,833	R. N. Isaacs	Tips of rods enameled to prevent wear of line.
May 18, 1858	20,309	Underwood & Bargis	Rod, with pulley set in tip.
Oct. 4, 1859	25,693	Henry Pritchard	Guides for lines on rods.
Dec. 24, 1867	72,667	J. H. Montrose	Rod, hinged like parasol-handle.

LIST OF PATENTS.

List of patents granted by the United States, &c.—Continued.

RODS—Continued.

Date of patent.	Number.	Inventor.	Subject of invention.
Mar. 17, 1870	100, 895	W. J. Hubbard..........	Jointed rod, screwed together, to prevent slipping apart, by male and female screws.
Sept. 26, 1871	119, 251	Thomas Tout..........	Rod, principally of wood, with lameneal of whalebone running longitudinally.

5. FLOATS, SINKERS, AND SWIVELS.

Date	Number	Inventor	Subject
Dec. 12, 1854	12, 060	J. W. Heard.............	Sinker, made hollow to contain shot, so that it may be adjusted to required weight.
Feb. 2, 1869	86, 609	J. A. Terrell	Float, made of glass.
Feb. 8, 1870	99, 572	James Ingram	Float, with ring and plugs in ends of it, so that it may be adjusted to line, without slipping over ends of line.
May 26, 1872	127, 218	Brown & Jarvis.........	Float, made of vulcanized rubber, for seines.
July 9, 1872	128, 885	E. Jewell	Float, for ready attachment to line, to avoid slipping over ends of line; spiral wire in either end.
Apr. 1, 1856	14, 587	Wooster Smith	Fish-hook and sinkers, used for cod-fishing.
July 7, 1863	39, 192	William Woodbury.....	Sinker, with spring inside for deep-sea fishing.
July 21, 1865	46, 453	E. F. Decker	Sinker, with guard-ring and swivel.
Sept. 25, 1866	58, 211	L. A. Burnham	Sinker, with lever, &c.
July 31, 1866	56, 857	J. A. Martin	Sinker, with lever, &c.
July 29, 1867	61, 625	J. A. Martin	Sinker, with lever, &c.
Dec. 10, 1867	71, 879	Martin Hiltz	Swivel, for anglers' use.
May 5, 1868	77, 628	L. D. Lothrop	Swivel, for anglers' use.
May 12, 1868	77, 774	W. H. Smith	Sinker, made in several pieces, to increase or decrease weight.
June 2, 1868	78, 546	E. F. Stacey.............	Nipper, or latch, to hold line.
Nov. 3, 1868	83, 681	Sewall Albee............	Jig, or sinker, three pieces, and method of attaching to hook.
Dec. 15, 1868	84, 885	Leach & Hutchins......	Mode of attaching hook to sinker.
Feb. 9, 1869	86, 786	F. Telgmanns...........	Sinker, sectional.
Aug. 3, 1869	93, 220	R. T. Osgood	Sinker, with spring and swivel, egg-shape.
Sept. 5, 1871	118, 772	H. Camp	Metallic line, with loops and reel.

6. PROJECTILES.

Date	Number	Inventor	Subject
July 29, 1841	2, 195	William Carseley	Spear.
Mar. 16, 1844	3, 490	Albert Moon............	Harpoon; harpoon contains bottle of explosive material, which operates to throw the flukes out when it strikes.
Sept. 19, 1846	4, 764	Oliver Allen	Lance.
Nov. 24, 1846	4, 865	Holmes & West.........	Lance or harpoon, moveable flukes.
Dec. 3, 1846	4, 873	Charles Randall........	Harpoon, moveable flukes.
Dec. 5, 1848	5, 949	Oliver Allen	Gun-harpoon.
June 4, 1850	7, 410	Robert Brown	Harpoon, mode of attaching line, } Gun-lances.
Aug. 20, 1850	7, 572	......do	Harpoon lance, mode of attaching line } Gun-lances.
Sept. 3, 1850	7, 610	C. F. Brown.............	Harpoon.
Nov. 19, 1850	7, 777	William Albertson	Harpoon.
May 6, 1851	8, 073	Charles Burt............	Exploding harpoon.
Mar. 30, 1852	8, 843	Sonnenburg & Richten..	Electric whaling-apparatus.
Apr. 6, 1852	8, 862	J. D. B. Stillman	Harpoon, moveable flukes and pulleys.
June 22, 1852	9, 044	C. C. Brand.............	Lance for killing whales. Re-issued August, 1856.
Aug. 19, 1856	15, 577	Nathan Schofield	Expanding spiral-winged projectile.
Mar. 10, 1857	16, 819	......do	Bomb-lance, with springs. Re-issued July 7, 1857.
Apr. 28, 1857	17, 173	Rufus Sibley	Bomb-lance, with moveable flukes.
May 26, 1857	17, 370	Grudehos & Eggers.....	Bomb-lance, with springs and moveable flukes.
May 26, 1857	17, 407	Rufus Sibley...........	Projectile.
May 19, 1857	17, 312	C. C. Brand	Projectile.
Oct. 20, 1857	18, 458	J. Q. Kelly	Harpoon.
Nov. 10, 1857	18, 568	H. Bates	Bomb.
Dec. 9, 1857	18, 824	N. Schofield............	Projectile.
Feb. 16, 1858	19, 363	H. W. Harkness.........	Harpoon and lance.
Aug. 17, 1858	21, 219	Rufus Sibley...........	Bomb-lance.
Aug. 24, 1858	21, 278	N. Schofield............	Harpoon-lance.
Nov. 2, 1858	21, 949	George Doyle..........	Harpoon.
Nov. 16, 1858	22, 054	A. F. & J. H. Andrews..	Bomb-lance.
May 3, 1859	23, 827	P. B. Comins	Bomb-lance.
June 14, 1859	24, 371	Robert Brown	Harpoon-bomb.
Aug. 9, 1859	25, 086	Isaac Goodspeed........	Projectile.
Dec. 11, 1860	30, 869	Theodore Briggs........	Harpoon-lance.
Jan. 22, 1861	31, 190	Thomas W. Roys........	Shoulder-gun, for harpoons, lances, &c.
July 16, 1861	32, 830	Goodspeed & Crawley ..	Guide for bomb-lance.
June 3, 1862	35, 474	Thomas W. Roys........	Rocket-harpoon.

REPORT OF COMMISSIONER OF FISH AND FISHERIES.

List of patents granted by the United States, &c.—Continued.

PROJECTILES—Continued.

Date of patent.	Number.	Inventor.	Subject of invention.
Apr. 21, 1863	38,207	M. Adams	Harpoon, with semi-revolving head.
Oct. 27, 1863	40,387	Oliver Allen	Bomb-lance, with perforated fire-proof diaphragm.
Feb. 21, 1865	46,437	Silas Barker	Exploding-harpoon.
Aug. 22, 1865	49,548	Ebenezer Pierce	Apparatus for killing whales.
Apr. 24, 1866	54,211	Roys & Lieliendahl	Rocket-harpoon.
Apr. 23, 1867	64,045	Robert E. Smith	Shooting-harpoon, grooved head, to receive the pivoted barb.
Dec. 3, 1867	71,763	Z. Kelley	Harpoon, with stops, springs, and catches.
June 9, 1868	78,675	......do......	Bomb-lance.
June 1, 1869	90,368	E. Pierce	Bomb-lance.
Dec. 7, 1869	97,693	J. P. Rechtens	Gun-harpoon.
May 7, 1872	126,388	Charles Freeman	Bomb-harpoons.

7. NETS AND POUNDS.

Mar. 21, 1838	....	Russell Evarts	Seine for deep-water fishing.
June 4, 1838	....	B. W. Hale	Seine for deep-water mackerel fishing.
Sept. 19, 1838	....	Cyrus Tracey	Seine.
Mar. 17, 1843	....	Harris Cook	Gill-net.
Apr. 25, 1843	....	John Downs	Form for making nets for taking eels.
Sept. 14, 1844	....	Carr, Shannon & Co.	Net or trap; place for bait similar to eel-pot.
Apr. 18, 1854	10,794	Charles De Saxe	Landing-net.
Apr. 27, 1858	20,125	Thomas Hall	Seine or net.
June 29, 1858	20,725	Benjamin Merritt	Seine for sea-fishing.
Apr. 8, 1862	34,887	F. Goodwin	Net or trap.
Apr. 25, 1863	39,676	W. Randolph	Net, to be anchored and used as trap.
June 19, 1866	53,635	Edw. A. Field	Net.
Aug. 7, 1866	56,917	Ferl & Larkin	Vertical deep-water fishing-net.
Nov. 6, 1866	59,429	William Maxwell	Net, double, with rigid mouth; can be anchored at any depth by floats and sinkers.
Feb. 26, 1867	62,481	C. C. Crossman	Net, attached to side of boat, so as to be lowered or raised.
Dec. 17, 1867	72,177	C. Drexel	Securing and feeding crabs.
Mar. 31, 1868	76,284	Daniel Will	Gill-net.
Apr. 7, 1868	76,387	Thomas Bell	Net-attachment for boats, with gauge to mast, to hoist or lower.
June 9, 1868	78,716	B. Arnold	Mode of making nets.
July 28, 1868	80,274	John Collins	Brace-seine.
Sept. 28, 1868	82,490	T. Cartwright	Set-net, to be anchored; the boat is attached about midway of the net, and a line is attached to small end of bag, and can be raised into boat and the fish taken out without disturbing the rest of the net.
Oct. 13, 1868	82,913	George D. Allen	Eel-pot.
Oct. 27, 1868	83,493	Smith Harper	Net.
Oct. 27, 1868	83,429	W. S. Wilcox	Pound net or trap.
Mar. 9, 1869	87,740	F. A. Wardmiller	Dip-net.
Feb. 8, 1870	99,713	P. C. Sabin	Purse-net, with bait-box, the net stretched on wires similar to umbrella, and, when ready to hoist, it is closed like a purse.
Apr. 4, 1871	113,292	J. E. Hammond	Fish-trap net.
Apr. 4, 1871	113,572	Benjamin Rider	Net-supporter.
Apr. 18, 1871	113,817	P. E. Tierman	Pound-net.
Aug. 15, 1871	117,957	L. H. Alexander	Net.
Nov. 14, 1871	120,974	R. Jeffrey	Seine.
Mar. 12, 1872	124,635	H. Smith	Seine, arranged so that portion of mouth may be below the surface.

8. OYSTER CULTURE AND GATHERING.

Mar. 2, 1858	19,516	Thomas Sheehan	Oyster rake or tongs.
Oct. 4, 1859	25,680	Thomas P. Sink	Reel or windlass, for hoisting oyster-dredges.
Oct. 11, 1864	44,634	George Jury	Reel for hoisting oyster tongs or dredges.
Oct. 2, 1866	58,426	Job Johnson	Oyster-rake.
Nov. 20, 1866	59,812	W. Belbin	Mode of hoisting oyster-dredge.
Mar. 17, 1868	75,550	Job Johnson	Oyster-tongs.
Apr. 14, 1868	76,697	Asa Barrett	Oyster-rake.
July 19, 1870	105,495	J. W. Sands	Grappling-tongs, (oyster.)
Sept. 27, 1870	107,740	E. Ward	Oyster-tongs.
Feb. 21, 1860	27,213	W. L. Force	Oyster-dredge.
May 20, 1862	35,324	J. H. Newcomb	Oyster-dredge.
May 5, 1863	38,436	Joseph Whitecars	Oyster-dredge.
Jan. 17, 1865	45,904	W. Belbin	Oyster-dredge.
June 5, 1866	55,223	Charles T. Belbin	Oyster-dredge.

LIST OF PATENTS.

List of patents granted by the United States, &c.—Continued.

OYSTER CULTURE AND GATHERING—Continued.

Date of patent.	Number.	Inventor.	Subject of invention.
Nov. 20, 1866	59, 812	W. Belbin	Oyster-dredge.
Jan. 4, 1867	65, 442	T. P. Sinks	Oyster-dredge.
Feb. 25, 1868	74, 857	......do	Oyster-dredge.
Apr. 7, 1868	77, 110	S. S. Shaw	Reel for oyster-dredge.
June 2, 1868	78, 509	C. T. Belbin	Oyster-dredge.
Aug. 18, 1868	81, 304	T. P. Sinks	Oyster-dredges.
June 19, 1869	85, 936	Daniel Kellers	Oyster-dredge chuck.
Apr. 27, 1869	89, 323	Thomas F. Mayhew	Oyster-dredge.
Nov. 30, 1869	97, 420	......do	Oyster-dredge.
Feb. 15, 1870	99, 900	R. O. & W. T. Howard	Reel for oyster-dredges.
Oct. 1, 1871	120, 463	T. P. Sinks	Oyster-dredge.
Nov. 28, 1871	121, 227	W. C. Baker	Oyster-dredge.
Nov. 28, 1871	121, 249	E. B. Lake	Oyster-dredge.
Jan, 2, 1872	122, 423	N. A. Williams	Oyster-dredge.
Jan. 16, 1872	122, 843	T. F. Mayhew	Reel or windlass for oyster-dredge.
Apr. 23, 1872	125, 964	T. W. Landon	Oyster-dredge.
June 11, 1872	127, 903	Benjamin F. Leyford	Oyster nurseries.
Aug. 20, 1872	130, 631	E. H. Frazier	Shell-oyster bucket.
Oct. 29, 1872	132, 668	J. A. Ketcham	Oyster-dredge.
Jan. 21, 1873	135, 167	Isaac Smith	Oyster-tongs.

9. PRESERVATION AND UTILIZATION OF FISH.

Date of patent.	Number.	Inventor.	Subject of invention.
Mar. 19, 1861	31, 736	Enock Piper	Preserving fish and meat.
Dec. 15, 1864	1, 618	James B. Herreshoff	Mode of treating fish-water for use in dyeing, &c. (Reissued December 15, 1864.)
Nov. 5, 1867	70, 435	George H. Herron	Improved mode of preparing fish for food. (Reissued January 16, 1872.)
May 19, 1868	78, 016	Benjamin Robinson	Improved process of obtaining gelatine from fish-heads.
Sept. 8, 1868	81, 987	William D Cutler	Improved method of preparing, desiccating, and preserving fish.
Oct. 27, 1868	83, 533	P. Nunan	Improvement in apparatus for preserving and freezing fish and meats, &c. (Drawing.)
Nov. 10, 1868	83, 836	William D. Cutler	Articles of food prepared from fish and potatoes.
Dec. 8, 1868	84, 801	Elisha Crowell	Improved article of prepared codfish.
Jan. 19, 1869	85, 913	William Davis	Improvement in freezing-box for fish. (Drawing.)
Jan. 19, 1869	86, 040	Thomas Sim	Improved compound for preserving fish.
Mar. 16, 1869	87, 986	Benj. F. Stephens	Improvement in putting up codfish.
Mar. 23, 1869	88, 064	J. Nicherson	Improved process of preparing fish for food.
May 25, 1869	90, 334	John Atwood, jr	Improved process of curing and putting up codfish.
June 8, 1869	90, 944	Havard & Harmony	Improved process of preserving meat, fowls, fish, &c.
Dec. 6, 1870	109, 820	D. Y. Howell	Device for freezing fish, meats, &c. (Drawing.)
Feb. 28, 1871	112, 129	Samuel H. Davis	Improvement in preserving fish by freezing.
May 21, 1872	127, 115	Isaac L. Stanley	Improved process of preparing fish for food.
Oct. 15, 1872	132, 316	Edward A. Pharo	Improvement in putting up salt mackerel and similar fish.
Dec. 15, 1868	84, 855	Edward E. Burnham	Preservation and improvement in fish or bait used in catching fish.
Feb. 11, 1868	74, 378	T. D. Kellogg	Improved method of preserving bait for fishing. (Drawing.)
Sept. 28, 1869	95, 179	R. A. Adams	Improvement in preserving fish.
Oct. 26, 1869	96, 288	George T. Thorp	Improved fish-bait.
Nov. 23, 1869	97, 145	R. A. Adams	Improvement in curing and preserving fish.
Mar. 29, 1870	101, 260	Silvanus Hamblin	Improved bait-mill for fishermen. (Drawing.)
Jan. 21, 1873	135, 113	Samuel A. Goodman, jr	Bait, composed of vegetable and animal matter; is mixed with ordinary bait.

10. FISH-CULTURE.

Date of patent.	Number.	Inventor.	Subject of invention.
Sept. 17, 1867	68, 871	Seth Green	Spawn-hatching mackerel.
Jan. 16, 1868	78, 952	W. H. Furman	Fish-breeding.
Mar. 16, 1868	78, 952	W. H. Tierman	Fish-breeding device.
Aug. —, 1868	80, 775	A. J. Smith	Transporting live fish.
June 11, 1872	127, 903	Benjamin F. Leyford	Oyster-nurseries.
July 11, 1871	116, 995	R. E. Sabin	Fish-nursery.
July 12, 1872	105, 176	A. S. Collins	Fish-spawning screen.
Oct. 22, 1872	132, 349	G. A. Brackett	Fishway.

REPORT OF COMMISSIONER OF FISH AND FISHERIES.

List of patents granted by the United States, &c.—Continued.

11. PATENTS GRANTED PRIOR TO 1834.

[The order shows that patents were granted to these persons, but the drawings and applications were destroyed by fire in 1836.]

Date of patent.	Number.	Inventor.	Subject of invention.
1795	....	Joseph Ellicott, Pa......	Catching fish.
1802	....	Nathaniel Robbins, N. J.	Mode of carrying fish in warm weather.
1809	....	Philip Groff, Pa.........	Seine.
1812	....	Samuel May, Pa.........	Seine.
1814	....	James Wells, jr., N. J...	Vessels and nets for fishing.
1818	....	James Drummond	Net.
1824	....	Daniel Gordon, Pa......	Catching fish.

No. 58,833.—RUSSEL N. ISAACS, New York, N. Y.—*Fishing Rod.*—October 16. 1866.—Enamel is applied to the guides through which the line passes on its way from the reel to the tip.

Claim.—The application to the metallic guides and tip of fishing rods an enamel, or covering of glass, porcelain, or any similar vitreous substance, to protect the line from friction and wear, substantially as described.

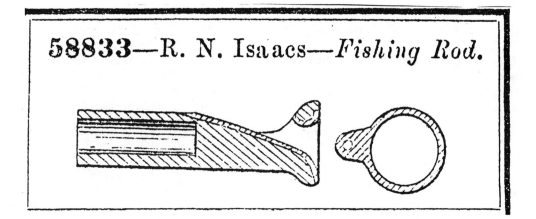

Bibliography

Baird, Spencer F. *Condition of The Sea Fisheries.* Washington, DC: Government Printing Office, 1873.

Bates, Joseph D. *Streamers and Bucktails: The Big Fish Flies.* New York: Alfred A. Knopf, 1979.

Bigelow, Paul. *Wrights and Privileges.* Athol, Mass.: Haley's, 1993.

Brown, Jim. *Fishing Reel Patents of the United States, 1838–1940.* Stamford, Conn.: Trico Press, 1985.

Brown, Jim. *A Treasury of Reels.* Manchester, Vt.: The American Museum of Fly Fishing, 1990.

Brown, John J. *American Angler's Guide.* New York: Burgess, Stringer, 1849.

Davy, Sir Humphrey. *Salmonia: or Days of Fly Fishing.* London: John Murray, 1828.

Demeritt, Dwight B. Jr. *Maine Made Guns And Their Makers.* Hallowell, Maine: Maine State Museum and Paul S. Plumer Jr., 1973.

"Ephemera" (Fitzgibbon, Edward). *A Handbook of Angling,* Third Edition. London: Longman, Brown, Green, and Longmans, 1853.

"Lake Brooks" (Harding, A. R.). *Science of Fishing.* Columbus, OH: A. R. Harding, 1912.

Henshall, James A. *Book of the Black Bass.* Cincinnati, OH: Robert Clarke, 1881.

Hogan, Austin, and Paul Schullery. *The Orvis Story.* Manchester, Vt.: The Orvis Company Inc., 1980.

Josselyn, John. *Account of Two Voyages to New-England.* Licensed by Roger L'estrange, 1673.

Keane, Martin J. *Classic Rods and Rodmakers.* New York: Winchester Press, 1976.

Kelly, Mary Kefover. *U.S. Fishing Rod Patents and Other Tackle.* Plano, Texas: Thomas B. Reel Company, 1990.

Leonard, J. Edson. *Flies.* New York: Barnes, 1950.

Liu, Allan J. (ed.). *The American Sporting Collector's Handbook.* Tulsa, OK: Winchester Press, 1982.

Marbury, Mary Orvis. *Favorite Flies and Their Histories.* New York: Houghton Mifflin, 1892.

Mather, Fred. *My Angling Friends.* New York: Forest and Stream, 1901.

Melner, Samuel, and Hermann Kessler. *Great Fishing Tackle Catalogs of the Golden Age.* New York: Crown, 1972.

Miller, Warren H. (ed.). *The Outdoorsman's Handbook.* New York: Field & Stream, 1916.

Munger, Albert J. *Those Old Fishing Reels.* Philadelphia: Century Graphics, 1982.

Munger, Albert J. *Old Fishing Tackle and Tales.* Philadelphia: Munger, 1987.

Murray, William H. H. *Adventures in The Wilderness; or Camp-Life In The Adirondacks.* Boston: Fields, Osgood, & Company, 1869.

Norris, Thaddeus. *The American Angler's Book,* New Edition. Philadelphia: E. H. Butler & Company, 1865.

Prouty, Lorenzo. *Fish: Their Habits and Haunts.* Boston: Cupples, Upham and Company, 1883.

Roosevelt, Robert Barnwell. *Superior Fishing.* New York: Orange Judd, 1884.

Schwiebert, Ernest. *Trout Tackle, Part Two.* New York: Dutton, 1978.

Scott, Genio C. *Fishing in American Waters.* New York: Harper & Brothers, 1869.

Schullery, Paul. *American Fly Fishing: A History.* New York: Lyons & Burford, 1987.

Shields, George O. (ed.). *American Game Fishes.* Chicago: Rand McNally, 1892.

Sinclair, Michael. *Bamboo Rod Restoration Handbook.* Grand Junction, Colo.: Centennial, 1994.

Sinclair, Michael. *Fishing Rods by Divine.* Grand Junction, Colo.: Centennial, 1993.

Smedley, Harold Hinsdill. *Fly Patterns and Their Origins,* Fourth Edition. Muskegon, Mich.: Westshore Publications, 1950.

South, Theophilus. *The Fly Fisher's Textbook.* London: R. Ackermann, 1841.

Spurr, Dick, and Gloria Jordan. *Wes Jordan: Profile of a Rodmaker.* Grand Junction, Colo.: Centennial, 1992.

Stoddart, Thomas Tod. *The Art of Angling, As Practised in Scotland.* Edinburgh: W. & R. Chambers, 1836.

Sutherland, Douglas. *The Salmon Book.* London: Collins, 1982.

Tolfrey, Frederic. *The Sportsman in Canada.* London: Newby, 1845.

Turner, Graham. *Fishing Tackle: A Collector's Guide.* London: Ward Lock, 1989.

Vernon, Steven K. *Antique Fishing Reels.* Harrisburg, Penn.: Stackpole, 1985.

Vernon, Steven K., and Frank M. Stewart III. *Fishing Reel Makers of Kentucky.* Plano, Texas: Thos. B. Reel Company, 1992.

Walton, Izaak. *The Compleat Angler.* Andrew Lang edition. London: F. M. Dent & Company, 1896.

Wells, Henry P. *Fly-Rods And Fly-Tackle.* New York and London: Harper & Brothers, 1885.

Wells, Henry P. *The American Salmon Fisherman.* New York: Harper & Brothers, 1886.

Index

(Italicized page numbers are references to illustrations of the makers' products)